NASA/SP—2010-216283

# A History of the Lightning Launch Commit Criteria and the Lightning Advisory Panel for America's Space Program

*Francis J. Merceret, Editor*
*NASA, John F. Kennedy Space Center*

*John C. Willett, Editor*
*Air Force Research Laboratory (Retired)*

*Hugh J. Christian*
*University of Alabama in Huntsville*

*James E. Dye*
*National Center for Atmospheric Research, Boulder, Colorado*

*E. Phillip Krider*
*University of Arizona, Department of Atmospheric Sciences*

*John T. Madura*
*NASA, John F. Kennedy Space Center*

*T. Paul O'Brien*
*Aerospace Corporation, El Segundo, California*

*W. David Rust*
*National Severe Storms Laboratory*

*Richard L. Walterscheid*
*Aerospace Corporation, Space Sciences Department, El Segundo, California*

National Aeronautics and
Space Administration

John F. Kennedy Space Center
Kennedy Space Center, FL 32899

**August 2010**

# Executive Summary

Since natural and artificially-initiated (or 'triggered') lightning are demonstrated hazards to the launch of space vehicles, the American space program has responded by establishing a set of Lightning Launch Commit Criteria (LLCC) and Definitions to mitigate the risk. The LLCC apply to all Federal Government ranges and have been adopted by the Federal Aviation Administration for application at state-operated and private spaceports.

The LLCC and their associated definitions have been developed, reviewed, and approved over the years of the American space program starting from relatively simple rules in the mid-twentieth century (that were not adequate) to a complex suite for launch operations in the early 21$^{st}$ century. During this evolutionary process, a "Lightning Advisory Panel (LAP)" of top American scientists in the field of atmospheric electricity was established to guide it.

This history document provides a context for and explanation of the evolution of the LLCC and the LAP. A companion document on the rationale is currently being prepared by the LAP to provide the physical, mathematical, and operational justification for the current LLCC.

Lightning that takes place naturally in thunderstorms was recognized as a threat to spaceflight early in the space age, and the early launch rules prohibited flight through thunderstorms. What had not been recognized until Apollo XII was launched in 1969, was that an ascending rocket could "trigger" lightning if it flew into highly electrified clouds, even in the absence of natural lightning. Apollo XII was struck twice by triggered lightning, and as a result, additional rules were added to forbid flying within 5 NM of thunderstorms or through certain non-thunderstorm clouds that experience showed had the potential of being electrified.

During the time between Apollo XII and the beginning of the Space Shuttle Program, several important research programs were carried out to learn more about lightning and cloud electricity in Florida, and additional instrumentation was installed at the KSC and Eastern Range (ER). The LLCC were modified for the Skylab program in 1973 and again for the Apollo-Soyuz Test Project in 1975. Expendable launch vehicles (ELV) had their own separate but similar rules. In 1979 the nascent Shuttle program adopted a set of LLCC that included constraints and waivers that were based on surface electric field mill measurements. The ELV rules did not adopt the field mill criteria.

In 1987, triggered lightning caused the loss of Atlas/Centaur 67 (AC 67) and its Fleetsatcom payload. A major cause of this accident was the failure of the ELV rules to include a field mill criterion. The Shuttle rules would have prohibited the launch of AC 67. In the aftermath of several investigations and a Congressional Hearing, a lightning "Peer Review Committee", a predecessor of the LAP, was established to guide NASA and the Air Force in the drafting of LLCC and the conduct of associated research. Several major research projects were undertaken, including the first "Airborne Field Mill Program" (ABFM I, 1990-1992) in which direct measurements of electric fields in Florida clouds were made from a specially instrumented aircraft. ABFM I resulted in several improvements to the LLCC, and the lessons learned ultimately led to an even more successful ABFM II campaign a decade later.

As new knowledge and additional operational experience has been gained, the LLCC have been updated to preserve or increase their safety and increase launch availability. All launches of ELV and manned vehicles now use the same rules, which simplifies their understanding and application and minimizes the cost of the weather infrastructure to support them. In the future, the processes and procedures that have evolved can be used to facilitate further improvements in both launch availability and safety as new research findings become available.

# Preface

Natural and triggered lightning is a demonstrated hazard to the launch of space vehicles, and the American space program has responded by establishing the "Lightning Launch Commit Criteria (LLCC)" to mitigate the risk. These LLCC are a complex set of rules with associated Definitions which must be satisfied before the launch of a space vehicle is permitted. The Definitions are an integral part of the LLCC and the term LLCC as used in this History is explicitly intended to include those Definitions. They apply to all Federal Government ranges including not only the well-known Eastern Range at Cape Canaveral, Florida and the Western Range at Vandenberg AFB, California, but also to smaller ranges such as the NASA range at Wallops Island, Virginia and the Air Force range at Kwajalein Atoll in the Pacific Ocean and others. In addition, these same LLCC have been adopted by the Federal Aviation Administration for application at state-operated and private spaceports.

The LLCC are developed and approved through a complex process, but the core science and recommendations for precise wording of the operative parts of the rules are provided by a "Lightning Advisory Panel (LAP)" consisting of American scientists working in atmospheric electricity and related disciplines including cloud physics and statistics. The LAP works closely with the operational personnel who must implement the LLCC in practice to assure that the rules are not only scientifically sound, but also realistic and practical.

The LLCC have evolved over the history of the American space program from relatively simple mid-twentieth century rules that proved inadequate to the complex suite that governs launch operations in the early part of the 21$^{st}$ century. Following the destruction of an Atlas-Centaur launch vehicle by triggered lightning in 1987, the LAP was established to guide the process. Expert guidance is required because there is always a tension between the essential need to fly safely by avoiding lightning strikes and the need to fly economically by avoiding unnecessary launch delays and scrubs. The LLCC have become complex because increased knowledge has permitted exceptions under certain conditions from what are otherwise broad prohibitions to flight. These exceptions reduce the number of occasions under which a launch will be scrubbed for violation of the LLCC when, in fact, it would have been safe to fly.

As the LLCC have become more complex, launch vehicle operators, range managers, and safety personnel have continuously requested briefings and discussions on the origin of the rules and the rationale behind them. This history document is designed to provide a historical context and explanation for the origin of the LLCC and the LAP. A companion rationale document is being prepared by the LAP to provide the scientific, mathematical, and operational basis for the current LLCC.

# Acknowledgements

Reconstructing more than 50 years of the history of weather support required a substantial effort not only by the authors and editors of this document, but also by colleagues and associates. Much of the history has not been formally documented before, and much of what existed at one time has now apparently been lost. This task would not have been possible without the assistance of many people who took time away from their busy workloads and deadlines to help us track down important events that are shrouded in the mists of the past.

We especially appreciate and recognize the contributions and assistance of the USAF 45th Weather Squadron at Cape Canaveral Air Force Station (CCAFS) and Patrick AFB in Florida and the NOAA Spaceflight Meteorology Group (SMG) at Johnson Spaceflight Center in Houston, Texas. We thank Bill Roeder and Billie Boyd of the 45th Weather Squadron and Michael Maier and David Chapman of the Eastern Range Technical Services Contractor (Computer Sciences Raytheon) for providing the history of lightning-related and other instrumentation at Kennedy Space Center (KSC) and CCAFS. We thank Tim Oram of SMG for his insights on weather support during the early years of manned spaceflight.

Bill Bihner and Jack Ernst (retired) at NASA headquarters provided recollections of the formation of the Peer Review Committee and its evolution into the modern Lightning Advisory Panel.

Launa Maier of the KSC Safety and Mission Assurance Directorate provided valuable recollections of the research programs and lightning infrastructure development during the years immediately following the AC 67 accident.

In addition to providing a figure and some references for the document, Jennifer Wilson of the KSC Weather Office handled the logistics of several face-to-face meetings of the LAP at KSC. Without these meetings dedicated to this History's organization and production, it could not have been completed.

Jennifer Rosenberger of the KSC Launch Processing Directorate did extensive reformatting and copy editing to prepare the original manuscript for public release in this NASA Special Publication series. We appreciate her diligence and attention to detail that substantially reduced the number of errors and inconsistencies in the presentation of the material.

Funding for the project was provided by the NASA Office of Safety and Mission Assurance (OSMA) Assurance Management Office (AMO). The authors and editors appreciate OSMA/AMO reviews of the final draft of this paper by Launa Maier and Tony Willingham.

# Notice

Mention of a proprietary product or service does not constitute an endorsement thereof by the Editors, the authors, or the National Aeronautics and Space Administration.

# Table of Contents

# Table of Figures

# List of Tables

# List of Acronyms

| | |
|---|---|
| 30WS | 30th Weather Squadron |
| 45WS | 45th Weather Squadron |
| 4DLSS | 4-Dimensional Lightning Surveillance System |
| AAT | ABFM Analysis Team |
| ABFM | Airborne Field Mill |
| AFCRL | Air Force Cambridge Research Laboratories |
| AFETR | Air Force Eastern Test Range |
| AGU | American Geophysical Union |
| AIAA | American Institute of Aeronautics and Astronautics |
| AMS | American Meteorological Society |
| AMU | Applied Meteorology Unit |
| ASTP | Apollo-Soyuz Test Program |
| ATC | Air Traffic Control |
| AWS | Air Weather Service |
| CaPE | Convection and Precipitation/Electrification Experiment |
| CCAFS | Cape Canaveral Air Force Station |
| CCFS | Cape Canaveral Forecast Facility |
| CG | Cloud to Ground |
| CGLSS | Cloud to Ground Lightning Surveillance System |
| CIF | Central Instrumentation Facility |
| CSR | Computer Sciences Raytheon |
| DRWP | Doppler Radar Wind Profiler |
| ELV | Expendable Launch Vehicle |
| ER | Eastern Range |
| ESMC | Eastern Space and Missile Center (early name for the ER) |
| ESSA | Environmental Sciences Services Administration (predecessor of NOAA) |
| FAA | Federal Aviation Administration |
| FSK | Frequency Shift Keying |
| IAMAP | International Association of Meteorological & Atmospheric Physics |
| IC | In-Cloud |
| ICAE | International Commission on Atmospheric Electricity |
| ID | Identification |
| IRCC | Interagency Review and Coordinating Committee |
| JSC | Johnson Spaceflight Center |
| KSC | Kennedy Space Center |
| LAP | Lightning Advisory Panel |
| LCC | Launch Commit Criteria |
| LCN | LCC Change Notice |
| LDAR | Lightning Detection and Ranging |
| LFCC | Lightning Flight Commit Criteria and Associated Definitions (FAA's version of the LLCC) |
| LLCC | Lightning Launch Commit Criteria (including their Associated Definitions) |
| LPLWS | Launch Pad Lightning Warning System |
| LUT | Launch Umbilical Tower |
| LWO | Launch Weather Officer |
| LWT | Launch Weather Team |
| MFFG | Median Filter First Guess |
| MSC | Manned Spaceflight Center (early name for JSC) |
| MSFC | Marshall Spaceflight Center |
| NASA | National Aeronautics and Space Administration |

| | |
|---|---|
| NCAR | National Center for Atmospheric Research |
| NMIMT | New Mexico Institute of Mining and Technology |
| NRL | Naval Research Laboratory |
| NOAA | National Oceanic and Atmospheric Administration |
| NSF | National Science Foundation |
| NWS | National Weather Service |
| OAT | Operational Acceptance Test |
| ONR | Office of Naval Research |
| PAFB | Patrick Air Force Base |
| PCM | Pulse Code Modulation |
| PI | Principal Investigator |
| PRC | Peer Review Committee (early name for the LAP) |
| QC | Quality Control |
| RF | Radio Frequency |
| ROCC | Range Operations Control Center (now the Morrell Operations Center [MOC]) |
| RTLP | Rocket Triggered Lightning Program |
| RTLS | Return to Launch Site (a Space Shuttle launch abort mode) |
| TBD | To Be Determined |
| TRIP | Thunderstorm Research International Program |
| SMC | Space and Missile Center |
| SMG | Spaceflight Meteorology Group |
| USAF | United States Air Force |
| VAFB | Vandenberg Air Force Base |
| VAHIRR | Volume Averaged Height Integrated Radar Reflectivity |
| WR | Western Range |
| WRWS | Western Range Weather System |
| WSO | Weather Support Office |

# Chapter 1   Introduction

Given the extraordinary record of accomplishments in the Earth and space sciences over the past 50 years, it is often difficult to remember the hazards that that have been overcome to make this progress possible. Lightning—both natural and artificially-initiated or 'triggered' discharges—is still the primary weather hazard to spaceflight operations.

When aerospace engineers are presented with a threat from natural and/or triggered lightning, they have only two options: (1) hardening the system to withstand the effects of nearby and direct strikes and (2) avoiding the hazard by flying only under safe conditions. Hardening a system against lightning involves using a design and materials that can withstand large transients in the electrical environment. This approach will have a high cost in terms of time, money, and weight and may not even be achievable. Avoiding lightning means maintaining access to information that can be used to warn when a hazard is present, and it also means taking appropriate actions when there is a threat. The Lightning Launch Commit Criteria (LLCC) and the associated Definitions are a set of rules that can be used to avoid both natural and triggered lightning hazards during space launches. The LLCC are not perfect, but they are based on experience that suggests they are safe and produce a minimal number of false alarms.

The present LLCC and Definitions are a coupled set of rules that should be evaluated as a single unit. They have evolved in the light of imperfect scientific knowledge about cloud electricity and lightning physics. As new knowledge and techniques have become available, the LLCC have been improved, and they have gotten safer and launch availability has increased. Much of the increased knowledge and advances in technology have been motivated and supported by the needs of spaceflight operations. In this History, we will trace the origin of the LLCC and the experiences that have led to the improvements. In a companion Rationale document, we will describe the structure and physical basis for the current LLCC, with the hope that future advances in science and technology, coupled with more experience and practice, will lead to even better LLCC.

## 1.0   Background

From the beginning of the American space program, lightning has been a concern. If lightning strikes an ascending launch vehicle it can inject currents in excess of 100 kA with rise times in the microsecond range. Even if the primary lightning current is conducted entirely, and, for the most part, harmlessly, by the vehicle's metallic skin, the resulting large electromagnetic fields can induce secondary currents in the interior structures leading to the destruction of the vehicle. Communications, guidance, navigation and control systems computers can be disrupted. Pyrotechnics designed to separate components in flight or as part of the range safety destruct system can also be triggered. When lightning strikes a space vehicle all effects are potentially adverse. The most likely result is loss of the mission, the booster and the payload.

In response to the threat of lightning and other weather hazards, weather observation and forecasting have been an integral part of ground support for launches at all American launch sites. In order to collect observations of sufficient density and sophistication, a specialized weather infrastructure is required at each site. To assure safety, mission success and consistency, a formal set of rules called "Lightning Launch Commit Criteria" (LLCC) and their associated Definitions prohibit launching unless certain weather constraints are met. For the newly-emerging private and state-operated spaceports, the Federal Aviation Administration applies similar rules that they call "Lightning Flight Commit Criteria" (LFCC). This history discusses the evolution of both the weather infrastructure and the LLCC from the beginning of American orbital spaceflight through the first decade of the 21st century. Just as importantly, it explores the organizations and people involved in designing and applying both the infrastructure and the LLCC.

This History focuses primarily on operations at Kennedy Space Center and Cape Canaveral Air Force Station (the Eastern Range), but the threat, and the responses to it, also affect Vandenberg Air Force Base (the

Western Range), Kwajalein Atoll (the Regan Test Range), NASA's range at Wallops Island, the Alaskan Space Authority range on Kodiak Island, and the new private ranges in New Mexico, Oklahoma and elsewhere. The LLCC/LFCC are applied at all of these places (with a very few local variations where necessary) and have been designed with all of them in mind. The focus on the east coast of Florida here results primarily from the availability of documentation and the fact that funding and administrative leadership for the Lightning Advisory Panel (LAP) and much of the research on lightning and cloud electricity conducted over the last 20 years has been provided by the Space Shuttle Program which launches only from KSC.

This work is organized largely but not strictly in chronological order beginning with the early days of America's orbital space program in the next chapter. Chapter breaks are focused on major events which, by coincidence, fall nearly on decadal boundaries. As Chapter 2 describes, until Apollo XII was struck by triggered lightning in 1969, the focus of lightning avoidance was on natural lightning. Chapter 3 describes the changes resulting from investigations and research following Apollo XII up to the beginning of the Space Shuttle era. Chapter 4 covers the Shuttle and unmanned vehicle programs through the destruction of Atlas-Centaur 67 by triggered lightning in 1987. Chapter 5 describes the changes resulting from investigations and research following AC 67 including the formation of the LAP and the first Airborne Field Mill Program (ABFM I) and closes out the 20th century. Chapter 6 picks up with the second Airborne Field Mill Program (ABFM II) which began in 2000 and takes us up to the present. In Chapter 7, the authors draw their conclusions and present their lessons learned. To keep the main body of the history from overwhelming the reader with detail, while assuring that the detail is available for those who might find it informative, eight appendices, some of them lengthy, are also provided. Because the material was drawn from a wide variety of sources covering about five decades, the units of measurement are not treated consistently. For ease of comparison, metric equivalents have been appended in parenthesis after non-metric units in most cases.

It will be helpful in understanding the significance of much of this history to know how weather support is presently provided. The remainder of this chapter will provide an overview of current (May 2010) weather support infrastructure and organization to the Shuttle Program and ELV operations at the Eastern Range (ER). A brief discussion of other Ranges is also provided. For similar reasons, a brief overview of the current Lightning Launch Commit Criteria (LLCC) is provided next.

## 1.1    A Summary of the 2009 Lightning Launch Commit Criteria (LLCC)

The current Lightning Launch Commit Criteria (LLCC), and their associated Definitions, Explanations and Examples are complex and are given in their entirety in Appendix I in several formats. To give the reader a feeling for the structure and content of the LLCC, a highly abbreviated paraphrase of each rule is offered here (in italics to emphasize that these are not the actual rules). Note that many details about the definitions of terms and the required measurement techniques have been omitted for clarity and simplicity. Therefore, these paraphrases do not adequately describe all the hazardous weather situations and must not be used for launch support.

Although the Definitions associated with these rules are not given here, they should be reviewed in Appendix I because they are an integral part of the LLCC, and because the logic and wording of the LLCC depend in a critical way on these definitions. It should also be noted that all of the LLCC must be satisfied and that each one requires clear and convincing evidence to trained weather personnel that its constraints are not violated. Under some conditions, trained weather personnel can make a clear and convincing determination that the LLCC are not violated based on visual observations alone. However, if the weather personnel have access to additional information, such as measurements from weather radar, lightning sensors, electric field mills, or aircraft, and if this information is within the criteria outlined in the LLCC, these data could allow a launch to take place when a visual observation alone would not.

### 1.1.1 Surface Electric Fields
(Appendix I, Section A1.11, 1.4A)

*Do not fly within 5 nm of any electric field mill that has shown readings in excess of 1 kV/m in the past 15 minutes. The field threshold can be raised to 1.5 kV/m if all clouds within 10 nm of the flight path are transparent or if they (a)have tops warmer than +5 °C and (b) have not been part of convective clouds with tops colder than -10 °C for at least 3 hours.*

### 1.1.2 Lightning
(Appendix I, Section A1.11, 1.4B)

*Do not fly within 10 nm of any type of lightning, or any convective cloud that has produced it, within the past 30 min. An exception is allowed if the cloud has moved beyond 10 nm and if an electric field mill within 5 nm of the lightning, and any other mills within 5 nm of the flight path, have shown less than 1000 V/m for at least 15 minutes.*

### 1.1.3 Cumulus Clouds
(Appendix I, Section A1.11, 1.4C)

*Do not fly within 10 nm of any cumulus cloud with a top colder than -20 °C, nor within 5 nm of any cumulus cloud with a top colder than -10 °C. Do not fly through any cumulus cloud with its top colder than +5 °C. An exception is allowed for a cumulus cloud with its top between ±5 °C if that cloud is not producing precipitation and if a field mill within 2 nm of that top, and any other mills within 5 nm of the flight path, have shown the vertical component of the electric field greater than −100 V/m, but less than +500 V/m, for at least 15 minutes.*

### 1.1.4 Attached Anvil Clouds
(Appendix I, Section A1.11, 1.4D1)

*Do not fly within 10 nm of any non-transparent, attached anvil for at least 30 minutes after the last lightning discharge occurs in the parent cloud or anvil, nor within 5 nm for 3 hours after the last lightning. Never fly within 3 nm of such a cloud. Two kinds of exceptions are allowed: 1) Flight is allowed up to 3 nm from an attached anvil at any time if it is colder than 0 °C everywhere within the prescribed distances of the flight path. 2) Flight is allowed through or within any distance of an attached anvil at any time if that anvil is colder than 0 °C everywhere and if its VAHIRR is less than 10 dBZ-km everywhere within 1 nm of the flight path.*

### 1.1.5 Detached Anvil Clouds
(Appendix I, Section A1.11, 1.4D2)

*Do not fly within 10 nm of a non-transparent, detached anvil for 30 minutes after the last lightning discharge occurs in the detached anvil (or in its parent cloud before detachment), nor within 3 nm for 3 hours after the last such lightning. Do not penetrate such an anvil for 3 hours after detachment, nor for 4 hours after it has produced lightning. Three kinds of exceptions are allowed: 1) Flight is allowed up to 3 nm from such an anvil during the first 30 minutes if the detached anvil is colder than 0 °C everywhere within 10 nm of the flight path. 2) Flight is allowed up to the edge of a detached anvil between 30 minutes and 3 hours after the last lightning if the radar reflectivity of all parts of that cloud within 5 nm of the flight path has been less than 10 dBZ, and if a field mill within 5 nm of the cloud, and any other mills within 5 nm of the flight path, have shown less than 1000 V/m for at least 15 minutes. 3) Flight is allowed through or within any distance of such an anvil at any time if the cloud is colder than 0 °C and if its VAHIRR is less than 10 dBZ-km everywhere within the prescribed distances of the flight path.*

### 1.1.6 Debris Clouds
(Appendix I, Section A1.11, 1.4E)

*Do not fly within 3 nm of a non-transparent debris cloud for 3 hours after it detaches or decays from its parent cloud and for 3 hours after it has produced lightning. Two kinds of exceptions are allowed to the 3 nm standoff requirement: 1) Flight is allowed up to the edge of a debris cloud at any time if the radar reflectivity of any part of that cloud within 5 nm of the flight path has been less than 10 dBZ, and if a field mill within 5 nm of the cloud, and any other mills within 5 nm of the flight path, have shown less than 1000 V/m, for at least 15 minutes. 2) Flight is allowed through or within any distance of such a debris cloud at any time if the cloud is colder than 0 °C everywhere within 5 nm of the flight path and if VAHIRR is less than 10 dBZ-km everywhere within prescribed distances of the flight path.*

### 1.1.7   Disturbed Weather
(Appendix I, Section A1.11, 1.4F)
*Do not fly through any non-transparent cloud associated with disturbed weather that has cloud tops colder than 0 °C and that, within 5 nm of the flight path, either contains moderate or greater precipitation or shows evidence of melting precipitation.*

### 1.1.8   Thick Cloud Layers
(Appendix I, Section A1.11, 1.4G)
*Do not fly through a non-transparent cloud layer that is thicker than 4500 feet and contains temperatures between 0 °C and -20 °C, nor through any non-transparent cloud layer that is connected to such a thick cloud layer within 5 nm of the flight path. An exception is allowed if the thick cloud layer is cirriform, is entirely colder than -15 °C, contains no liquid water and has never been associated with a convective cloud.*

### 1.1.9   Smoke Plumes
(Appendix I, Section A1.11, 1.4H)
*Do not fly through any cumulus cloud that develops from a smoke plume for 60 minutes after it has detached from that plume.*

### 1.1.10   Triboelectrification
(Appendix I, Section A1.12, G417.23)
*Do not fly through any cloud (transparent or not) that is colder than -10 ℃ at vehicle velocities less than 3000 ft/s unless the vehicle has been treated or hardened against surface discharges.*

## 1.2   The Current Weather Infrastructure at the Eastern Range (including KSC)
The Eastern Range benefits from an extensive suite of weather instrumentation including surface meteorological towers, surface electric field mills, rain gauges, weather radar systems, weather balloon systems, Doppler wind profiling radars and more. A complete description of each system is presented in Appendix VIII. In this section we discuss the general types of instrumentation and their application to the LLCC and to weather support in general at the Range. The material is organized in the same manner and order as Appendix VIII. Thus, to find the detail applicable to section 1.2.x below, consult Appendix section A8.x where x ranges from 1 to 6.

### 1.2.1   Atmospheric-Electrical Sensors
The Range's suite of lightning and electric-field sensors permits the detection and location of electric-charge centers aloft and lightning within clouds, between clouds, and from cloud to ground. It also permits an assessment of the magnitude and polarity of the lightning current in strokes to ground. The LLCC have many sections explicitly referring to these measurements. In addition, the lightning and electric-field sensors are used by the 45th Weather Squadron (45WS) forecasters for the issuance and cancellation of lightning watches and warnings for ground facilities at KSC/CCAFS and aircraft operations at the Shuttle Landing Facility at KSC.

Also, if a lightning strike occurs in close proximity to a launch complex or payload processing facility, it is sometimes required to retest the vehicle and/or payload to assure that no damage has been caused by the direct or induced electric currents from the nearby strike. The ability to precisely locate the strike points and to estimate their amplitudes informs the decision-making process about whether potentially costly retesting is required.

Among the most unique and important of the atmospheric-electrical sensors at KSC/CCAFS is the large-area network of ground-based electric-field mills and the associated display that is known as the Launch Pad Lightning Warning System, or LPLWS (see Appendix VIII, A8.1.1). This system is unique in that it is the only observing system that is capable of giving advance warning of electrified clouds *before* the first lightning flash. It is used not only to issue lightning warnings for ground operations but also to provide relief in several of the current LLCC under otherwise questionable conditions.

Together, the suite of atmospheric electrical sensors at the KSC/ER is without question the most comprehensive and the most accurate that exists anywhere in the world today.

### 1.2.2    Weather Radar

Weather radar is used extensively, both in determining whether the LLCC are satisfied, and for forecasting the weather. The LLCC have sections explicitly referring to radar-based criteria (See Appendix I) for determining the locations, edges and heights of clouds and for assessing the threat of triggered lightning. Significant relief from the restrictive nature of the rules is available through the use of a radar-derived quantity called the Volume Averaged Height Integrated Radar Reflectivity (VAHIRR) (See, e.g., 1.1.4 above). In addition, weather radar is essential to the issuance and cancellation of weather watches and warnings for protection of personnel, ground systems and facilities at KSC/CCAFS and aircraft operations at the Shuttle Landing Facility at KSC and the Air Force airfields at CCAFS and PAFB.

### 1.2.3    Surface Observations

Surface observations by meteorological instruments and trained human observers are used together with weather radar to determine quantities used by the LLCC including the altitude of the cloud base, the presence of thunder and whether or not the clouds are transparent. These data are also used to assess compliance with ground winds and visibility constraints applicable to launches and ground processing operations. They also support the issuance and cancellation of weather watches and warnings for protection of personnel, ground systems and facilities at KSC/CCAFS.

### 1.2.4    Upper Air Observations

Measurements of winds, temperature, pressure and relative humidity throughout the atmosphere from just above the surface to altitudes well above 20 km are essential for day to day weather forecasting. In addition, wind measurements over these altitudes are used for generating the steering commands used by launch vehicles on ascent and assessing the resulting aerodynamic loads that the vehicle will experience in flight. Finally, winds at the level of cirrus anvil clouds can be used to estimate when a thunderstorm anvil will form upstream that may cause a violation of the anvil LLCC.

Weather reconnaissance aircraft are used to determine the height and thickness of clouds required by the LLCC more precisely than radar is able to do. Although not available for day-to-day operations, on day of launch aircraft support may be provided. (See section A8.4.3)

### 1.2.5    Other Observations and Forecast Assets

A variety of tools is available to assist forecasters in making effective use of the large amount and variety of information provided by the weather instrumentation. These include systems for displaying the information in graphical or tabular form, and automated or semi-automated tools for forecasting the threat of specific weather events.

### 1.2.6 *Numerical Weather Prediction*

The use of computer models has become an integral part of modern weather forecasting, and has lead to substantial improvements in the accuracy and lead time of forecasts. The 45WS uses the models run by the NOAA National Centers for Environmental Prediction as well as models run by the U.S. Navy and European governments. In addition, it has access to a high resolution local model run by the National Weather Service Office in nearby Melbourne, FL. These models all provide guidance which the forecasters may weigh and evaluate to inform their forecasts.

## 1.3 The Organization of Weather Support to the Shuttle Program and the Eastern Range

All operational weather support to launch and ground processing operations at KSC and CCAFS is provided by the 45th Weather Squadron housed at Range Weather Operations in the Morrell Operations Center at CCAFS. Operational support for Shuttle landings (at any location including KSC) and on orbit operations is provided by the Spaceflight Meteorology Group (SMG) which is operated by the National Weather Service at Johnson Space Center (JSC) in Houston, TX. Weather support for Shuttle is split between SMG and 45WS because they have different customers and different mission requirements. SMG serves the Flight Directors at JSC and is co-located with them, while the 45WS serves the Launch Director and Shuttle Processing Directorate, both located at KSC. The primary responsibility for Shuttle weather support transfers from 45WS to SMG as soon as the Space Shuttle Vehicle leaves the launch pad and returns to 45WS upon "wheels stop" on landing.

In 1991 NASA, the Air Force and the National Weather Service collaborated to establish the Applied Meteorology Unit (AMU) co-located with Range Weather Operations at CCAFS and managed by the KSC Weather Office. The AMU provides technology development, evaluation and transition services to the 45WS, SMG and the NWS office at Melbourne, FL as well as directly to the Shuttle program. Many of the highly specialized tools used for weather support to both manned and unmanned operations on the Eastern Range and at KSC were developed or transitioned to operations by the AMU (Bauman et al., 2004).

The KSC Weather Office (KSCWO) has an agency-wide responsibility to assure that weather support to NASA spaceflight missions and operations, manned or unmanned, is adequate. This involves making sure weather requirements are well formulated and well stated by the NASA programs and properly understood by the weather support providers. The KSCWO manages or provides oversight to KSC-owned weather infrastructure, conducts or manages weather research in support of operations, and manages KSC budget line items directly related to weather support. Finally, the Office serves as a liaison between the weather support service providers and NASA managers, performing the functions of a staff meteorology office to NASA programs operating at KSC or CCAFS and the KSC Center Director.

## 1.4 Infrastructure and Organization at Other Ranges

The Western Range Weather System consists of 26 fully instrumented wind towers, six 915 MHz Boundary Layer Doppler Radar Wind Profilers/mini Sound Detection and Ranging System sites, a modernized Advanced Weather Interactive Processing System, a WSR-88D NEXRAD weather radar, a Lightning Location and Protection System, and two Automated Meteorological Profiling and Real Time Meteorological Profiling Systems. This entire suite of 173 instruments is used, in conjunction with contracted weather aircraft, to evaluate Lightning Launch Commit Criteria during launch operations. The Western Range does not have a network of electric field mills.

# Chapter 2    The Early Years (through 1969)

## 2.0    Background

U.S. Government-supported research into rocket-propelled vehicles began during World War II and accelerated with the capture of German V-2 rockets at the end of the war. The V-2 was far more capable than anything the U.S. had developed, and a special launch facility was built at White Sands, New Mexico to support test flights of the captured weapons. The New Mexico site proved inadequate for experiments with multi-stage versions of the V-2 since these long range flights would overfly populated areas.

After a number of studies involving considerations of everything from orbital mechanics to the logistics of transporting launch vehicle components to the launch site, the Joint Long Range Proving Ground was established in 1949 by the Department of Defense on the site of the old Banana River Air Station at Cape Canaveral, Florida. In 1951 this was renamed the Air Force Missile Test Center and in 1964 it was again renamed, this time to the Air Force Eastern Test Range. [Source: http://en.wikipedia.org/wiki/Eastern_Range retrieved 25 February 2010.] The Range consisted not only of the assets located at Cape Canaveral, but also of a network of down-range tracking stations and ships including stations at Antigua, Ascension Island and Argentia, Newfoundland. The best known tracking ship was the specially designed Vanguard which supported many of the Apollo launches and also made a number of goodwill stops at ports where the public was invited to tour the ship.

On 29 July, 1958, President Dwight D. Eisenhower signed Public Law 85-568, creating the National Aeronautics and Space Administration, also known as NASA. In July 1962, the agency established its Launch Operations Center adjacent to the Air Force test facility at Cape Canaveral on Florida's east coast, and renamed it in late 1963 to honor the president who put America on the path to the moon. The source for the material on the history of KSC in this section was <http://www.nasa.gov/centers/kennedy/about/history/timeline/1950.html accessed 25 March 2010>. Also see Benson and Faherty (1978).

On 24 Aug 1961 NASA announced that it intended to expand the Cape Canaveral facilities for manned lunar flight and other missions requiring advanced Saturn and Nova boosters by acquiring 80,000 acres of land north and west of the Air Force Missile Test Center facilities at the Cape. The U.S. Army Corps of Engineers was designated to act as real estate acquisition agent for NASA, and the Lands Division of the Justice Department was designated to handle the legal aspects.

In July 1963 construction of the Vehicle Assembly Building began. At the time of its construction it was the largest building in the world, with an enclosed volume of 129,482,000 cubic feet. At a height of 525 feet, 10 inches to the top of the finished roof of its high bays it was constructed with 98,590 tons of steel and 65,000 cubic yards of concrete. Two years later, construction of the first stretch of the Crawlerway, between the Vehicle Assembly Building and Launch Pad 39A, was completed. Consisting of two 40-foot (12m) wide lanes separated by a 50-foot (15.2m) median, the Crawlerway was designed to support the 17,000,000 pound (7,727,300 kg) load of a Transporter carrying a Mobile Launcher and Apollo-Saturn V.

In January 1966 the first of three Mobile Launchers was moved into the Vehicle Assembly Building, which was essentially complete. Weighing in at 10,500,000 pounds, the Mobile Launchers functioned as erection platforms for the Apollo/Saturn Vs and launch stands for the vehicle at the pads. These same refurbished Mobile Launchers are used in the Space Shuttle program and are called Mobile Launch Platforms. In May of that year, the first Apollo/Saturn V, a facilities test model, was rolled out of the Vehicle Assembly Building atop a 6,000,000 pound Transporter, a public sign that the Apollo Program was well under way. Each of the two Transporters, built by the Marion Power Shovel Co., had a load capacity of 12,000,000 pounds, and when loaded, could traverse the Crawlerway between the VAB and the pad at a top speed of one mile per hour.

Today, these same "Crawler-Transporters" are used to move the Space Shuttle stacks atop their Mobile Launcher Platforms out to the pads for launch.

From the earliest days of aviation it became obvious that safe flight through the atmosphere depends critically on understanding the weather and avoiding its hazards. Although the flight of a spacecraft through the atmosphere is much briefer than that of an airliner, it is even more vulnerable to weather threats. What is less obvious is that ground processing of launch vehicles and their payloads is also extremely weather sensitive.

The major concerns for launch operations are winds and lightning. During ascent, the wind exerts both forces and torques on the rocket. In order to fly the desired trajectory, the vehicle must adjust its steering to account for the effects of the wind. In extreme cases, the vehicle may not have sufficient steering capability (called "control authority") to make the required trajectory adjustments. As a result, the payload could end up in an unacceptable orbit. In some cases, flying the required trajectory may result in aerodynamic stresses that exceed the strength of the vehicle, resulting in its destruction. Even the surface winds at the launch pad are a concern. As soon as the rocket lifts off, and before it has risen above any structures on the pad such as umbilical towers, strong winds could push the rocket into these structures causing destruction of both the vehicle and the pad.

Lightning is a concern for both launch and ground processing operations. During ascent, the absence of nearby natural lightning does not indicate the absence of a threat. If the flight path passes through electrically charged clouds, the rocket can initiate a lightning discharge that would not have occurred naturally. Avoiding such "triggered lightning" is the major function of the Lightning Launch Commit Criteria. A lightning strike can destroy a launch vehicle. The loss of Atlas Centaur 67 (see section 4.3.2 below) is an example. Natural lightning striking near the launch pad can damage electronic circuits in a payload, booster or ground processing equipment even if the lightning does not attach directly to the pad structure. This is because the electromagnetic fields from nearby strikes can induce harmful currents in nearly any network of electrical conductors. See the discussion of Gemini 2 in section 2.3 below for an example.

There are other weather concerns that can impact ground processing or launch, including heavy rainfall, extreme temperatures and reduced visibility, but these will not be discussed further here. The important point is that the weather infrastructure, weather support organizations and processes, and weather-related constraints to operations described in this History are all required for safe operation of any spaceport. The observations and lessons learned at KSC, the Eastern Range and other locations described here should be applicable to any spaceport anywhere.

## 2.1    Infrastructure

The early weather support infrastructure at the Eastern Range included not only trained weather observers, weather balloons and wind towers, but what for the time was an advanced technology, weather radar to detect and track precipitation. During the 1950s and 60s, the Range used a 3cm radar designated CPS-9 (Boyd et al., 2003). Its short wavelength made it quite sensitive and capable of detecting very light precipitation and even non-precipitating clouds. However, its short wavelength also caused serious attenuation in moderate to heavy precipitation (*ibid*).

To support the 1960s manned spaceflight program, NASA constructed a 150 meter tall tower instrumented at seven levels to measure wind speed and direction, temperature and relative humidity. This provided vertical profiles of the properties of the atmosphere from the surface to a height near that of the top of the Apollo-Saturn launch vehicle, thus enabling the entire environment to which the vehicle was exposed on the launch pad to be characterized.  This tower was located northwest of Shuttle launch complex 39 (LC39) and east of the north end of the current Shuttle Landing Facility (SLF). Additional 18 meter towers were located at LC39 (Johnson, 1976). When the SLF was built for the Shuttle program, three 10m towers were located respectively near the north, center and south ends of the runway.

In addition to these towers dedicated to launch and landing support, NASA built a "mesonet" of surface wind towers on or near KSC while the Air Force built a similar network of 14 towers on and near Cape Canaveral Air Force Station. The surface tower networks were used for forecasting local weather such as thunderstorm initiation and for support to ground processing operations and facilities. As discussed in Chapter 4, Section 4.1, these two networks were later combined. Observations of current weather, cloud type and coverage, visibility and current weather were made by human weather observers (see Appendix VIII, section A8.3.2).

Prior to the Apollo 12 incident, eight radioactive probes and eight corona current sensors were used to measure the atmospheric electric field (or potential gradient). The principle of operation of the radioactive probes assumes that a portion of the ions of one sign created by a radioactive source close to a conductor will be carried away by the field and a portion of the other sign will be returned to the sensor, depending upon the polarity of the field, until an equilibrium current flow is established. The output can be used as a measure of the field intensity and was recorded on strip chart recorders. The corona current sensor consisted of a 4-foot long whip antenna that was connected to a microammeter. This detector measured the current that flowed to or from the antenna in a large electric field, again roughly in proportion to the ambient field. All information was available in the weather office in real time. The primary disadvantages of both of these sensors were that they exhibited a non-linear response to the external field, and they were sensitive to precipitation, wind, and rapid field changes in a thundery environment. The sensors were also mounted close to buildings and on towers rather than on uniform terrain that was cleared of vegetation.

Ground-based assets did not provide the only weather observations on launch day. Weather reconnaissance aircraft were used at least as far back as the Apollo program. The exact landing site for Apollo 11 was retargeted based on aircraft confirmation of satellite data that indicated that a thunderstorm with tops to 50,000 feet (15.2 km) was in the vicinity of the original landing target [Source: The Rescue of Apollo 11, Noel A. McCormach, Center for the Study of National Reconnaissance, accessed at <http://libertyyes.homestead.com/Hank-Brandli-25.html>]. A variety of aircraft provided by NOAA and the DoD were used for launch support to Apollo, the Apollo-Soyuz Test Project and Skylab <http://www.srh.noaa.gov/smg/apollo.htm>. These practices have continued into the present and reconnaissance aircraft routinely support both manned and unmanned spaceflight.

Preventing damage due to a lightning striking a spacecraft or ground support equipment at the launch pad is critical. If lightning strikes near a launch pad while a vehicle is present, it is highly desirable to know the exact location and intensity of the strike. The need to retest sensitive electronics can depend on the polarity, magnitude and rise time of the lightning current. The most important infrastructure solution for the Apollo launch pads was an insulated, folding mast mounted atop the hammerhead crane that capped the launch umbilical tower (LUT). This mast held up two overhead wires that were grounded several hundred meters from the mobile launch platform. A lightning rod on top of the mast extended above the vehicle sufficiently so that the lightning current would travel down one or both of the catenary wires and go into the ground without harming the vehicle itself. The hammerhead crane was the highest thing on the launch platform, and was capable of being extended over the top of the Saturn V vehicle (Benson and Faherty, 1978, Ch. 13).

This solution was only partially effective. Although no discharge ever struck the vehicle directly during the Apollo program, lightning did strike the LUT many times. To determine if a diverted or nearby strike may have induced sufficient voltages or currents to potentially damage the vehicle, payload, or ground support equipment, sensors were installed where the catenary wire went to ground and around the pad. In addition, three automatic TV cameras were installed around the periphery of the pad to help verify that lightning did not strike the vehicle and to estimate the strike point either by showing actual location or by triangulation.

## 2.2 Organization

Weather support for all phases of Project Mercury (1960 – 1962) from ground processing through launch to landing was managed by the Project Mercury Weather Support Group (PMWSG) of the US Weather Bureau's (USWB) National Meteorological Center (NMC) at Suitland, MD. In addition to the Suitland Office, there were PMWSG Units at Patrick AFB, FL and the National Hurricane Center (NHC) in Miami, FL.

Several high level reorganizations and changes in nomenclature took place between 1962 and 1966 as the space program grew beyond Mercury. The USWB became the Environmental Science Services Administration (ESSA) and the PMWSG became the Spaceflight Meteorology Group (SMG). The Suitland and NHC Units remained in place, but the Patrick AFB Unit was moved up the coast to Cape Canaveral/KSC and renamed the Cape Unit, while a Manned Spacecraft Center Unit was established in Houston, TX. A description of weather support during this period may be found in Nagler (1966). Under the 1963 Webb-McNamara agreement between NASA and DoD, the USAF provided standard weather observation and forecasting services to all Range users, and NASA provided specialized support through SMG.

During the Apollo era, ESSA became the National Weather Service (NWS) but the organization of SMG and its various units remained relatively stable. The NWS did pick up some responsibilities for coordinating weather support to NASA and Department of Defense (DoD) unmanned test ranges in the US and the Pacific.

## 2.3 Operations

Although a major focus of the LLCC today is on natural and artificially-initiated or *triggered* lightning, America's space program was concerned almost immediately with natural lightning and its destructive potential due to a steady stream of lightning damage reports beginning with Mercury and Gemini programs. Some of the research performed and the techniques developed to avoid and mitigate damage from natural lightning during ground processing helped later when the space program became aware of the even greater risk posed by rocket-triggered lightning during ascent.

*Gemini 2 & 5* On 17 Aug 1964 while the Gemini 2 launch vehicle (GLV-2) was being tested on launch complex 19, several observers reported a lightning strike at or near the pad. All testing was halted for a thorough investigation of the event. Inspection revealed no physical markings of any kind but disclosed a number of failed components, mostly in ground equipment with some in GLV-2. This indicated that complex 19 had not been hit directly. The damage was attributed to indirect electromagnetic effects. [Source: <http://history.nasa.gov/SP-4002/p2b.htm accessed 24 Mar 2010>] The strike, coupled with other problems, resulted in postponement of GLV-2 until 19 Jan 1965. Subsequently, Gemini-Titan-5 was also postponed when a similar near miss affected the onboard computers, resulting in a two day delay (Grimwood et al., 1969)

*Apollo I* Natural lightning immediately impacted the Apollo program. On May 25, 1966, the first Apollo vehicle, a full-scale mockup of a Saturn V moon mission vehicle, identical in size and weight to the operational vehicle, rolled from the Vehicle Assembly Building to the Launch Pad. Two days later lightning struck the LUT on the Mobile Launcher (ML) initiating a series of failures which ultimately caused an object to contact Apollo's second stage (Durret, 1976).

The damage caused by the Apollo I accident immediately raised interest and concern about lightning and its hazards, especially during Apollo rollout and pad processing, and led the Apollo program to develop a weather infrastructure with unique sensor systems, configurations, and data exploitation techniques to mitigate the potential damage from natural lightning. This unique infrastructure proved to be even more valuable in a few years when NASA unexpectedly encountered another kind of lightning—*rocket triggered lightning*.

*Apollo XII* Through 1969, the only lightning launch constraint was "do not launch through a thunderstorm." This inadequate constraint led to a rude awakening and near disaster. Any complacency with lightning was shattered on 14 Nov 1969 during the ascent of Apollo XII, America's second manned attempt to land on the moon. As launch time approached, a southward moving cold front was passing thru central Florida producing cloudy skies and rain showers. Radar indicated *cumulus congestus* cloud tops in the Cape area ranging from 18,000to 23,000 feet (5.5 – 7.0 km) well above the 12,400 foot (3.8 km) freezing level. Although no lightning was reported in the Cape area before the launch, potential gradient measurements from the eight sensors (installed in the mid to late 1960s) were varying significantly thus implying rapidly and highly fluctuating electric fields aloft over the launch area before T-0. For example, readings at Site 8, adjacent to Launch Complex 39, varied from +6 to -11 (units unknown) in the 20 minutes before launch (NASA, 1970, *pp12-17*). MSC Mission Control did not want to delay the launch, especially with President Nixon and wife Pat in attendance. Technically, the only applicable LLCC was not to fly into thunderstorms and it was not violated because no thunder was being heard. Nevertheless the decision was extremely unwise given the information available which suggested strongly that the environment was electrically active.

Thirty six seconds after liftoff at 1122 EDT, Apollo XII was jolted. As the Saturn V rocket passed through 6500 feet (2 km) a lightning strike disconnected the service module's fuel cells that produced power for the command module. This frightening event was followed sixteen seconds later by another lightning strike at an altitude of 14,500 feet (4.4 km). This strike upset the Command Module's navigation system, causing temporary loss of the system's stabilizing platform and denial of critical flight data to Mission Control at MSC. Flight Director Gerry Griffin considered an abort while experts on the affected systems considered alternatives. A young engineer recalled a similar anomaly during a training simulation a year earlier and recommended resetting a switch to restore communications. Based on this recommendation Mission Control told the crew; "Flight, try SCE (Signal Condition Equipment) to Aux". Astronaut Alan Bean positioned the SCE switch, which was located near his seat, as instructed. That restored flight telemetry to JSC's computers. Fortunately, the Saturn V booster's systems, including navigation and communications, had not been affected by the lightning strikes.

Because nobody in the Apollo program had anticipated that the vehicle might be struck by lightning under non-thundery conditions, or understood that lightning discharges could be initiated just by flying a vehicle and its exhaust plume into a highly electrified cloud, the first efforts were simply to understand what had happened to the spacecraft and whether the astronauts should be allowed to continue their mission. This responsibility fell to the Mission Evaluation Team at MSC and its Head, Donald D. Arabian, who quickly determined that permanent damage to Apollo XII was minimal and that it was OK for the astronauts to continue their mission.

Figure 2.3.1 shows a sketch of the clouds that were present at the time of the launch of Apollo XII and the lightning strikes that were triggered by the vehicle together with many of the weather measurements that were considered in the analysis of the incident (taken from Brook et al., 1970).

Appendix VI describes the timeline of events that followed the Apollo XII lightning incident and lists the scientific panels and reviews that provided a basis for a multi-center NASA Report in February 1970 (NASA, 1970). The MSC report describes the weather conditions that were present at the time of the launch, the relevant instrumentation, and shows photos of the CG strokes that were recorded near the launch pad. It also describes the ways that the lightning current affected various components of Apollo XII and the mechanisms whereby this current caused transient upsets and damage. The report ends with a conclusion that the best way to avoid a similar incident in future Apollo missions is to avoid natural lightning or highly electrified clouds during launch rather than attempting to harden the vehicle. In order to avoid these hazards, several new Lightning Launch Commit Criteria (LLCC) were adopted that were based largely on the weather conditions under which the ESSA forecasters at KSC had previously observed high electric fields. These LLCC were used for the first time during the launch of Apollo XIII and continued through the launch of Apollo XVII and the subsequent Skylab missions.

Post analyses of the Apollo 12 incident revealed a new danger, triggered lightning, that had not been previously assessed in space program meetings on procedures and constraints, but the hazard itself was already known to the aviation community [Fitzgerald (1967; 1970)].

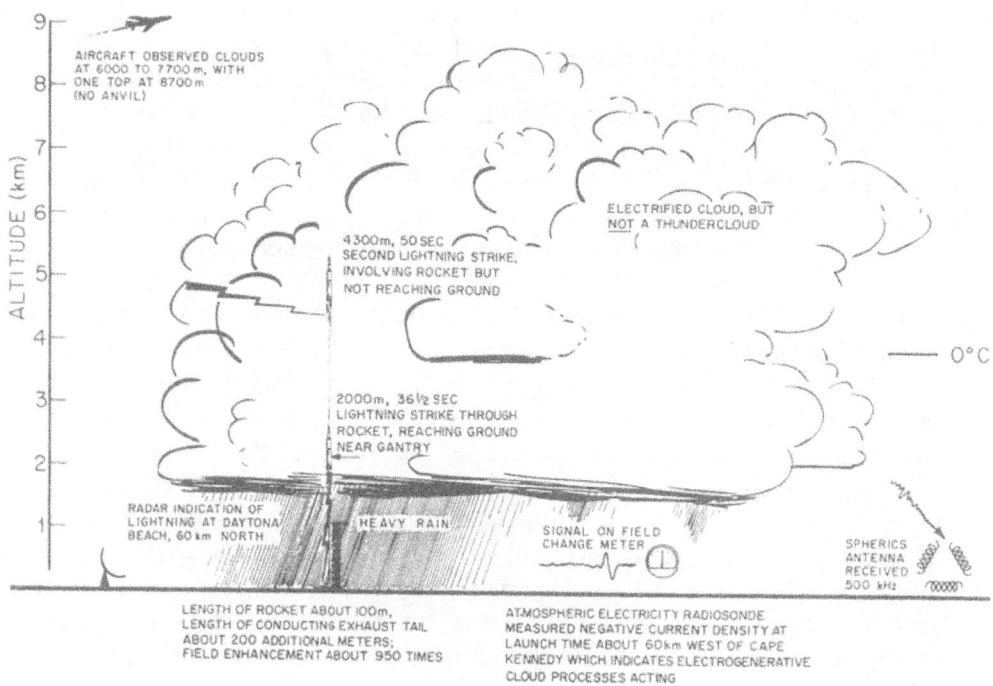

Figure 1 — Electrical phenomena—Apollo 12 launch, November 14, 1969

**Figure 2.3.1 Sketch of meteorological conditions during the Apollo XII incident**

## 2.4    Research

As we shall see in the next chapter, the Apollo XII incident spurred a great deal of lightning research in the aerospace community. There had been some previous research. Measuring electric fields aloft, especially inside a cloud, is an extremely challenging undertaking which is difficult even today after significant advances in both science and technology. Nonetheless, as early as 1957, Clark (1957) measured fair weather electric fields as a function of altitude using four rotating-vane field mills mounted symmetrically on the wing tips and the top and bottom of the fuselage of a P4Y aircraft. Similarly, in the 1960s Fitzgerald (1965, 1977) used rotating-vane field mills mounted on C-130 and F-100 F aircraft in conjunction with a U-2 high altitude aircraft to measure the electric fields simultaneously above and in thunderstorms at KSC. He used symmetric field mill mounting and electronic compensation to minimize the effects of aircraft charging on the measurement of the ambient field. For one cloud Fitzgerald (1965) studied, the strongest electric field was associated with the region of highest radar reflectivity with the vector field apparently pointing toward (or away from) the centers of charge.

When considering these early works, it is important to remember that through the early 1970s, the aircraft measurements were being recorded on strip-chart recorders or on analog magnetic tape recorders carried on the aircraft or sometimes telemetered to the ground; the radar data were recorded photographically; and measurements of the cloud microphysics were made by subsequent analysis of particles that had impacted on formvar replicator films, oil-coated glass slides, a moving belt of soft metal foil, or "snow sticks" as well as viewing the particles impacting upon the windscreen. The enormous progress that has been made in our understanding of the electrical structure of thunderstorms and the associated lightning over the past 3 to 4 decades is mainly due to the advances that have been made in analog and digital electronics, electronic data storage capabilities, and display technology.

The early efforts to examine the electrification of clouds in the lower atmosphere were not limited to passive examination of existing terrestrial clouds. On 25 April 1962 Saturn C-1 SA-2 lifted off from Cape Canaveral in the second suborbital test flight of the Saturn I configuration. The payload, designated "Project Highwater", was 86 tons of water ballast that was deliberately dumped into the ionosphere at an altitude of 105 km to determine the effects of a sudden release of a large volume of water at high altitude. The water cloud continued to rise on an inertial trajectory to an altitude of 145 km and expanded to several km in diameter. A similar test was conducted on 16 November 1962 during flight SA-3. Both tests created "lightning-like" radio disturbances that were recorded on the ground. [Source: http://en.wikipedia.org/wiki/Project_Highwater accessed 22 February 2010]

# Chapter 3   The Aftermath of Apollo XII (1969 – 1980)

## 3.0     The Apollo XII Investigations

The Apollo XII lightning investigation team concluded that although the Saturn V design contained safeguards against electrical discharges, lightning could damage spacecraft electronics such as solid-state devices. Nevertheless the team opposed any spacecraft modifications due to impacts to schedule, cost, and capability (due to added weight) but instead recommended changing the LLCC to significantly lower the risk of being struck by lightning again. The following month, atmospheric electricity scientists, while participating in the December 1969 AGU meeting, designed new rules which attempted to identify/avoid all hazardous weather associated with atmospheric electricity (Benson and Faherty, 1978).

NASA estimated the new Lightning Launch Commit Criteria (LLCC) would increase the probability of a February delay to 18% from its existing 10%, and the probability of a summer scrub to 18% from its existing 3%. The new LLCC were used for Apollo's remaining moon missions, and until 1988 for Expendable Launch Vehicle (ELV) missions (NASA, 1970).

### Post Apollo 12 Lightning LCC (LLCC) 1970

*Space vehicle will not be launched if nominal flight path will carry vehicle:*
Within 5 statute miles (sm) of a cumulonimbus (thunderstorm) cloud;
Within 3 sm of anvil associated with a thunderstorm;
Through cold front or squall line clouds which extend above 10,000 feet;
Through middle cloud layers 6000 feet or greater in depth where the freeze level is in the clouds;
Through cumulus clouds with tops at 10,000 feet or higher.

The Apollo 12 lightning incident brought into clear focus many of the scientific questions that were still not understood about the small- and large-scale processes that create high electric fields inside both thundery and non-thundery clouds, and also about the physics of lightning discharges. Because of this attention and NASA's urgent "need-to-know," program managers like James Hughes in the Office of Naval Research (ONR), Ronald L. Taylor in the National Science Foundation (NSF), and others in ESSA and NASA were able to secure funding to enhance their existing investigations and support new studies of the salient phenomena. Measurements of the optical temperature of the Apollo 15 exhaust plume showed peak values that were near 2370 °C (Pifer and Krider, 1972), and the presence of such temperatures in the plume, combined with trace amounts of sodium in the fuel, had major implications for artificially-initiating or "triggering" lightning (Krider et al., 1975). In section 3.4, we will review additional studies that were undertaken at KSC by a number of investigators both inside and outside NASA as part of the Thunderstorm Research International Program (TRIP).

In subsequent years, as new techniques became available for research - particularly the development of large-scale integrated circuits for both analog and digital electronics - huge advances were made in computers, memory technology, and communications. These advances in turn led to major improvements in the ability of scientists to conduct experiments in atmospheric electricity, display the results, and analyze the data. As new knowledge became available, the LLCC were revised, updated, and improved to increase both launch availability and launch safety.

Among the scientific advances that were made in the early- and mid-1970s, largely as a result of the Apollo 12 lightning incident, were the development of (a) an improved ground-based method for detecting and displaying cloud electric fields and (b) improved methods for detecting, locating, and displaying both intracloud and cloud-to-ground (CG) lightning discharges. These advances were tested at KSC and were the basis for many of the subsequent improvements in the LLCC.

## 3.1 Infrastructure

Except as noted below in this section, the weather infrastructure at KSC/CCAFS remained the same as that presented in Section 2.1 during the period covered by this chapter.

During the 1970s, the CPS-9 3 cm weather radar at the Eastern Range was replaced with a 5 cm AN/FPS-77 located atop the Range Control Center on Cape Canaveral, but this location resulted in serious RF interference to sensitive spacecraft (Boyd *et al.*, 2003). This location also placed many of the launch sites within the radar "cone of silence" centered on the radar site. That deficit was remedied by using data from the National Weather Service WSR-57 radar at Daytona Beach, FL about 50 miles (80 km) north of Cape Canaveral, and the FPS 77 radars at KSC and Patrick AFB about 30 miles (50 km) south of the Cape.

By the mid 1970s, a network of ground-based electric field mills had been established and designated the Launch Pad Lightning Warning System (LPLWS), a name that continues until this day although the details of the sensor design have changed. For more details and a picture, see Appendix VIII, section 8.1.1. If forecasters observed the surface fields increasing to above fair weather values of ~200-500 V/m, and other data indicated the fields may continue to rise and exceed 3000 V/m, they would issue a warning for lightning within 5 nm (9 km) as soon as needed to provide at least a 30 minute lead time before the first strike. The 3000 V/m surface threshold magnitude was based on theoretical and empirical data which related surface fields to the fields aloft required to initiate lightning. The field mill system could also be used to estimate charge center locations and accompanying fields, and indicate lightning was present in the local area, but could not locate lightning impact points.

During the early years of the LPLWS the calibration methodologies had not matured and data from this period was not always quantitatively precise to the degree it is today. The Secretariat of the Thunderstorm Research International Program (see Section 3.4.4 below) was notified of some concerns regarding the accuracy of the data in a letter dated 15 September 1976 (unpublished). The authors, C.C. Moore and Stephen Friberg, well-respected scientists in the field of atmospheric electricity, had conducted a series of comparisons between LPLWS sensors and a portable, research grade field mill. They reported their results and made some recommendations for improved calibration procedures.

The use of surface field measurements to infer field strengths aloft becomes less reliable for clouds either above 3050 meters or not over the field mill network (Arabian, 1976, page 2). A valuable ally was a Lear jet stationed at Patrick AFB equipped with field mills, that could fly where required and directly measure the fields aloft. The Learjet's 45,000 foot (13.7 km) service ceiling, 464 knot (240 m/s) cruising speed, 6800 ft/min (35 m/s) climb rate, and ability to accommodate three passengers plus instrumentation made it an ideal platform for helping improve lightning warning accuracy.

## 3.2 Organization

The time between the end of the Apollo program in the mid to late 1970s and the beginning of the Shuttle Program early in the 1980s was a time of transition for weather support to manned spaceflight. The NWS responsibilities for coordinating weather support to NASA and Department of Defense (DoD) unmanned test ranges in the US and the Pacific that were assigned in the late 60s had ended by 1975.

By 1978, SMG operations were consolidated at JSC, and support for ground processing and launch activities at the Kennedy Space Center was assumed by the US Air Force at Cape Canaveral Air Force Station. SMG remained responsible for support to on-orbit and landing operations world-wide including Return to Launch Site (RTLS) landings at KSC. This division of labor has remained in place until the current time and appeared to have been adopted by NASA for the Constellation Program that was scheduled to replace the Shuttle

Program. The status of Constellation Program components such as the Orion crew capsule is uncertain at the time this is being written. An in-depth description of Shuttle weather support is found in Bellue et al. (2006).

## 3.3 Operations

### 3.3.1 Apollo XIV – Apollo XVI
Apollo XII was not the end of Apollo's lightning troubles. Apollo XIV was delayed 40 minutes by thunderstorms near the launch pad. Apollo XV fared even worse. While it was at the pad, several lightning strikes damaged two spacecraft sensors and ten ground support equipment systems in June and July 1971. Finally, lightning struck the pad twice in March 1973 but Apollo XVI remained undamaged thanks to mitigation techniques adopted after Apollo XV.

### 3.3.2 Skylab
Five months after the last Apollo moon mission, Apollo XVII, the first Skylab mission launched on 14 May 1973. Skylab 1 was an unmanned mission that was followed by four manned Skylab Station missions. The Skylab LLCC were identical to the new post Apollo 12 LLCC, except Skylab combined the first two criteria into one rule.

**Skylab (1973) LLCC (Arabian, 1976)**
*Space vehicle will not be launched if nominal flight path will carry vehicle:*
- *Within 5 sm of a cumulonimbus (thunderstorm) cloud or within 3 sm of an associated anvil;*
- *Through cold front or squall line clouds which extend above 10K feet;*
- *Through middle cloud layers 6000 feet or greater in depth where the freeze level is in the clouds;*
- *Through cumulus clouds with tops at 10,000 feet or higher*

**Skylab 2**: Natural lightning problems during ground processing operations, similar to those which had threatened Apollo's XIV, XV, and XVI, continued with the remaining three Skylab missions. Lightning struck Skylab 2's launch facility structure damaging four spacecraft sensors in May 1973. Then thunderstorms, which had continued from the previous day into the night and morning hours, threatened Skylab 2's countdown. However by launch time they were far enough west to allow a safe launch.
**Skylab 3**: Lightning struck the launch pad damaging ten spacecraft sensors, eight ground sensors and three ground service equipment (GSE) systems in June 1973.
**Skylab 4**: Among other equipment, a 200KA strike damaged the guidance and navigation inertial measurement unit, a signal conditioner and four spacecraft sensors in August 1973 (Arabian, 1976).

### 3.3.3 The Apollo-Soyuz Test Project (ASTP)
The ASTP deserves special attention and is discussed at length here. The triggering of lightning by Apollo 12 in 1969 alerted the community to the threat of launch vehicles (and their ionized exhaust plumes) triggering lightning when launched into clouds containing strong electric fields. This in turn, raised questions about which clouds produce these fields and where these fields are located inside those clouds, particularly at the Kennedy Space Center where thunderstorms and lightning are frequent. Plans had been made for an earth-orbiting rendezvous of an Apollo vehicle with a Soyuz vehicle to take place in July 1975. Both launches were planned to take place on the same day. The Soyuz vehicle would launch first from the Soviet Union, and if that was successful, the Apollo would launch 7 ½ hours later from KSC. Because of the low Earth orbit, the time-window for a successful Apollo launch was only 5 - 8 min (Arabian, 1976), and it needed to take place at 3:50 PM local time on July 15, the diurnal maximum for lightning at KSC in a month of peak lightning activity.

Given the international politics of that era, NASA felt intense pressure that this Apollo launch not be delayed unless it was absolutely necessary (Kanter, 1975). At that time, a large network of surface electric field sensors was in place at KSC to identify impending lightning hazards to space vehicles, but "very little data were available to correlate the field intensities aloft to those measured at ground level". Consequently a research program was conducted jointly by KSC and NOAA using four aircraft to make airborne electric field measurements at different altitudes in clouds over the surface network (Arabian, 1976). The four aircraft were the New Mexico Institute of Mining and Technology (NMIMT) Schweizer powered glider instrumented with rotating-vane field mills, a NOAA T-29 aircraft instrumented with 2 cylindrical field mills, one on the nose and one atop the fuselage (Kasemir, 1972); a NRL/NOAA S-2D instrumented with 2 cylindrical field mills, and a NASA C-45 (NASA 6) instrumented with 2 cylindrical mills. The C-45 was also equipped to dispense chaff if it was necessary to reduce the electric fields over the launch path just before the Apollo launch. The results from the aircraft study "indicated that, below an altitude of 3050 meters, the airborne data compared favorably with the ground instrumentation. Above this altitude, however, the ground-based sensors could not be relied upon to provide an accurate measurement of the electric field intensities aloft" (Arabian, 1976). Therefore airborne measurements of electric field were deemed necessary to support the Apollo launch.

In order to be able to measure the electric field along the entire portion of the flight path of the Apollo vehicle where it might be subject to lightning strikes, four aircraft in addition to those used in the KSC/NOAA study were used: A NASA/Ames Learjet and a NASA/JSC T-38, both instrumented by the Stanford Research Institute with rotating-vane field mills, an Air Force C-130 instrumented with rotating-vane mills by Air Force Cambridge Research Laboratory (AFCRL) and a RF-4C from Kirtland Air Force Base instrumented by the Air Force Weapons Laboratory. A detailed flight pattern with exit and entry points for all aircraft was worked out for the hour immediately preceding the launch. The flight pattern and altitude assignments of the eight aircraft are shown in figure 3.3.3-1.

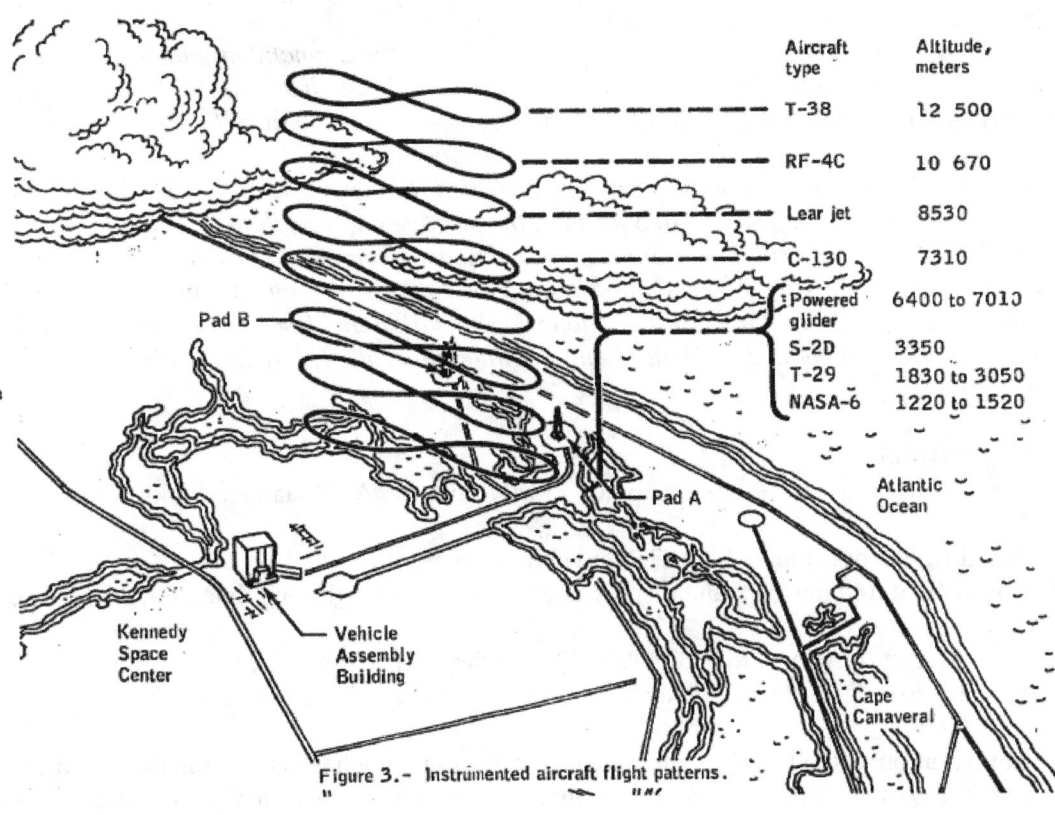

Figure 3.- Instrumented aircraft flight patterns.

**Figure 3.3.3-1  Flight plan to determine electric fields prior to the Apollo-Soyuz Test Project**

Note:  [Reproduced from Arabian, 1976]

For the 10 days prior to July 15 there had been thunderstorms every afternoon with much rain and lightning, some hail and even one tornado, but the evening before the scheduled ASTP launch the prevailing winds changed from southwest to southeast. This greatly reduced the potential for thunderstorms at the time of the launch. Although there were showers in the early afternoon on July 15 (Kanter, 1975), the launch proceeded on schedule at 3:50 PM local time (Arabian, 1976).

The conclusions drawn from these studies and from the multiple aircraft instrumented for the ASTP launch were that launch delays due to the presence of clouds could be avoided by measuring the electric field aloft along the flight path just prior to launch using multiple instrumented aircraft and that further research was required to characterize the electrical properties of various cloud types (Arabian, 1976).

The ASTP launch rules were as follows (Heritage, 1988, page 4-4):

**Apollo-Soyuz Lightning LLCC 1975**
*Space vehicle will not be launched if the nominal flight path will carry vehicle:*
   A. *Through a cumulonimbus (thunderstorm) cloud;*
   B. *Within 5 sm of a cumulonimbus (thunderstorm) cloud or within 3 sm of an associated anvil. This rule may be relaxed at the discretion of the Launch Director if the electric field at the launch pad is less than 1 kV//m with a very narrow launch window*
   C. *Through cold front or squall line clouds which extend above 10K feet;*
   D. *Through middle cloud layers 6000 feet or greater in depth where the freeze level is in the clouds;*
   E. *Through cumulus clouds with tops at 10,000 feet or higher;*
   F. *Rules C, D, and E above may be relaxed at the discretion of the Launch Director when electric field measurements in the launch pad area are stable and measure less than 1 kV/m. [Editor's note: It seems likely that the 1 kV/m threshold was only for a foul-weather polarity field. Explicit use of the absolute value of the electric field was not made in the operational rules until 1991.]*
   G. *Rules C, D, and E above may be further relaxed provided that airborne and ground electric field measurements are less than or equal to 3 kV/m at the surface and less than or equal to a vertical profile of electric field, EC(H), where EC(H) varies linearly from 3 kV/m at the surface to 15 kV/m at 25K feet and remains constant above that altitude. This rule may be applied only if vertical field measurements along the flight path in a 3-mile area are within the described envelope and there are no rapid fluctuations of about 3 kV/m at about 1 minute intervals within the 3 mile area measured by the ground mills.*

### 3.3.4    The Viking Missions
The two launches of Viking spacecraft that took place shortly after the ASTP launch took advantage of three legacies of ASTP.
   1. The modified LLCC developed for ASTP and shown above.
   2. The improved field mill network and display and the improved lightning detection capabilities at KSC/CCAFS.
   3. Three of the aircraft (NASA 6, the NOAA T-29, and the NRL/NOAA S-2D) used during ASTP for the airborne measurements of electric field aloft remained for the Viking launches.

Although the documentation of the LLCC at the time of the Viking launches is unfortunately incomplete, we are including a discussion of the Viking missions because it demonstrates historically the extent to which personnel in the space program at that time attempted to use new measurement techniques and scientifically based approaches to provide safety while still not being overly restrictive. Although there were airborne measurements of electric fields prior to launch, as for the ASTP, the aircraft were required to depart the launch area no later than T minus 5 minutes (Arabian, 1976, p. 9). This left a 5 minute time-gap without airborne electric field measurements. However, there were continuous measurements of the surface electric field from the field mill network, so Heinz Kasemir and colleagues at NOAA used the surface measurements to develop a probabilistic technique for forecasting the movement of the electric field contours that were measured at the surface. (W. D. Rust, personal communication).

There was only a short time to do the research and develop the techniques for interpreting the surface electric field measurements. The main objective was to devise some sort of observationally-based prediction technique that could forecast when high electric fields moving into the launch area would result in natural or triggered lightning. For the ground-based data, the criterion was the probability that a 1-kV/m contour line would reach a five-mile radius around a launch pad, which was the demarcation distance of a no-go for launch. The electrical state of the atmosphere in the pad area was to be evaluated continuously, but with specific emphasis on both ten minutes and five minutes before the scheduled launch. As far as we are aware, no documentation remains that describes the details of this technique.

It was decided to support the upcoming Viking launches using this probability prediction technique. The NOAA/Environmental Research Laboratory/Atmospheric Physics and Chemistry Laboratory in Boulder, Colorado, provided an advisor, W. David Rust, who had experience in measuring atmospheric electric fields. The main responsibility of the NOAA advisor was to inform the KSC weather office if a no-go recommendation for launch was needed given the electrical state of the atmosphere at the time. (W. D. Rust, personal communication)

**Viking 1** was under pressure to launch on-time due to the goal of landing on Mars on the 4th of July 1975 as a major bicentennial event. The launch window was only seconds long due to the planetary trajectory, and this window was on a summer afternoon at the peak of normal thunderstorm activity. The vehicle was launched on 20 August 1975 through thick middle clouds with temperatures approaching the freezing level (Durret, 1976). Even though there were thick mid-level clouds, the launch was permitted because the electric field measurements aloft and at the surface were at acceptable levels. (Arabian, 1976, p. 14)

**Viking 2** was launched from pad 41 on 9 September 1975 through opaque clouds that were estimated to be at a height of 2450 meters (8100 feet) (Arabian, 1976, p. 14). Large convective storms had developed offshore and were moving from the east toward the launch site. There were also thunderstorms over land to the west of the launch site. Just a few minutes before the scheduled launch, it was not clear whether a 1-kV/m contour of the surface field measurements would reach the 5 mile demarcation distance and create a 'no-go' condition for launch. The NOAA advisor used the rudimentary prediction technique and recommended go-for launch, and it was successful. The situation that existed shortly after the time of launch is depicted in Figure 3.3.4-1 below. In the figure, the 1-kV/m threshold is the outer most contour of the electric field. Additional contours are probably at 1 kV/m intervals. Electric field polarities are reported using the potential gradient sign convention, i.e., E = -grad V. It is likely that for the ASTP and Viking launches, only foul weather polarity fields (negative in this plot) were considered when applying the LLCC. Thus the single positive contour line near pad 41 did not delay the launch. Outlines of the radar reflectivity from a base-level scan (reflectivity factor unavailable) are shown as white clouds and the radar-derived cloud tops are as indicated.

About a minute after launch, lightning occurred off the coast nearby. Shortly thereafter, the pad 41 area went red and remained so for a significant time. Both the Viking 1 and Viking 2 launches are examples of the usefulness of real-time electric field measurements aloft and at the surface for optimal launch support.

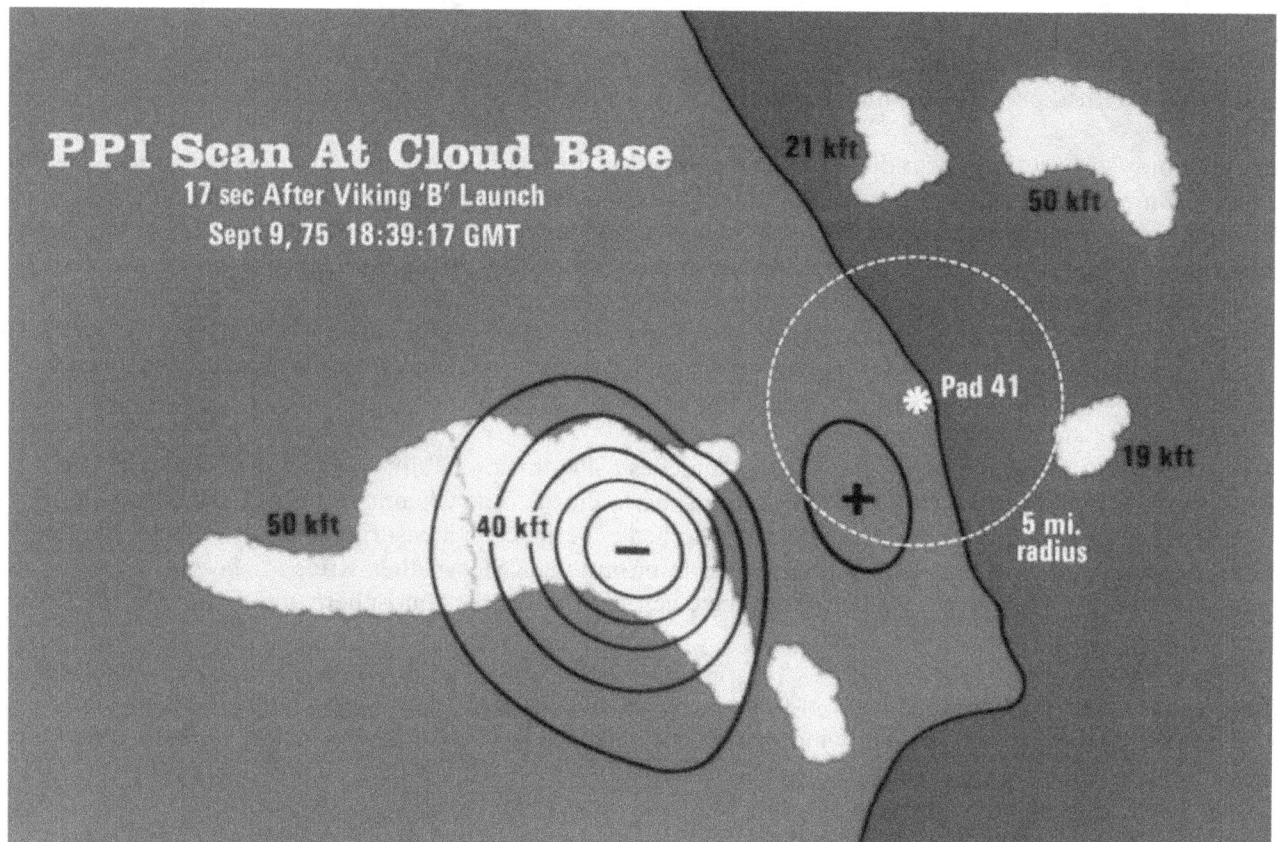

**Figure 3.3.4-1  Contours of the electric field at the ground shortly after the launch of Viking 2**

### 3.3.5  *Atlas Centaur 38 (AC 38)*

Although the manned spaceflight program continued to refine its launch criteria to safely avoid triggered lightning, the ELV community, for unknown reasons, continued to rely only on the launch criteria developed immediately (just one month) after Apollo 12. AC 38 illustrates a close call with the near catastrophic consequences (later realized in the destruction of AC 67) that can result from using faulty LLCC or faulty implementation procedures. In preparing this history, only one AC 38 source document could be located, a memo by James Nicholson that discusses the weather assessments during AC 38 [Source: *Memo for Record, Date: 28 May 1976, Subject: "Consultative Support for Atlas-Centaur Launch 13 May 76", Signed: James R Nicholson.*]. Mr. Nicholson is personally known to several of us as a very professional and credible source. His memo is summarized below.

Background: At the time of AC 38 (13 May 1976), the Air Force provided weather support for CCAFS and the NOAA/National Weather Service (NWS) provided weather support for KSC. The memo was written by a NWS forecaster based on the telephone conversations he and another NWS forecaster had with the Air Force Launch Weather Officer (LWO). The Post Apollo 12 Lightning LCC (LLCC) which governed AC 38 were provided at the end of section 3.0 above.

Significant points in the NWS memo:
1. AF and KSC field mills were different with different scales. Field mills from the two networks had never been collocated to calibrate the differences. Thus NWS did not understand how to compare Range electric field mill readings with KSC readings to determine if the AF sensors were exceeding the NASA LLCC.
2. Although the AF field mill network was operational, the KSC field mill network was down which denied the AF important upstream electric field data.

3. The NASA 6 aircraft was not available for weather reconnaissance.
4. Cumulus began growing late morning into towering Cumulus tops 22,000 feet (6.7 km) by 1145 EDT, and a thunderstorm near Cocoa at 1230 EDT top 44000 feet (13.4 km) drifting slowly ENE and eventually over into Port Canaveral. Convection over Merritt Island eventually shifted from line just W of KSC to along I95 sending Anvil clouds and Mammatocumulus over KSC and CCAFS from 1600EDT thru the 1828 EDT launch. At 1800 EDT radar indicated clouds over LC 39 had bases ~15000 feet (4.6 km) with tops to 33000 feet (10 km).
5. When the Range informed NWS at T-20 minutes (1740EDT) that the launch would occur at the scheduled time (1800 EDT), NWS expressed concern about: A. Anvil clouds overhead; B. Layer of Mammatocumulus which in their past experience had been associated with disturbed cloud electric fields; and C. Radar observations of convection. Thus NWS recommended not launching.
6. AF LWO said electric fields were disturbed but 'seemed to be quieting down'.
7. 1755EDT: Range delayed launch due to weather.
8. 1820 EDT: AF LWO recommended the Range proceed with launch
9. NWS replied radar observations indicated vertical thickness of echoes 'seemed to be decreasing in vicinity of Launch Pad 36. Memo did not mention if NWS had withdrawn their recommendation not to launch.
10. 1821 EDT: NWS detected lightning strike at 040 degrees. Based on thunder, strike was estimated at 10-12 miles distant.
11. 1828 EDT: Successful launch T-0 Observation: 3/10 CB @3000ft (914 m), 5/10 AC @13,000 ft (4 km), 5/10 AC @18,000 ft (5.5 km).
12. LWO tells NWS: A. AC 38 entered clouds ~20,000 feet (6.1 km). B. AF field mills were reading 12 kV/m which he guessed would translate to ~3 KV/m for KSC mills (thus significantly exceeding the NASA threshold of 1kV/M).
13. 14 May 1976 (next day): Weather Commander shows NWS a running time plot of field mill observations (a strip chart). NWS assesses plot as 'indicating disturbed atmospheric (electrical) conditions as one would expect from visual observations of clouds'.
14. Weather Commander mentions a TV camera filming the launch showed a lightning flash descending part way to ground, about 6 seconds after vehicle entered the cloud base.

### 3.3.6  *F11 Europa II*

During the Apollo era, the Europeans were having their own problems with hazardous electrical discharges that had a different cause—triboelectrification. Although this History is based on the American space program, it cannot be completely separated from the world at large, and the Europa incident described here illuminates a significant oversight in the American program at the time. F11 Europa II was launched on 5 November 1971 during benign weather. At T+105 seconds and at 27 km altitude, the launcher lost its attitude control because the guidance computer in the rocket's third stage had ceased to function. Other electrical anomalies lasting about three seconds occurred at the same time. Less than a minute later, the resulting stresses on the first and second stages caused them to explode. This catastrophic accident led to the cancelation of a major European cooperative program.

The resulting investigation concluded electrostatic discharge caused the accident and identified three different mechanisms that could build up excess charge on a booster:

1. Electrostatic induction could cause significant charge to accumulate on the vehicle before launch if the environmental field is high, for example due to a nearby electrical storm. The vehicle will then carry the charge aloft where it will discharge in the lower atmospheric pressure.
   <u>Conclusion</u>: *Ruled out* because this model does not explain the observed series of discharges; and because the weather was benign.
2. The vehicle could become charged by friction due to collisions with snow, ice, and/or cloud particles.

Conclusion: *Ruled out* because of the good weather, and because encountering ice and snow particles at 27 km is rare.

3. The vehicle could become charged because negative and positive ions in the hot booster exhaust have different mobilites and will cause a charge separation and drain current to the metallic nozzle wall. Conclusion: *Accepted* as the cause based on thorough testing and the agreement of results with models and theory. For a thorough explanation of the causes of this accident see Taillet (1974).

Despite the Europa failure and the fact that electrostatic charging was cited as the cause of two Minutemen failures, two Titan guidance anomalies, and the likely cause of two Scout failures (Andrus *et al.*, 1969), the American Space Program did not introduce electrostatic charging and triboelectrification into its LLCC until the 1988 post AC 67 revision. The 1988 Triboelectrification rule recommended by the LAP only addresses collisions with cloud particles (as in Cause 2 above). The LLCC prohibited flight thru clouds at altitudes colder than $-10\,°C$ unless the vehicle has been treated for a resistivity of no more than $10^{-9}$ ohms per square. Nanevicz (1973) established $6 \times 10^{-9}$ as the threshold and the Heritage Committee reduced it to just $10^{-9}$ ohms per square to allow for drift in surface resistivity and other uncertainties.

# 3.4 Research

### 3.4.1 Attempts to Reduce Lightning
The early 1970s was a period when much effort, both research and operational, was being put into cloud modification in order to increase precipitation and/or reduce the damage from hail storms. One area that was also being pursued was an attempt to reduce lightning by seeding clouds with conducting chaff needles. Holitza and Kasemir (1974) and Kasemir et al. (1976) reported on studies to reduce the electric field and lightning in small thunderstorms in eastern Colorado and Wyoming. In principle, the sharp points on the chaff would go into corona in strong electric fields and thereby reduce the fields inside the storm. Both seeded and control clouds were investigated. For these studies, a B-29 aircraft was instrumented with two cylindrical field mills (Kasemir, 1972), one on the nose and one on the top of the fuselage over the wings to measure the electric fields in the storms. The initial experiments suggested that the electric field at cloud base decayed about five times faster in seeded storms than in control storms (Holitza and Kasemir, 1974). Additionally, the number of lightning events decreased to one-third or less of those in the control storms (Kasemir et al., 1976).

### 3.4.2 Russian Research
The early development of LLCC for hazards in stratiform clouds was hampered by the lack of published data inside those clouds in the western scientific literature. In contrast, and ultimately fortunately, there were a large amount of in-storm electric field data in several reports from scientists in the USSR. It was difficult for the US space program to rely upon the USSR data because American scientists had no first-hand knowledge about its quality. Because the senior Russian scientist, I. M. Imyanitov, was favorably perceived by western scientists in atmospheric electricity, however, it was ultimately decided to use the Russian data. At some time during the process of interpreting and using the data, there was a controversy over the maximum electric field strength within stratiform clouds. The issue was finally resolved when it was discovered that the units of electric field, e.g., volts per meter, were incorrect in some of the translated text. Once everyone understood that there was a translation error, it was decided to correct the units and the data for the LLCC.

The following are short descriptions of various stratiform cloud genera from the Glossary of Meteorology (Glickman, 2000):
- Stratus – a gray cloud layer with a rather uniform base.
- Stratocumulus – predominantly stratiform in the form of a gray or whitish layer or patch, which nearly always has dark parts and is non-fibrous (except for virga)
- Altocumulus – a gray or bluish (never white) sheet or layer of striated, fibrous or uniform appearance, and may be precipitating.

- <u>Nimbostratus</u> – gray colored and often dark, rendered diffuse by more or less continuously falling rain, snow or sleet and not accompanied by lightning, thunder or hail.

Over two decades, I. M. Imyanitov and his colleagues studied stratiform clouds in the USSR, mainly at three locations with different latitudes: Leningrad (St. Petersburg) at approximately 60 °N latitude, Kiev at 50 °N, and Tashkent at 41 °N. Stratiform clouds were examined from the perspective of vertical profiles of the electric field obtained with instrumented airplanes. The data that Imyanitov and colleagues summarized came from about 900 spiraling ascent soundings up through clouds to 6 km MSL. Imyanitov et al. (1971) reported each data point to be an average of 100 m vertically from a climb rate of about 4 m s$^{-1}$, and 3 km horizontally from a horizontal speed of about 50 m s$^{-1}$. The observations in the USSR showed that in general the maximum electric field increases as the cloud genus moves from stratus (St) to stratocumulus (Sc) to altostratus (As) to nimbostratus (Ns). Furthermore, within a genus, thicker clouds tended to have larger maximum electric field. The average maximum electric field for these genera (at 60 °N) was larger in the summer than in the winter by factors ranging from about one to seven. A summary of the Russian observations of the electrical structure of stratiform clouds is also available in MacGorman and Rust (1998, sections 2.1-2.3). What is clear from the Russian data is that stratiform clouds can be electrified, sometimes highly for clouds with thicknesses of 2000 m or greater, and thus, they do pose a significant threat to space vehicles flying in them or in their vicinity.

### 3.4.3    *Post-ASTP Airborne Experiments*
After the ASTP mission, because of the unique capabilities of the Learjet to climb with the tops of growing cumulus clouds and to fly at or above anvil altitudes, arrangements were made for the NASA/Ames Learjet instrumented by Stanford Research Institute to remain at Patrick Air Force Base for the summer of 1975 to make high altitude electric field measurements. The Learjet was flown around developing thunderstorms cells to study the fields in the vicinity with particular emphasis on the region of the anvil. Most of the measurements started when a cell was very young and followed the development throughout the growth period and continued until it began to disintegrate (Arabian, 1976). Measurements were gathered from ten different storm cells.

The Learjet had flight capability to 45,000 ft (13.7 km), cruise speed of 240 m s$^{-1}$ and climb rate of ~2000 m min$^{-1}$. Four rotating-vane mills were installed on the NASA/Ames Learjet and were calibrated by measuring the potential at the location of each mill on a model of the aircraft in an electrostatic cage. Appendix C of Arabian (1976) gives a general discussion of the measurement of electric fields from aircraft as well as a brief discussion of the calibration of the Learjet. The procedure is most probably an early version of Kositsky et al. (1991).

Isolated storm cells were selected for study so that measured electric fields were not affected by nearby storms. They also wanted to avoid unsafe flying conditions and wanted to insure level flight for making the measurements (Arabian, 1976). The position of the aircraft measurements could be related to ground-based radar observations. Apparently in-cloud measurements were not made. From a limited analysis of the data from these storms, Arabian (1976) stated that the high electric fields observed in the vicinity of the anvils were most likely associated with charge concentrations in the thunderstorm main vertical body. An example taken from Arabian (1976) is shown in figure 3.4.3-1 below. Because the aircraft was flying in clear air it is likely that the electric fields observed above the anvil are less than those that would have been observed inside the anvil because of the effects of screening layers.

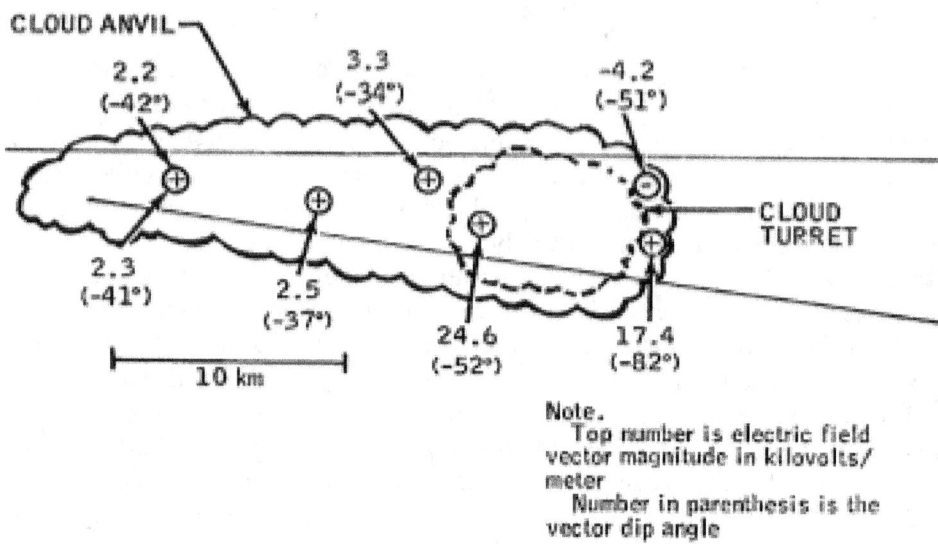

Figure C-10.- Electri field vectors above an anvil at 12.5 km altitude.

**Figure 3.4.3-1 Measurements of electric field around a thunderstorm**

Note: Measurements are from the NASA Ames Learjet [Reproduced from Arabian, 1976]

In the weeks following the ASTP launch, a NOAA study was undertaken using the C-45, the S2-D and T-29 aircraft to investigate how effectively electric fields might be reduced by the release of chaff into Florida thunderstorms (Rust et al., 1977). The sample size was small, but the initial results were not promising, probably because Florida storms are larger and more complex than the small storms that were studied in Colorado and Wyoming by Holitza and Kasemir (1974).

### 3.4.4    *Thunderstorm Research International Project (TRIP)*

As a result of the Apollo XII triggered lightning event, NASA encouraged and supported a number of collaborative experiments at KSC involving several different groups of atmospheric scientists . These efforts were primarily intended to define the meteorological conditions under which rockets would trigger lightning and to understand better the characteristics of Florida lightning. They culminated in an intensive series of ground and airborne measurements that were conducted prior to and during the Apollo/Soyuz Test Project (see section 3.3.3). After those experiences, NASA was well aware of the mutual benefits of collaborative efforts and issued an invitation to the atmospheric electrical community to participate in a new series of investigations at KSC during the summers of 1976, 1977, and 1978. In December, 1975, this opportunity was enthusiastically grasped by the scientists in atmospheric electricity who attended the Fall Annual AGU Meeting in San Francisco.

A very loose organization for the coordinated studies emerged during and after that AGU meeting. There were three overall components [Pierce, 1976]:

1.  *Principal Investigators (PIs).* Each *PI* brought his own equipment, furnished his own financial support for personnel, etc., and made his own measurements.
2.  *Technical support group from KSC.* This group assisted the *PIs* by providing sites for the experimental equipment and some basic facilities at those sites and elsewhere. The facilities included shelter,

electric power, air conditioning, timing, telephone, and other linkages. Thus, data obtained by different investigators/techniques could be compared and the characteristics of different physical parameters could be correlated.

3. *Secretariat.* The secretariat comprised Dr. L. H. Ruhnke and Dr. E. T. Pierce. Dr. Ruhnke's main function was to act as an interface facilitating communications between the PIs and the KSC technical support group. Dr. Pierce's role was to describe the project(s) to the outside scientific community and to coordinate the interpretation and presentation of the scientific results. The organization was essentially cooperative, i.e. there was no formal structure and neither the Secretariat nor the KSC support group had any directive powers over the PIs. However, the PIs were expected to accept that the generous site support provided by KSC was limited, that no individual investigator could claim an excessive share of this support, and that any conflicts between PIs over, for example, experimental compatibility, must be resolved cooperatively.

The International Commission on Atmospheric Electricity (ICAE) is a topical commission within the International Association of Meteorological and Atmospheric Physics (IAMAP). The ICAE had been advocating coordinated thunderstorm investigations for several years, and all three officers of the ICAE were already participating in the work at KSC. Prof. John Latham (President, ICAE) was a Principal Investigator, and Dr. Lothar Ruhnke (Secretary, ICAE) and Dr. E. T. Pierce (Honorary President, ICAE) were the Secretariat. During the Fifth International Conference on Atmospheric Electricity held in Garmisch-Partenkirchen, Germany, 2-7 September 1974, the ICAE passed a unanimous resolution supporting a new thunderstorm project at KSC [Dolezalek and Reiter, 1977, p. 809]. Because of this international approval and likely participation and because "program" represented the informality of the organization better than "project," the preferred description for the cooperative thunderstorm studies at KSC was the Thunderstorm Research International Program, which had a convenient acronym, TRIP.

Perhaps the greatest experiment within the TRIP was the concept of a coordinated study itself. The basic idea was that the combination of results obtained by different investigators, studying the same storms at the same time and in the same place, would provide much more valuable scientific information than the identical investigations conducted independently on different storms and in different locations. Clearly, such collaboration could only happen if there was a host organization (like KSC) that needed the information and could provide technical support such as power, telephone, timing, etc. Also, there had to be enthusiastic investigators who were ready and willing to share their data freely on a voluntary basis.

The first TRIP experiments were held at KSC in the summer of 1976 and involved about 70 scientists in 19 different research groups [Pierce, 1976]. Subsequent campaigns were held at KSC during the summers of 1977 and 1978, and at the New Mexico Institute of Mining and Technology in Socorro in 1979-1981. Publications describing the results of these investigations have been summarized in numerous review articles and books on atmospheric electricity and related topics such as Uman et al. (1975); Mason and Moore (1976); Rust and Krehbiel (1976); Uman (1976); Dolezalek and Reiter (1977); Gaskell et al. (1978); Uman et al.,(1978); Cotton (1979); Illingworth and Krehbiel (1981); Hallett (1983); Lhermitte (1979); Lhermitte and Krehbiel (1979); Gaskell and Illingworth (1980); Latham (1981); Volland (1982); Lhermitte and Williams (1983), Uman (1983); Golde (1977), Uman et al. (1978); Parker and Kasemir (1982); Volland (1982), Uman and Krider [1982}; Volland (1982); Illingworth (1985); Williams (1985); Krider and Roble (1986); Uman (1987); Latham and Dye (1989); Uman and Krider (1989); Williams (1988, 1989) and others.

### 3.4.5   *Sensor Development*
As new techniques became available for research - particularly the development of large-scale integrated circuits for both analog and digital electronics - huge advances were made in computers, memory technology, and communications. These advances in turn led to major improvements in the ability of scientists to make measurements in atmospheric electricity, display the results, and analyze the data. As new knowledge became

available, the Lightning Launch Commit Criteria (LLCC) were revised, updated, and improved to increase both launch availability and launch safety.

Among the scientific advances that were made in the early- and mid-1970s, largely as a result of the Apollo 12 lightning incident, were the development of (a) an improved ground-based method for detecting and displaying cloud electric fields and (b) improved methods for detecting, locating, and displaying both intracloud and cloud-to-ground (CG) lightning discharges. These advances were tested at KSC and were the basis for many of the subsequent improvements in the LLCC.

### 3.4.5.1 Launch Pad Lightning Warning System (Field Mill Network)

The LPLWS is basically a large-area network of electrostatic field sensors (field mills) that measures the vertical component of the electrostatic field, E, produced by cloud charges aloft, even if those clouds are not producing lightning. The sensors are termed 'mills' because they employ a rotating and grounded metallic shutter to alternately cover and uncover a set of insulated stators that respond to E. The amplitude and phase of the AC current flowing to and from the stators is proportional the amplitude and polarity of the local electric field. When an electrified cloud forms overhead or moves into the region from somewhere else, the E field will increase in magnitude and often change polarity. In most cases, E will be large close to the cloud charges and small farther away, so 2-dimensional maps of E will show approximately where the cloud charges are located.

If the cloud electric field is measured in conjunction with a weather radar that senses precipitation, the onset of the electric field can be compared with the onset and type of precipitation, the rate of echo growth, and the time-evolution of the cells, thereby improving both the nowcasting (i.e. detection) and forecasting of high fields aloft. When lightning discharges occur, maps and analyses of the lightning-caused changes in E, or $\Delta$E, can be used to determine an approximate location (in 3-dimensions) of the change in the cloud charge (Maier and Krider, 1986; Koshak and Krider, 1989; Krider, 1989).

The unique features of the LPLWS are its large area (approximately 20 x 30 square kilometers), the number of field mills (25 to 33), and the fact that each sensor is mounted (and calibrated) in the same way on uniform sites that are cleared of vegetation. A network such as this minimizes the frequency of false alarms, such as might be caused by a single sensor having an incorrect reading, and also any 'failures-to-warn' that might occur if the positive and negative cloud charges are not vertically aligned or if the field seen by a single sensor is masked by intervening space charge. Further details on the implementation and calibration of the LPLWS at the KSC-ER are given in Appendix VII by Michael W. Maier.

Among the first scientific studies that were based on the field mill network were analyses of the electrostatic field changes produced by Florida lightning by Jacobson and Krider (1976) and the overall behavior of E under both small and large storms at KSC by Livingston and Krider (1978). These studies showed that the charge centers inside Florida storms are located at altitudes where the air temperatures are below freezing, i.e. where the environmental temperatures are between about -10 °C and -20 °C, and that the time-average values of E are often surprisingly small when averaged over five minute intervals, even under active storms. The finding that the lightning charges are located at subfreezing temperatures lent support to the idea that a non-inductive ice-ice collision process is the dominant microphysical mechanism in cloud electrification (Saunders, 1988; 2008), and the LLCC still rely on this assumption.

### 3.4.5.2 Cloud-to-Ground Lightning Surveillance System (CGLSS)

KSC has made major contributions to the development of two complementary systems for detecting and locating lightning, the Cloud-to-Ground Lightning Surveillance System (or CGLSS) and the Lightning Detection and Ranging (or LDAR) system. The CGLSS utilizes a network of gated, broadband electric and magnetic field sensors (Krider and Noggle, 1975) to detect the waveform signatures that are characteristic of return strokes, the high-current components of CG flashes (Krider et al, 1976; Herrman et al., 1976). When a proper signature is detected (in the time-domain) at two or more known locations (the antenna sites), the coincident times-of-arrival and magnetic directions can be used to compute the point where a return stroke strikes the ground (Krider et al., 1980; Cummins et al., 1998; 2006).

According to a time-domain antenna theory developed by M. A. Uman and his collaborators shortly after the Apollo 12 incident (see for example Uman et al., 1975), the initial peak of the electromagnetic pulse that is radiated by a return stroke is proportional the peak current in the stroke, multiplied by the speed of the stroke propagating up the leader channel, and divided by the distance to the stroke. [Note: this theory is sometimes called the simple 'Transmission-Line Model' or TLM because it assumes the current pulse propagates up a straight channel, without distortion, and at a constant speed.] Since the CGLSS measures the peak field and can compute the stroke location, and since the stroke velocities are known and roughly constant, the CGLSS can also provide an estimate of the peak current in the stroke and its polarity.

The first CGLSS system was installed at the KSC-ER between 1 June 1979 and 12 July 1979. It was a prototype consisting of three medium-gain magnetic direction-finders (DFs) and was installed as part of the Federal Evaluation of Lightning Tracking Systems (FELTS). This system was subsequently purchased in February 1981 with joint funding provided by NASA and the Air Force. Later upgrades were performed as described in Chapters 4 and 5.

### 3.4.5.3 Lightning Detection and Ranging (LDAR) System

The first LDAR system was developed at KSC by Carl Lennon and associates after a design described by Proctor (1971). It contained seven broadband VHF radio receivers that were deployed at the sites shown in Figure 3.4.5-1 and were precisely time-synchronized, initially using microwave communications links and later GPS timing. Each site received VHF radiation at 66 MHz, logarithmically amplified the signal, and then transmitted the time and key signal parameters to a central station where the source locations were computed [Lennon and Maier, 1991; Maier et al., 1995]. The 3-D locations of the sources of lightning VHF pulses are computed using the differences in the times-of-arrival of the signals detected at the different receiver sites. Since the main sources of VHF radio emissions are the processes associated with air breakdown, the LDAR system detects primarily the in-cloud portions of CG flashes, leader processes, and intracloud discharges. Today each LDAR receiver site operates automatically and is powered by batteries that are recharged by solar panels. The first LDAR system was developed by the KSC Instrumentation and Measurements Branch, and the current one is operated and maintained by the ER Technical Services Contractor. The 45 WS receives and evaluates the LDAR data 24 hours a day, 7 days a week. For further details on the evolution of the LDAR system and other instrumentation at the KSC-ER, see Boyd et al. (1995), Harms, et al.(1997; 1998; 2001) and Roeder et al.(1999).

**Figure 3.4.5-1  KSC LDAR Sites**

### 3.4.5.4  U.S. National Lightning Detection Network (NLDN)

Data from the U.S. National Lightning Detection Network (NLDN) (Cummins et al., 1998, 2006; Cummins and Murphy, 2009) have been used to detect and track CG lightning flashes beyond the range of the CGLSS and LDAR systems since the early 1990s. The NLDN sensors are similar to those used in the CGLSS except that they have higher gains and larger distances between the sensors. The coincident data from two or more sensors are collected and processed in real-time by a network control center in Tucson, Arizona, and the GPS time, location, and polarity of each lightning stroke, together with an estimate of its peak current, are provided to the KSC-ER in real-time. For further information about the NLDN and its history, see Cummins and Murphy (2009).

# Chapter 4   The Space Shuttle Era Begins (1981 – 1987)

## 4.0   Background

The Space Shuttle Program initially attempted to design a vehicle to withstand lightning and thus avoid the restrictiveness of the LLCC. However testing and analyses were not yet complete when Shuttle began its flight test phase in 1981, thus Shuttle initially adopted many of the post Apollo 12 LLCC (Heritage, 1988, para4-4, p 4-6.), although with modifications. The first Shuttle-specific LLCC (1979) read as follows (Heritage, 1988, Table 4-3; Kapryan (1978)]:

### Original Space Shuttle LLCC 1979

*Space vehicle will not be launched if nominal flight path will carry vehicle:*
- *Through a cumulonimbus (thunderstorm) cloud;*
- *Within 5 sm (7.8 km) of a cumulonimbus (thunderstorm) cloud or the edge of an associated anvil that is within 5 sm of its radar cell. At the discretion of the Launch Director the anvil may be penetrated if the static electric field is less than 1kV/m and the 5 sm rule is maintained.*
- *Through or within 5sm of any other clouds where radar shows virga or precipitation and tops extend (or will extend at the time of launch) above the -10C temperature altitude and four or more field mills within the launch site have changed more than 500 V/m.*
- *Through single layer clouds in the dissipation stage and where these clouds have activated the delta electric field contours within a period of 10 minutes before launch. The maximum electric fields associated with subject cloud as measured on the ground must not exceed 2 kV/m. The electric field meter readings for double layer clouds cannot exceed 500 V/m.*

Additional knowledge about cloud electrification accumulated over the years due to research and operational experience. During the extensive program-wide reviews of weather support that followed the Challenger accident described in section 4.3 below, an improved set of Shuttle LLCC were adopted. These read as follows (Heritage, 1988, Table 4-4):

### 1986 Shuttle Lightning Launch Commit Criteria (LLCC)

*Do not launch if vehicle path is:*
- *Within 5 nm of a cumulonimbus cloud or the edge of associated anvil cloud;*
- *Within 5 nm of any convective cloud whose top extends to -20C isotherm with virga/precipitation;*
- *Through any cloud where precipitation is observed;*
- *Through dissipating clouds in which the electric field network has detected lightning within 15 minutes prior to launch;*
- *Through any cloud if ground level electric field at launch site is greater [in magnitude] than + or - 1000 V/m.*

These included three major changes from the 1979 version:
1. The wording was simplified.
2. The surface field mill threshold was made consistent at 1000 V/m.
3. The threshold isotherm for convective cloud tops was changed from -10 °C to -20 °C based on research such as that of Cobb (1975).

## 4.1   Infrastructure

Except as noted below in this section, the weather infrastructure at KSC/CCAFS remained the same as that presented in Section 3.1 during the period covered by this chapter.

Early in the Shuttle program the wind tower network that NASA maintained at KSC and the similar network maintained by the Air Force at CCAFS were combined as a single system belonging to the Air Force to reduce duplication of effort and cost. The combined network expanded to 29 and later to 47 locations (Billie Boyd, 45WS, private communication). Although a few sites have been decommissioned, this is essentially the configuration of the wind tower network today. Most of these towers are 16.5m (54ft) tall and instrumented for wind at two levels, 12 and 54 feet, and for temperature at two levels, 6 and 54 feet. In addition to the 150 meter tower mentioned above there are three other tall towers, each 62m (204 ft) tall instrumented at 4 levels [Eastern Range Handbook, 2007].

The CGLSS system was upgraded during this period. In August 1983, a low-gain system was added to obtain more accurate locations near the KSC-ER launch complexes. By February 1984, the system contained two low-gain direction-finders (DFs) that were located at the Ti-Co Airport (28.5N 80.8W) and on Merritt Island (28.4N 81.3W), and three medium-gain sensors, one co-located with the low-gain sensor on Merritt Island, and the other two were at the Orlando and Melbourne Airports (Harms, et al., 2001).

## 4.2    Organization

On 1 October 1978, the Air Force assumed full operational weather support for launch and ground processing operations at KSC and the Eastern Range, whether manned or unmanned. SMG retained responsibility for on-orbit and landing operations at all locations including KSC. Following the Challenger accident in 1986, NASA funded additional Air Force support for the Shuttle program including a dedicated full-time civilian launch weather officer. Although there have been some organizational name changes, the overall structure and content of weather support has not changed significantly since then. Support is currently provided by the 45th Weather Squadron within the 45th Space Wing at Cape Canaveral Air Force Station. Although the structure and content of the support has remained relatively stable, the infrastructure and instrumentation has changed substantially and is discussed later in this History.

## 4.3    Operations

### 4.3.1    Challenger

On a cold, clear morning in January 1986, the Space Shuttle Challenger and its crew of seven, including school teacher Christa McAuliffe, ascended above launch complex 39 at Kennedy Space Center to the cheers of spectators on site and around the world, including tens of thousands of school children watching on television. About 70 seconds into the flight, after the routine call "go at throttle up" had been issued, the vehicle exploded and the crew fell to their deaths in the Atlantic Ocean just a few miles off the Cape Canaveral seashore. Extreme cold weather, high wind shear aloft, and defective design of the O-ring seals at the solid rocket booster segment joints combined to permit the super-hot booster flame to burn through the booster case and into the external tank containing tons of explosive rocket fuel for the Shuttle's main engines. The defining operational incident during the opening years of the Shuttle Program had nothing directly to do with lightning, but the destruction of the Challenger and the ensuing investigations (e.g., Rogers, 1986) exposed generic weaknesses in weather support procedures and processes. The procedures and processes used to avoid lightning hazards were not immune to these weaknesses.

In the early 1980s, weather was the single most frequent cause of Space Shuttle launch scrubs and delays. Delays and scrubs were costly both in terms of labor hours and for expendables such as propellants consumed during each countdown. In an attempt to reduce these delays and scrubs, NASA had convened a "blue ribbon" panel of experts in November of 1985 to assess NASA weather support activities and plans. The panel, formally called the Space Shuttle Weather Forecasting Advisory Panel and informally called the Theon Panel after its chairman, Dr. John Theon, had circulated preliminary recommendations just prior to the Challenger accident. These were reviewed and revised by the panel following Challenger. In October 1986, it published its official recommendations (Theon, 1986, 1988). The mission of the Theon Panel and its recommendations are important for this History because they address infrastructure, technological, and organizational issues directly relevant to the development and application of the LLCC.

The Theon Panels specific tasks included the following:
- *Review current weather observation, analysis, and forecasting facilities, systems, and procedures which provide the weather forecasts to Shuttle managers for Shuttle ground processing, launch, and landing operations;*
- *Review future plans for these facilities, systems, and procedures;*
- *Advise NASA Associate Administrator for Space Flight on current and planned capabilities for providing state-of-the-art operational weather forecasts;*
- *Recommend processes and/technologies to the Associate Administrator to improve weather forecasts for the planning and execution of Shuttle operations.*

The resulting set of recommendations (as cited in National Research Council, 1988, Appendix C, pp59-63.) was extensive:
- Establish a Weather Support Office at top level of Shuttle operations to plan, organize, focus and direct weather support activities. The Office Chief should be a senior atmospheric scientist or senior technical manager with strong operations background;
    - Establish a small, highly qualified, well-trained, and dedicated team of forecasters who provide weather support for operations. Forecasters should be integral part of Shuttle team and remain for extended periods (5-10 years). Ensure their continuity and their devotion to the task;
    - Establish a Technique Transition Unit at each operational site. Purpose: Ensure forecasters have latest research results and tools; are trained to use tools effectively; and act as an interface between the R&D community and the forecast team;
- Quantify weather hazards;
- Re-examine models used for wind-loading calculations in view of the upcoming installation of the new KSC wind profiler and its ability to detect the rapid wind profile changes which undoubtedly occur under many weather situations;
    - Establish an external Shuttle Weather Advisory Panel of experts;
    - Integrate applications into MIDDS weather display systems at JSC, CCAFS, EAFB, and VAFB;
- NEXRAD (Doppler weather radar): Integrate into MIDDS and develop Cape-specific applications;
    - Improve the Mesonet weather tower network;
    - Develop Thermal Protection System (tile) rain erosion measurement and aircraft instrumentation;
- Review/develop local short/long-term mesoscale forecast models and Shuttle specific applications;
    - Maintain KSC artificial intelligence study at a modest research level.

Because of the cost involved in implementing many of these recommendations, the NASA Administrator forwarded the report of the Theon Panel to the National Research Council to request a "second opinion". While the NRC was considering whether to review the Theon report, NASA and the Department of Defense suffered another major accident, the loss of AC 67 described in the next section below. As a result, the NRC agreed to review the Theon report but in the context of the whole American space program rather than just that of the Shuttle program. The NRC reinforced and expanded upon the recommendations of the Theon Panel and many of them were adopted as described in Chapter 5, section 5.0.2 below.

### 4.3.2 *Atlas Centaur 67 (AC 67)*
Late in the afternoon of 26 March 1987, while the NRC was still considering whether to review the Theon Report as requested by NASA, an Atlas-Centaur designated AC 67 lifted off into a dark, precipitating

stratiform overcast from launch complex 36B at Cape Canaveral Air Force Station. The payload was a Fleet SatCom satellite for the US Navy. NASA was responsible for managing the launch. Just under 50 seconds later a cloud to ground lightning strike was photographed at the pad and two seconds after that the vehicle began to break up. Destruction of AC 67 was completed by the Range Safety Officer. The NRC responded to NASA by agreeing to review the Theon Panel report, but in the broader context of the whole American Space Program rather than just the Shuttle Program. A discussion of panels convened, recommendations and actions pursued as a result of AC 67 is presented in detail in Chapter 5.

The LLCC used for AC 67 were very similar to the post-Apollo XII rules described in Section 3.0 and read as follows (Heritage, 1988, p. 4-9, Table 4-5)

### Atlas/Centaur 67 Lightning Launch Commit Criteria (1987)

*Flight path of the vehicle should not be:*
  A. *Through a thunderstorm/cumulonimbus cloud.*
  B. *Within 5 miles of thunderstorm/cumulonimbus cloud or 3 miles of associated anvil top.*
  C. *Through cold front/squall line associated clouds with tops 10,000 ft or higher.*
  D. *Through middle-level cloud layers, 6,000 ft or greater in depth, when the freezing level is in the clouds.*
  E. *Through cumulus clouds with the freezing level in the clouds.*

## 4.4    Research

### 4.4.1   KSC Rocket Triggered Lightning Program (RTLP)

This program was organized and operated by the KSC Engineering Directorate (DE) rather than by the KSC Weather Office. The NASA/DE RTLP lead engineer was William Jafferis, who was extremely enthusiastic about the project. Mr. Jafferis was convinced, even before the AC 67 disaster in 1987 (see above and Section 5.0), that not enough was known about lightning in general and triggered lightning in particular to safely conduct space launches in the Cape Canaveral area. Throughout its nine-year history (1983 through 1991) he managed and directed the program with funding from the Shuttle Program at KSC, supplemented by private sources such as the Electric Power Research Institute, Florida Power and Light, and others. By its conclusion, he had built KSC into the world's leading center for rocket-triggering research. Three LAP members (Krider, Willett, and Rust) participated in the RTLP at various times during its life. In addition to numerous scientific papers and technical reports, an article about the program was published in the New York Times in the aftermath of the AC 67 accident (Wilford, 1987).

Summertime rocket-triggering experiments by the RTLP were begun in 1983 south of Melbourne, FL (Richmond, 1984). They were moved some 50 km north in 1984 to a site just east of the Vertical Assembly Building (VAB) at the NASA Kennedy Space Center (KSC). In 1985 triggering operations were moved again, about 15 km further north to the west bank of the Mosquito Lagoon, where they continued through 1991. The Mosquito Lagoon facility eventually comprised two instrumented triggering locations (the original "land pad," and a newer "water pad," described below), two control and instrumentation trailers (one a discarded Chesapeake & Ohio railroad caboose), a "headquarters" building dubbed the Atmospheric Sciences Field Laboratory (ASFL), and an un-energized, 448 m long, electric-power distribution line running approximately between the land pad and the ASFL. Diesel generators were provided for operations during storms, when commercial power in the area was unreliable. A tethered balloon (aerostat) was even operated during 1989 and 1990 to support instruments aloft and to trigger lightning from an elevated and ground-isolated platform. In August of 1991, however, an accident with injury involving the unintended launch of a 2.75" Folding-Fin Aircraft Rocket by French investigators brought the program to an end [Mishap Investigation Team Report,

1991]. Earlier that same season, the aerostat had blown away and been lost at sea, which did not help the program's standing with KSC management.

Thereafter, triggering activities by some of the participants and funding sources moved to Camp Blanding in Starke, FL, under the management of the University of Florida, where they continue and flourish to this day. Unfortunately, this move, coupled with the loss of Shuttle Program funding, eliminated any motivation to address the operational needs of the space program. The research at Camp Blanding became even more focused on the concerns of the electric power industry (although this focus has broadened recently because of significant long-term funding from the National Science Foundation).

Fundamentally, the demise of the KSC RTLP resulted from the contemporary reorganization of weather infrastructure throughout NASA that had been recommended by the Theon and NRC panels. The new NASA Weather Support Office (see Section 5.2.0) led by Jack Ernst at NASA Headquarters and its counterpart at KSC (see Section 5.2.2) led by Jan Zysko, assumed control of the Shuttle weather-related "research" funding to ensure that it was directed towards narrowly defined, operational needs. In February, 1989, Jim Nicholson (NASA/KSC/PT, the conduit for this funding) conducted a blue-ribbon panel review of the RTLP to get recommendations on how it could be restructured to support Shuttle operations more directly. This panel consisted of the following experts on lightning and atmospheric electricity:  E. Philip Krider (University of Arizona, Chair), Jean-Louis Boulay (ONERA, France), Charles B. Moore (NMIMT), Richard E. Orville (State University of New York at Albany), Martin A. Uman (University of Florida, Gainesville), and John C. Willett (AFGL). Its report [Review of and Recommendations for the Rocket Triggered Lightning Program (RTLP) at the NASA Kennedy Space Center, revised August 1, 1989] identified several experimental efforts that would be of great value to the operational community and that could be carried out by the RTLP. It also recommended some changes in the organization of the RTLP, its funding, and the objectives to achieve these ends. Unfortunately, the recommended changes never came to pass, and this more or less guaranteed the termination of the RTLP when the above-mentioned accidents occurred.

French technicians from the Centre Etudes Nucleaires de Grenoble (CENG), who described the campaigns in a series of technical reports (e.g., Eybert-Berard et al., 1986; Barret, 1986; Eybert-Berard et al., 1988, 1989; Barret et al., 1990; Barret et al., 1991), provided some of the infrastructure, although KSC also developed its own operational capability. Two different triggering platforms were used at the Mosquito-Lagoon site. A "land pad" was constructed near the shore of the lagoon in 1985 to facilitate engineering tests on isolated conducting bodies and other hardware, and its configuration is quite complicated (see Leteinturier et al., 1990). A simpler "water pad" was constructed about 30 m offshore in 1987 to enable more precise electromagnetic measurements (Leteinturier et al., 1991).

The triggering techniques used in the RTLP have been described in detail by Laroche et al. (1985, 1989a, and 1991). Two basic rocket-and-wire methods were used at KSC; in both, a spool of fine Kevlar (TM)-wrapped copper wire was raised by a meter-long, plastic-bodied, black-power-fueled rocket. (This approach of raising the spool was pioneered at the RTLP in 1983. Previously a spool of cotton-wrapped steel wire had been mounted on the ground, and the rocket lifted only the free end of the wire.) The rockets could reach altitudes of about one kilometer in a several seconds. The "classical" technique (sometimes called LRSG) involved unspooling a continuous, grounded wire as the rocket ascended. This method has the advantage of allowing a direct measurement of the lightning current at the base of the wire (lightning channel). The "altitude" technique (LRSA -- also pioneered at KSC in 1987) is similar, except that the rocket unspools a few hundred meters of non-conducting line before the conducting (un-grounded) wire was paid out. Altitude triggering was developed to simulate better how lightning strikes airborne vehicles and to study the leaders that develop early in the discharge between the lower end of the triggering wire and ground. A later innovation in the altitude technique was the addition of a short length of grounded wire below the non-conducting segment (LRSAG), to facilitate studying the ground-striking process and allow direct measurement of currents in this "lightning rod." The key scientific results from these experiments are given in Section 5.4.3.

### 4.4.2    *Research on Lightning Strikes to Aircraft*

More or less contemporaneously with the KSC Rocket-Triggered Lightning Program, several experimental studies were performed on the physics and phenomenology of lightning strikes to aircraft, and these have been reviewed by Uman and Rakov (2003), which is the basis for most of the following material. We will return to this subject in more detail in the separate Rationale document because of the fact that most lightning strikes to airborne vehicles are triggered by the vehicle itself after it flies into a high electric field.

Our understanding of the initiation and characteristics of lightning strikes to aircraft is derived primarily from airborne studies involving four instrumented aircraft, an F-100F, an F-106B, a CV-580, and a C-160. The F-100F project, termed Rough Rider, could have been described in Chapter 2, but it is covered here to keep it together with the other three. Rough Rider was operated by Don Fitzgerald of the Air Force Geophysics Laboratory in collaboration with the FAA between 1964 and 1966 and is described by Fitzgerald (1967) and by Petterson and Wood (1968). The F-100F, a single-engine jet fighter, penetrated thunderstorms to measure turbulence and to obtain photographic, shock wave, and electric current records of lightning. Data were recorded for 49 lightning discharges on rather primitive instrumentation by today's standards.

The NASA F-106B, a delta wing, single-engine jet fighter, was operated by the NASA Langley Research Center. It flew about 1500 thunderstorm traversals at altitudes ranging from 5000 to 40,000 feet (1.5–12 km) and was struck by lightning 714 times between 1980 and 1986 (Pitts et al., 1988). Statistics were compiled for aircraft surface electric and magnetic field derivatives and for lightning currents and current derivatives flowing through the aircraft. Detailed information on the instrumentation and data obtained can be found in numerous publications listed by Uman and Rakov (2003).

The USAF/FAA CV-580, a twin-engine turboprop transport aircraft, was instrumented as described by Rustan (1986) and others and was flown in 1984 and 1985. Numerous sensors measured the time derivative of the surface electric-field intensity, the rate-of-change of the surface current density, the "slow" electric-field change (with a lower-frequency response of 1 Hz), and the current injected into the tail and wingtips. In addition, five electric field mills were mounted at various locations on the fuselage. The latter instrumentation yielded some of the first, reasonably credible measurements of the ambient fields and aircraft charge immediately prior to lightning strikes (Bailey and Anderson, 1987; Anderson and Bailey, 1987).

The C-160 research aircraft that was operated by ONERA in the Transall field programs in France during 1984 and 1988 was a twin-engine aircraft similar to the CV-580 but somewhat larger. For the 1988 study the C-160 was instrumented specifically for investigation of the initial processes of lightning attachment, with a network of five electric field mills, numerous capacitive antennas and current shunts, and a high-speed video system (Lalande et al., 1999, and others). The field-mill system was again used to determine the ambient field and charge on the aircraft just prior to lightning strikes.

The key results from these experimental programs are the following:

1.  About 90% of lightning strikes to flying aircraft, and by inference to spacecraft in the troposphere, are triggered by the aircraft (Lalande and Bondiou-Clergerie, 1997).
2.  The maximum current derivatives ($dI/dt$) measured on aircraft can be as large as those in lightning return strokes at the ground (see Section 5.4.3), although the peak currents are not as large and the rise times of the fast-rising portions are shorter (Pitts *et al.*, 1987).
3.  Current pulses that have large $dI/dt$ values and high repetition rates occur, primarily during the early stages of lightning attachment. This observation has led to addition of a so-called "multiple burst" waveform to the aircraft test standards (*e.g.*, Plumer, 1992).

# Chapter 5   The Aftermath of Atlas Centaur 67 (1987 – 1999)

## 5.0     The AC 67 Investigations

The AC 67accident closely followed the Challenger accident, thus it generated national attention. The result was a series of investigations by NASA; the Air Force; the Navy; the National Research Council; the media; standing conferences including AIAA, AMS, AGU, and AF/NASA; Congress (including formal House Hearings); and three Flight Rules workshops with JSC, MSFC, KSC, AF, and consultants participating. Following are dates when Reviews were directed and either held or published:

| | |
|---|---|
| Oct 86 | NASA Theon Panel publishes report pre-AC 67 (See section 4.3) |
| 27 Mar 87 | Admiral Truly, Hq NASA/M, Office of Space Flight, directs formation of AC 67 accident board chaired by Jon Busse, Goddard Space Flight Center MD (GSFC) |
| 11 May 87 | Hq Air Force directs AF Space Division to form Lightning Review Committee to investigate AC 67 incident – results in the "Heritage Committee" report on 31 August 1988 |
| 9 June 87 | NASA/Busse Preliminary Report |
| 15 July 87 | NASA/Busse Final Report |
| 4 August 87 | Congressional Hearings Chaired by Representative Bill Nelson |
| 5 August 88 | National Research Council publishes *Meteorological Support for Space Operations--Review and Recommendations* – the result of the NRC's review of the Theon Panel report cited above. |
| 31 August 88 | Air Force Lightning Review Committee, chaired by Hugh Heritage, publishes final report: Launch Vehicle Lightning/Atmospheric Electrical Constraints Post AC 67 Incident |

Each of the three expert panels formed to determine the cause(s) of the accident and to recommend ways to prevent such events from happening in the future is described separately below. In addition to these panels of technical experts, there was a Congressional investigation and hearings before Congressional Subcommittees. One of these is also described in detail below.

### 5.0.1   Busse Panel

NASA Headquarters appointed the Atlas/Centaur-67/FLTSATCOM F-6 Investigation Board chaired by Jon R. Busse. This Board in turn invited a Sub-Panel of experts on lightning and electromagnetic pulse to investigate the weather and the probable causes of the AC 67 incident and to make recommendations. The members of this Sub-Panel are presented in Table 5.0.1-1.

The Sub-Panel published their results in a NASA Report (Christian *et al.*, 1987). The same authors also published a peer reviewed article in the Journal of Geophysical Research (Christian et al., 1989).

The Final Report of the Atlas/Centaur-67/FLTSATCOM F-6 Investigation Board (hereafter termed the Busse Report) was approved on May 11, 1987, and was officially dated July 15, 1987. The basic conclusion of the Busse Report was that the breakup of the Atlas/Centaur-67 vehicle was caused by large dynamic loads during flight as a result of an excessive angle of attack. This in turn was caused by an unplanned, hard-over booster engine gimbal command being issued by the Centaur Digital Computer Unit (DCU).

The most credible mechanism for causing the booster engine hard-over command was the DCU reacting to an external electrical transient. All other mechanisms are improbable. The electrical transient was caused by a triggered lightning strike on the vehicle. The strike occurred on the vehicle nose fairing and coupled into the ground bus network through multiple paths, including the wiring from the booster external equipment pod and the nose fairing accelerometer wiring. These transients entered the DCU and caused the alteration of a single word in the DCU memory associated with the computation of the Atlas engine yaw commands.

| Panel Member Name and Title | Panel Member Affiliation |
|---|---|
| Bruce D. Fisher<br>Chairperson | Research Engineer, NASA Langley Research Center |
| Keith E. Crouch<br>Vice President | Lightning Technologies, Inc. |
| Dr. Vladislav Mazur<br>Research Physicist | NOAA National Severe Storms Laboratory |
| Dr. Lothar H. Ruhnke<br>Head, Atmospheric Physics Branch | Naval Research Laboratory |
| Dr. Hugh J. Christian, Jr.<br>Research Scientist | NASA Marshall Space Flight Center |
| Dr. Rodney A. Perala<br>Vice President | Electro Magnetic Applications, Inc. |

**Table 5.0.1-1  Members of the expert Sub-Panel of the Busse Panel**

In the opinion of the Board, the most probable cause of the mission failure was launching the AC 67 vehicle into atmospheric conditions conducive to triggered lightning and in violation of the established criteria used to avoid potential electrical hazards.

Chapter II of the Busse Report contains a number of Determinations, Findings, Observations, and Recommendations, among which are:

*Determination:  The vehicle was launched into adverse weather conditions of rain, clouds, and intense electric fields. The conditions existing at launch time were in violation of the criterion stating that the flight path of the vehicle should not be through middle-level cloud layers 6,000 feet or greater in depth, when the freezing level is in the clouds. (p. II-1)*

*Finding:  Before the AC 67 launch, there were a significant number of indications that generally the weather was unfavorable and that specifically there was a lightning hazard. Yet the real import of these indications escaped the launch management team because of imprecise communications, lack of awareness, or both.*

*Recommendation:  It should be emphasized to all management personnel, whether directly or indirectly involved in the launch decision process, that they have a responsibility to exercise awareness,*
*judgment, and leadership. Management must be constantly alert to variations to planned launch activities. They must provide a questioning and challenging attitude towards the validity of inputs being made. (p. II-2);*

*Finding:  The electric field mill data just prior to launch indicated the potential for lightning near the AC 67 pad. While the field mill data is not a part of the stated potential electrical hazards criteria, weather personnel were aware of the field mill data and the possible hazard to launch.*

*Recommendation:  An electric field mill criterion should be established for Expendable Launch Vehicles (ELVs) to formally establish electric field strength measurements as an element of determining lightning potential.*

*Recommendation:  The electric field mill system should be transitioned to operational status. (p. II-3)*

and

*Finding:  The weather criteria currently used for ELVs by NASA will not avoid all situations for known risks of lightning strikes.*

*Recommendation:  The (weather) criteria should be revised to take into consideration the additional knowledge and measurement techniques of lightning phenomena that have been developed since the present ELV criteria were established in 1970. The Shuttle and ELV weather criteria should converge where practical. (p. II-3).*

Appendix H of the Busse report (Volume II, Book 3) should be noted. It includes a detailed analysis of measurements that were obtained during the AC 67 incident (pp. H-3 to H-137), an analysis of the weather and applicable LLCC at the time of the incident (H-137 to H-139), an evaluation of the lightning detection systems that were operating at the time of the incident (H-141 to H-143), an analysis of the susceptibility of the AC 67 vehicle to lightning, and several recommendations for hardening the vehicle against lightning disturbances.

Attachment H-1 to Appendix H describes the electric field mills that were operating at the KSC-ESMC at the time of AC 67 incident and the use of network of such sensors to detect and locate electrified clouds and lightning. It also reviews and compares four other lightning detection systems that covered the launch area at the time of the incident (see pp H-181 to H-234).

Attachment H-2 to Appendix H contains a description of the electric field mill calibration procedures that were used at the ESMC and the stability of those sensors over time (pp. H-243 to H-254), a model of the physics of triggering lightning in an electrified environment, including the effects of the vehicle geometry and its exhaust plume (pp. H-257 to H-279), and a comparison of electric fields measured on the ground with those measured aloft (H-280 to H-296). Attachment H-2 also contains a report of a lightning discharge that was observed during the launch of A/C-38 (H-297 to H-310).

### 5.0.2  The National Academy of Sciences
The National Academy of Sciences (NAS) appointed a "Panel on Meteorological Support for Space Operations" that was chaired by Dr. Charles L. Hosler of the Pennsylvania State University. The other members of the NAS Panel were:

Gregory S. Forbes, Pennsylvania State University
Joseph B. Klemp, National Center for Atmospheric Research
E. Philip Krider, University of Arizona
John A. McGinley, National Oceanic and Atmospheric Administration
Peter S. Ray, Florida State University
Leonard W. Snellman, University of Utah

The panel was assisted by John D. Perry and Fred D. White of the NAS staff. The Hosler panel had a number of meetings with people and organizations that were involved in space operations, and in 1988 it published a Report (National Research Council, 1988, hereafter NRC) that highlighted the importance of weather in space operations and made five specific recommendations to improve weather support at the KSC-ER. These were:
1. Quantify more rigorously the relationships between the magnitudes of weather variables and the hazards they pose to space vehicles. Flight rules and launch commit criteria should be based on these relationships. (NRC, p3.)
2. New and improved instrumentation must be used to detect weather conditions and phenomena that are hazardous to space operations. (NRC, p4.)

3.  A number of emerging techniques for weather analysis and forecasting and decision making must be actively pursued. (NRC, p5.)
4.  Give clear and unambiguous authority to the Weather Support Office and give it sufficient budgetary authority to ensure an integrated and coordinated meteorological support program for all phases of manned and unmanned spaceflight. (NRC, p6.)
5.  An Applied Research and Forecasting Facility (ARFF) should be established at Kennedy Space Center to promote the development and application of new measurement technology and new weather analysis and forecasting techniques to improve weather support for space operations, to provide forecaster education and training, to coordinate field programs involving the meteorological community and to conduct an active visiting scientist program. (ibid).

Many of these recommendations were at least partially implemented or completely implemented in modified form. The Lightning LCC have been substantially improved based on two field programs, ABFM I (section 5.4.1 below) and ABFM II (section 6.4 below) consistent with recommendation 1. The improvements in lightning instrumentation described in section 5.1 follow from recommendation 2. The creation of the Applied Meteorology Unit (see section 5.2.3) implemented recommendation 5 for the purpose of achieving recommendation 3. Recommendation 4 was at least partially implemented by providing Weather Support Office with sufficient funding and authority to operate the AMU and conduct detailed annual reviews of weather support to the Space Shuttle program.

### 5.0.3    *Aerospace Corporation-Heritage*

The Air Force Systems Command, Space Division, asked the Aerospace Corporation to study the weather conditions surrounding the AC 67 incident and the existing launch vehicle weather constraints and to provide recommendations for improvement. To satisfy this request, the Aerospace Corporation appointed a 'Lightning Review Committee' (LRC) that had 5 members from the Aerospace Corporation and 6 outside experts with expertise in lightning and atmospheric electricity. All members are listed in Table 5.0.3-1 taken from the Heritage Report (Heritage, 1988), page xvi.

This group operated as both a Team and a Committee. It held numerous meetings and discussions, and each member was asked to contribute to the Team in his own expert area. As a convened body, the group was required to consider, act, report, and recommend, and thereby was a Committee. Each committee member, including the chairman, had one equally weighted vote. Meetings were conducted with a provision for open, informal discussion of issues so that all committee members wishing to express their views could be heard.

The LRC was provided adequate time for discussion so that 'unanimous approval' of issues could be optimized. On occasions when this approach was not successful, a member vote was taken, and a simple majority was required to approve or reject an issue. The majority contained participating members from two subsets, the subsets being: (1) Aerospace employees and (2) outside consultants. Members with strong minority views were encouraged to submit perspective views, and they were allotted an Appendix of the LRC Report to express them. All items distributed for review by the LRC, other than during meetings, were given a one week suspense date. The member receipt of the review material, without subsequent response, was considered for voting purposes to be a vote in favor.

| Name | Company | Title |
|---|---|---|
| Bruce Edgar | Aerospace Corporation | Scientist, Space Test Program |
| Hugh Heritage | Aerospace Corporation | Chairman -Lightning Review Team, Manager – Electrical Systems, Space Launch Operations |
| Robert H. Holzworth II | University of Washington | Professor of Geophysics, Adjunct Professor of Physics |
| William Jafferis | NASA/KSC | Advanced Projects |

| Name | Company | Title |
|---|---|---|
| Harry Koons | Aerospace Corporation | Senior Scientist, Space Plasma Physics |
| E. Philip Krider | University of Arizona | Professor and Director -Institute of Atmospheric Physics |
| Glenn Light | Aerospace Corporation | Project Engineer -Space Surveillance and Tracking Systems |
| Joseph E. Nanevicz | SRI International | Deputy Director -Electromagnetic Sciences Laboratory |
| W. David Rust | NOAA | National Severe Storm Laboratory |
| Martin A. Uman | University of Florida | Professor of Electrical Engineering |
| Richard Walterscheid | Aerospace Corporation | Research Scientist, Atmospheric Dynamics |
| Harry Wilson | Aerospace Corporation | |

**Table 5.0.3-1  Lightning Review Committee**

The LRC published its Report as Aerospace Report No. TOR-0088(3441-45)-2 and it is dated 31 August 1988. It contains an Executive Summary, 11 Chapters and an Appendix. The Chapter titles are:

Chapter 1.  CLOUD ELECTRIFICATION
Chapter 2.  NATURAL AND TRIGGERED LIGHTNING
Chapter 3.  LAUNCH VEHICLE ELECTRIFICATION -SELF INDUCED, CLOUD INDUCED
Chapter 4.  RELEVANT LAUNCH RULES
Chapter 5.  RELEVANT INSTRUMENTATION
Chapter 6.  INSTRUMENT RECOMMENDATIONS
Chapter 7.  RECOMMENDED LAUNCH CONSTRAINTS AND RATIONALE
Chapter 8.  TRAINING AND EDUCATION
Chapter 9.  RECOMMENDED RESEARCH
Chapter 10. SUMMARY AND CONCLUSIONS
Chapter 11. GLOSSARY, TERMS, AND DEFINITIONS

The primary protection method adopted by the LRC was to hold (delay) the launch under any of the following four conditions from paragraph 7.2 of the Report:

a.  When natural lightning has recently occurred in the vicinity of the launch complex or along the planned flight path (see Constraint I in paragraph 7.2).
b.  When meteorological measurements indicate that electric fields may be present with sufficient intensity to cause the vehicle to trigger lightning (Constraint II). This heading comprises seven cloud-based rules similar to those in the current LLCC, plus exceptions to several of those rules based on electric-field measurements aloft along the flight path.
c.  When atmospheric electrical measurements at the ground indicate that electric fields may have sufficient intensity to cause the vehicle to trigger lightning (Constraint III).
d.  When atmospheric conditions are such that the launch vehicle interacting with cloud particles, precipitation, or electric fields might charge and produce electrostatic discharges capable of affecting vehicle subsystems (Constraint IV).

The specific launch constraints recommended by the LRC, assuming that the primary hazards are vehicle-triggered lightning, natural lightning, and triboelectrification (due to interactions of the vehicle with its environment during the ascent phase of the mission) are described in Chapter 7 of Heritage (1988) and are presented in our Appendix I, section A1.8, but not reproduced here because of their length.

The LRC concluded that 'The primary atmospheric electrical hazard to launches is vehicle triggered lightning; that is, lightning initiated by the launch vehicle itself in situations where natural lightning otherwise would not occur. Launch vehicles, by virtue of their enhancement (concentration) of the ambient electric field, can initiate lightning in electric fields that are tens to hundreds of times smaller than what would be required to initiate natural lightning. Electric fields at the ground greater than about 1 kV/m, and all fields aloft greater than about 5 kV/m (or even lower at high altitudes), indicate that there is a potential for vehicle-triggered discharges. Fields of this level and higher can be generated by common thunderstorms and also by non-thunderstorm clouds, such as nimbostratus, cirrus anvils, other types of layer clouds, and cumulus clouds in various stages of development. In fact, the triggered strikes to both Apollo 12 and Atlas/Centaur-67 occurred in stratus clouds associated with fronts. Additional hazards to the launch vehicles include direct strikes from natural lightning and the effects of vehicle electrification.' (p. 10-1)

### 5.0.4    *The Subcommittee on Space Science and Applications Hearing*

This section consists of notes paraphrasing key portions of the transcript of the hearing before the House Subcommittee (Committee on Science, Space and Technology, 1987). Unless the text is in quotation marks and explicitly designated as a direct quotation, it is not a verbatim reproduction of what was said. The proceedings are summarized in great detail because they provide significant insight into the mindsets of those responsible for assessing the weather and those responsible for assessing the weather's potential impact on the mission's safety and success. The authors of this History have added notes and comments throughout the section to emphasize or expand on key issues. In many cases, the initials of the specific author of the note or comment are provided.

Although the payload was owned by the U.S. Navy, the Department of Defense had contracted with NASA to provide the launch services and with the Air Force provided the operational weather support. As a result, much of the heat from the Congressional inquiry fell on NASA and the Air Force. In the transcript below the Congressmen are identified by their party affiliation and state in parentheses following their names. Key witnesses whose identity may not be stated explicitly in the proceedings below include the following:

- Jon R. Busse – Chairman of the NASA panel that investigated AC 67
- Brig. Gen. George E. Chapman – Commander of Air Weather Service
- Maj. Gen. Donald L. Cromer – Vice Commander of Air Force Space Division
- Lt. Gen. Forrest McCartney – NASA KSC Center Director
- George A. Rodney – NASA Associate Administrator for Safety, Reliability, Maintainability and Quality Assurance
- Adm. Richard Truly – NASA Administrator
- Lt. Col. John Warburton – Commander of the 45th Weather Squadron during the AC 67 launch
- James Womak – NASA/KSC Acting Director, ELV operations

Following many of the paraphrases below, the page number where the testimony may be found in Committee on Science, Space and Technology (1987) is noted in parentheses.

Mr. Nelson (D FL), *Opening Remarks*:  Purpose of Hearings--Learn what happened to cause accident and how to prevent future accidents. 'NASA management seemed to be improving after Challenger, was it real and lasting, or was NASA slipping'?

Mr. Walker (R PA), *Opening Remarks*: Purposely keep man in loop because the best general purpose computer ever devised rested between the ears of every human being. Every member of launch team has a responsibility to stand up and say this doesn't make sense, aren't we doing something dumb here? Thus want everyone to be open and honest. Over 2000 launches occurred from CCAFS last 30+ years with virtually no mission failures due to weather.

Mr. Busse was the first witness and began by summarizing his written statement including the conclusions and recommendations from his investigation. (*See previous section, NASA Busse Report, for details.*)

Chairman Nelson asked Busse if his panel determined why NASA and ELV criteria for electrical hazards were different. Busse replied the Shuttle criteria were revised (for Shuttle) in late 1970s but ELVs did not feel revisions were necessary but there was nothing in writing.

Chairman Nelson asked Busse if field mills were used during countdown. Busse replied yes, there's evidence in transcripts of Cape Weather discussing field mill readings with launch managers during countdown. Data is available in two forms: continuous readings on strip charts for selected mills; and also on a contour display which takes data from all operational mills and generates contours of constant electric field intensity. Busse added that only 21 of the 34 field mills were operating during countdown. (p 18)

Chairman Nelson questioned the rationale for the 1000 V/m criterion. Bruce Fisher (NASA Langley engineer involved in F106 research programs) replied it was based on the accumulated research in the scientific literature from balloon borne experiments with field mills, airborne field mill research, and especially the experiments done at KSC's rocket triggered lightning site. A more detailed written explanation was later provided by Dr. Hugh Christian of NASA/MSFC and Mr. Dwight Surter of NASA/JSC: Cape foliage goes into Corona at about 3-5000V/m and creates screening layers between the surface and the charged clouds above. Rocket triggered lightning experiments suggest lightning can be triggered when fields reach 3000 V/m. 1000 V/m was chosen as the constraint threshold to provide an added level of safety from 3000 V/m and because it's below the value at which the Cape foliage begins to go into corona and produce space charge and screening layers. (p 21)

*(Note: The LAP is currently preparing a companion document to this History explaining the scientific, technical and operational rationale behind the current LLCC. The Rationale document is being prepared as part of the same project as this History. The Rationale will contain additional detail and explanation based on latest research findings and theory. )*

Chairman Nelson then asked, 'since 3000 V/m is necessary to trigger lightning, how much greater is the risk of triggered lightning if the fields were reading 6000-8000 V/m as they were prior to AC 67 launch'? No panel members could answer. (p 23)

Chairman Nelson asked if you looked outside that day at time of launch, what kind of weather would you have seen? Busse and Lt Col Kim (USAF, Weather commander at Egland AFB, representing AWS) replied it was raining with cloud bases at 4500 Broken and 8000 feet (1.4 and 2.4 km) Overcast. There was thunder from thunderstorms which had been moving east all day from the Florida West Coast to the Florida East Coast. (ibid)

Chairman Nelson questioned why there was no aircraft used. Busse replied an F-16 aircraft was supposed to come from McDill AFB in Tampa but the weather was so poor in Tampa the aircraft could not take off. Mr Bauman added no aircraft were available from Patrick AFB so they decided to just rely on Pilot Reports. (pp 23-24)

When discussing the 6000 foot cloud thickness LCC, Congressman Walker concluded there was great uncertainty in the triggered lightning LCC and that much may have been learned since the AC 67 LLCC were written in 1970. Bruce Fisher agreed.

*(Note: The brief dialogue was a good justification for the formation and continuing contributions of the Lightning Advisory Panel!) (p 25. See also discussion on p 28 for additional justification for a standing Lightning Advisory Panel)*

Rep. Skaggs, member of Chairman Nelson's Science, Space and Technology Committee, was concerned that the analyses and discussions seemed to be compartmentalized around specific LCC. He asked whether someone was tasked to look at the data and situation more holistically. He did not receive an answer to his question and this concern was expressed in other ways throughout the Hearings.

*(Note: Today's LLCC address his concern, at least partially, in the Common Sense Rule, and the launch day LWT processes require the presence of a Senior Weather Officer, preferably the weather unit's Commander or Director of Operations). (pp 26-27)*

Throughout the Hearings the subject repeatedly arose of what the LWO (Launch Weather Officer) intended when he said 'weather is GO for launch' vs. what the Launch Director and his team thought the LWO intended. The LWT (Launch Weather Team) intended 'GO' to mean the LWT believed all the weather LCC had been met, but the LWT also believed the final determination whether the LCC were met was the Launch Director's (LD) responsibility based on the briefings and data provided him by the LWO and his LWT. However, the LD believed the LWT meant GO with no caveats for reviewing the data further beyond the questioning of the LWO by the LD during their discussions and weather briefs over the net.

Mr. Womack: 'When we got the GO for launch, I assumed that was a GO for criteria and we met the criteria, period --in retrospect I should have questioned them more closely on the criteria that day'. (pp 29-31,37, 38, 52, 53)

When Rep Buechner asked Mr. Bauman, who'd been the Launch Director for 30 Delta launches, whether he'd have launched that day, he relied 'probably Yes' because 'weather had always been accurate before.' (p33)

Cost of AC 67 vehicle and payload ~ $163M (in 1987 $). (p 35)

Mr. Busse: Burden of proof ought to be on proving the weather is good enough rather than on proving the weather is bad.

*(Note: This concept of operations is documented in today's LCC, and has been the case since the major 1988 LLCC revisions, which require the LWT to be 'clearly convinced LLCC have not been violated' i.e. are Green.'*

Mr. Busse: Field mills are not a cure all for understanding all weather problems or conditions that might exist with respect to lightning. (p 37)

Mr. Womack: Decision to launch A/C vehicle required concurrence of three officials: Launch Vehicle Manager, Mission (payload) Manager, and Launch Director, in addition to Range who evaluate readiness of Weather, Range Safety, Pad Safety, and Vehicle Tracking. (pp 43-44)

LtC. Warburton: 'We believed the constraints were met, and we recommended the launch proceed based on the weather.'(p 48)

LtC. Warburton: It was an overrunning or frontal type of day vs. a summer thunderstorm type of day. (p 49)

Mr. Walker: Someone who was there described the rain coming down was like a monsoon rain--it was difficult to drive cars near launch site or on the Cape. (p 53)

Rep. Walker to Launch Director Womack: Were you aware of the field mill readings.?

Mr. Womack: We discussed field mills in general, the fact they were being monitored, but I received no numerical numbers. (p 56)

Rep. Packard to Warburton: Is there a margin of safety in weather criteria?

LtC. Warburton: I don't believe there's much margin.

Rep. Packard: Should there be?

LtC. Warburton: Depends on risk or necessity of launch. Some launches must occur. Some launches don't need to occur, and there may even need to be different launch constraint for those. (ibid)

Mr. Womack: General McCartney had advised us that we did not have to launch that day, so be very aware. (p 59)

LtC. Warburton: During one of the 51L [Shuttle Challenger. Ed.] attempts the weather team thought we'd met constraints; John Young [An astronaut. Ed.] in airplane thought we'd met constraints; but the Launch Director (Bob Sieck) called the situation too marginal and believed we should scrub. We've had that type of experience in the past. (ibid)

Chairman Nelson: A/C decision seems to have been made in midst of poor communications and blurred lines of responsibility--frustrating that Atlas launch team was not well coordinated after all the years of enormous and successful experience. Seems the forecasters should have been aware of their role in the launch decision. The Launch Director should have been more critical of the GO recommendation from the forecasters, and especially as was testified here today, that there was no pressure, no urgent need to launch on that particular day. So it seems from what we've heard so far the launch decision process was flawed with perhaps some of the same communications mistakes that were uncovered by the Rogers Commission [The Commission that investigated the Challenger accident. Ed.], namely that individuals overseeing the launch had to prove they were not ready to 'go,' rather than prove they were ready for launch. (p 60)

Adm. Truly: I've gone to Nat'l Research Council and asked for a blue ribbon study--to get some advice where we might put future funding and focus our efforts and do it in a coordinated way so we don't duplicate efforts already underway. (p 62)

Adm. Truly: Revisions to the weather LCC have been initiated. Each will be clearly documented, along with rationale, and the required sensors and instrumentation to assure it is met. (p 65)

*Note (JCW): Although the Heritage Committee [Heritage, 1988] did provide rationale for their proposed rules, subsequent changes have not been so justified, providing the motivation for the present pair of reports.*

Maj Gen. Cromer: The Weather Office (Det 11) has developed both checklists for each LCC that state the data and mandatory instrumentation required to evaluate each criterion, and new CCTV displays that show the current and T-0 forecast for each constraint. In addition, a Senior Weather Detachment representative will be collocated with the Mission Director to provide face to face advice on marginal weather conditions and more in-depth insight into weather and resultant risks.

*Note (JTM): Both NASA and the AF later evaluated this recommendation with respect to the KSC and the Eastern Range and declined it because they did not want to separate the weather person from his/her weather data and team, especially given the very dynamic weather patterns in America's 'lightning capital. However, the Western Range has a different concept of operations that suits their geographical layout and weather patterns better and they do station a weather officer with the Mission Director. (p 79)*

Chairman Nelson to BG Chapman: To what degree is the 1000V/m criterion a matter of discussion among AF weather experts?

Gen. Chapman: Criterion not discussed with me. But I'd like to add that it is an operational criterion that is very common throughout the AF and the Army or those commands that I support.

*Note: Rep. Nelson was apparently referring to the field mill criterion in the Shuttle LLCC (see Appendix I Section A1.6) since the ELVs did not have such a rule at the time. It's doubtful BG Chapman understood Rep. Nelson's question correctly. The General was likely referring to electrical constraints other commands have for different types of operations that are also sensitive to electrical discharges. (p88)*

Chairman Nelson to Cromer: Where are you applying the field mill LCC?

Gen. Cromer: Field Mill System only exists at Patrick (CCAFS and KSC). We can't apply it at Vandenberg because there's no system there. We'll deploy it where appropriate. I don't know that we have enough prevalent conditions for induced lightning on West Coast to apply field mill data there. (p 89)

*(Note: Although Field Mills existed at a few locations around the world, the Field Mill network at CCAFS was probably the only system in existence used for operations)*

Chairman Nelson to Chapman: Communication between your weather specialists and the Launch Director broke down. Neither completely communicated to the other their impression of the current situation. What do you intend to do to implement from what we've learned from our mistakes?

Gen. Chapman: We will emphasize what we always do throughout our support to all our NASA, Air Force and Army customers: go directly to them and ensure we understand their requirements. We will then document our procedures in detailed checklists, as General Cromer mentioned, to ensure we haven't missed anything. Checklists will be a very important part of our process as they are throughout the Air Force. (p 90)

Chairman Nelson: The AF specialists were surprised to learn the count was proceeding. I wonder if the rigid chain of command authority, the checklist mentality would be an impingement upon expressing a difference of opinion?

Chapman: No sir, I don't think so. We have a questioning attitude at all times.

Chairman Nelson: How will you tell your specialists to communicate their level of confidence? Chapman: Difficult question to answer. In each case there is a judgment call. We will do as our customer wants and as Admiral Truly recommended and require clear and convincing evidence.

*Note: Level of confidence is now expressed in Probabilities for <u>Forecasts</u>, but for final evaluations/assessments of observed weather conditions, the Launch Weather Team must be '<u>Clearly Convinced</u>' the constraints have been met (i.e. are Green) otherwise they are Red. (p 90)*

Chairman Nelson to Truly and McCartney: Did the Launch Director or the weather office have the final responsibility to ensure the weather requirement were met. Truly and McCartney to Nelson: Clearly the Launch Director had the final responsibility.

*(Note: On the Launch Weather Team, the Senior Weather Officer has the final responsibility for the assessment communicated to the Launch Director. However, as stated above all LWT members must be 'clearly convinced LLCC are met, otherwise LLCC are Red) (p 92)*

Chairman Nelson asked Truly about a Nowcasting Facility he'd heard about. Truly referenced the NRC assessment and report on weather support to America's Space Program which recommended a technology transition facility co-located with forecasters. Truly said NASA was in process of assessing the NRCs recommendation. (p 94)

*(Comment (JTM): That 'Nowcasting Facility' does not do forecasting but does Technology Transition based on customer taskings. It is called the Applied Meteorology Unit, was formed in 1991, is co-located with the AF forecasters at CCAFS, and has been providing very valuable tailored weather support products to both manned and unmanned space programs ever since-- see Sections 1.3 and 5.2.3.)*

Gen. McCartney: It was clearly a matter of everyone there in the launch decision team, the Launch Director, the Launch Vehicle Manager, and the Mission Director felt it was a clear situation where the mission would not be jeopardized by launching. In retrospect, it was not the right decision to make and as the Senior Manager there, if I had any indication, I, like Admiral Truly, would have stopped it immediately. We had discussed all afternoon, well before the launch, there would be no waiver of any criteria. The criteria were there, they would be followed, and there was never any intent to waive any of those criteria. So it was clearly a situation of everyone felt the criteria were adequate, it was being met, and the decision was made to go. In retrospect it was the wrong decision. (pp 92-93)

*(Comment (JTM): The exchange continued. The weather personnel thought they'd communicated to the launch team that the situation was 'close' and 'hairy' and it was the launch team's responsibility to make the final assessment whether the weather criteria were met; while the launch team, knowing the weather was a concern and having discussed weather all afternoon, remembers hearing weather personnel say 'the criteria are met' and 'weather is GO for launch.' Thus weather personnel were surprised the launch proceeded, and the launch team was surprised to learn they should have realized it was barely acceptable to launch according to the criteria, since critical airplane and weather radar data were missing, and rawinsonde data was questionable during rain.) (pp 93-94).*

Chairman Nelson read the following from a transcript of communications between Cape Weather and the launch team 4 minutes before launch: "We are monitoring our field mills right now and the (electric) potential is up east and right about the SLF which is just outside the 5 mile limit. Over the pad itself we don't show any potential for lightning, and to the east of the pad, which is the path of the missile, it looks good and we haven't had any lightning within 10 nm of the Cape." (p 94)

*(Comment (JTM): Two concerns: 1. The statements "(Electric) potential is up east", and later: "to the east of the pad, which is the path of the missile, it looks good" appear to contradict. That aside, there are no field mills or other measurements over water, thus one cannot state 'there's no potential.' 2. Reference the statement: "Over the pad itself we don't show any potential for lightning": At the time, the potential near the pad was in process of transitioning from high negative fields to high positive fields, that is, the field was passing through low values on its way to high values of the opposite polarity. It's just an indication of the atmosphere's disturbed state and no guarantee the electric potential isn't high along the flight path. If one just views an electric field mill value in isolation and not its temporal trends, and the spatial distribution of electric fields surrounding the mill and along the flight path, one will be misled regarding the danger. The field mill display system available to the weather team includes three important displays: 1. A display of individual field mill values; 2. A display based on readings from all the other field mills, of contours connecting values of equal potential for a spatial view of the surrounding fields; and 3. A display, of 'strip charts, with plots of the time history of each field mill's values over time. This allows the Launch Weather Team to determine if each field mill has been varying little, at fair weather values, or is fluctuating widely between positive and negative electric polarities.)*

Chairman Nelson to Truly: Why does NASA use field mills for Shuttle but not for ELVs?

Adm. Truly: I do not have an answer but in hindsight it was a mistake. (p 95)

*(Comment (JCW): This is really the key question.)*

Adm. Truly then addressed an earlier suggestion by the AF Det 11 Weather Commander that their weather Go/No Go calls might be more conservative for the manned Shuttle missions than for the unmanned ELV launches. Truly said, "I don't think NASA has any launch today that would require a triggered lightning potential to be different for an expensive satellite that is going to launch from the Cape or anywhere else, as compared to a Shuttle." Nelson agreed and said that while there may be differences with the way they are put together and in their redundancies, there should be no differences in the way weather is applied and evaluated. (p 97)

Chairman Nelson to Cromer: An aircraft was not available. What will you do in the future?

Gen. Cromer: If we can not verify a clear line of sight along the flight path and no aircraft are available, we will cancel the launch.

*(Comment (JTM): While an aircraft can provide critical data under some weather situations that would make the difference between whether the launch weather team is clearly convinced the LLCC have been satisfied, there are many situations where the nature of the clouds along the flight path, their type, thickness, temperature levels, history, etc, does not require data from an aircraft because proper data is available from other sources.) (ibid)*

Gen. McCartney to Nelson: Air Force weather has greatly improved their displays. The presentation of data is a significant improvement to what decision makers have had before. (pp 97-98)

Gen. McCartney and Cromer to Nelson: A Senior Weather Officer will be available in the blockhouse, eye ball to eyeball with the launch director and launch team rather than on the net, so they can question him or her and assist them in interpreting the weather data they are receiving.

*(Comment (JTM): Western Range: Weather Officer is located with the launch team. Eastern Range: Shuttle management considered proposal but decided Weather Officer should be co-located with all the weather data, especially since the weather can change so rapidly. They wanted all training on how to interpret data done before countdown supplemented over net if needed.) (p 98)*

George Rodney--Director Safety, Reliability, Maintainability, and Quality Assurance (SRM&QA) at NASA HQ—opening statement: I chose not to go to the launch and also did not send any of my senior managers because I considered it a low risk event of a highly mature system and crew. This is a good example of a Lesson Learned. "We are continually in a high risk business. There isn't any such thing as a low risk operation in our business, and we can never be complacent with that thought." (p 100)

Mr. Rodney--Written Statement: How can such an accident be possible in the new NASA safety conscious environment at NASA? Three factors: A. Many successful launches conducted at KSC in conditions that were not blue sky and sunny. The veteran AC 67 launch team may have been less than appropriately sensitive to the hazards imposed by marginal conditions and may have been led to depend unquestionably on the LCC to protect the vehicle from weather induced hazards. B. Veteran launch teams like the AC 67 team have can-do attitudes they can handle any anomalies that arise in a normal countdown. While hindsight shows resolution of some AC 67 deviations was inappropriate, the team members believed they were reviewing, processing and resolving these deviations correctly and in a manner consistent with past successful launches. C. I believe the

ELV side of NASA has not received comparable SRM&QA emphasis when compared to Shuttle. I believe these underlying factors contributed to what the (Busse) board described as a lack of awareness of the "...indications that the weather was unfavorable and that specifically there was a lightning hazard." The additional constraints, clarification of procedures, and oversight measures that the new launch decision process will impose, will counteract such unconscious group predispositions and group dynamics. (pp 103-104)

*(Comment (JTM): 'Group predispositions and group dynamics' were also threats to the weather support process, especially with the new, post AC 67 lightning LCC. The new LCC were much more complicated and thus prone to misinterpretation, confusion, and thus propensity to being dominated by the loudest voice(s). Consequently, the Senior Weather Officer position was created to oversee the entire process and ensure group dynamics did not bias the evaluation. The Senior Weather Officer ensures all roles and responsibilities are properly executed, all LCC are properly evaluated, and the evaluation results communicated according to the new rules/guidelines).*

Chairman Nelson to Rodney: Everybody now embraces the idea you should use field mills as another measure for expendable rockets, in addition to the Space Shuttle using them. We had to learn through a painful experience that's what we are going to do. What's your conclusion as to why somebody wouldn't have concluded that anyway, without having to go through and lose an Atlas?

Mr. Rodney: We had planned to bounce the Shuttle against the ELVs and other programs, but it hadn't occurred at the time of AC 67. (p 108)

Chairman Nelson: So since NASA had gotten into a mindset they were now having to get rid of ELVs, that they really didn't worry about the launch criteria, having it updated to what otherwise would be using more recent technical data, that they didn't have when that criteria was established years ago. Is there truth to that? (ibid)

Rodney: Yes sir, and it was coming off a very acceptable successful ELV launch rate. (ibid)

Chairman Nelson: Thank you all for a most enlightening hearing.

*(Comment (JTM): Important issues related to accident the Hearing failed to consider:*
*Prior to launch, the Duty Forecaster issued a Lightning Advisory for lightning within 5nm based on the very high 8000 V/m Field Mill readings. The Advisory should have prevented launch. However the forecaster's shift ended before launch which allowed the forecaster's successor to cancel the Lightning Advisory and the countdown to proceed. When the Lightning Advisory was canceled, the Field Mill readings were still 8000V/m and remained so at launch.*

*The Lightning Advisory discussed above was in effect when the 'GO' was given for launch. Weather radar was not operational due to upgrade—major data source for evaluating LLCC. When discussing validity of using balloon data as a substitute for the missing aircraft, data in evaluating whether clouds extended above the freezing level and violated the Thick Cloud LLCC, the Hearings failed to consider that the balloon blows downstream with the wind. Consequently the moisture data the balloon was measuring may not have represented the moisture present and whether clouds were likely present above the freezing level along the flight path.*

*Rain, heavy at times, was reported throughout CCAFS and KSC. This, and the presence of nearby thunderstorms, was not consistent with either a drying atmosphere aloft or with cloud tops below the freezing level which were necessary in order to declare the Thick Cloud LLCC was Green. The weather team helped justify their 'Go for Launch' decision based on the decrease of several nearby field mills from very large values to near zero. Hearing participants failed to mention that fact by itself was insufficient unless it included an evaluation of the electric field contours and field mill strip charts. This would have clearly demonstrated to the Congressmen that the observed electric field decrease to zero was just part of a larger very disturbed electric field which was swinging from large positive values through zero to large negative values (and vice versa . In fact, the electric field did continue changing polarity by thus passing through zero to 8 kV/m near T-0.*

*Many major signs of potentially dangerous electric fields aloft were present simultaneously: high surface electric fields, locally heavy rain, upstream thunderstorms and lightning with upper winds capable of advecting charge laden debris cloud over CCAFS. The Apollo XII incident in 1969, AC 38 in 1976 (discussed earlier in this report) and AC 67 in 1987, though years apart, are startling in terms of the similarities between the weather conditions, faulty or ambiguous commit criteria and faulty evaluation processes for applying the criteria. The Apollo XII and AC 67 incidents were triggered lightning events, and the AC 38 incident was probably also such an event. The fact that faulty processes, so ingrained and numerous, do not occur overnight, likely indicates a failure of higher headquarters (HHQ) to fulfill their oversight and insight responsibilities. There was no indication of effective HHQ inspections, staff assistance visits, simulations, etc.)*

## 5.1   Infrastructure

Except as noted below in this section, the weather infrastructure at KSC/CCAFS remained the same as that presented in Section 4.1 during the period covered by this chapter, but as described below, a number of improvements were made.

### 5.1.1   LPLWS Upgrade

In 1988 NASA established requirements for an upgrade to its network of field mills which the Range called the Launch Pad Lightning Warning System or LPLWS. A NASA Requirements Document (79K32611) was released by the joint NASA/ESMC LPLWS Task Team. The Phase I requirements were (Michael Maier, CSR, private communication):

- Add new field mill sites to provide coverage of all areas of interest
- Conduct survey to verify location of all field mills
- Replace the aging Tektronix monitor and strip chart recorders in the CCFF
- Improve exposure of all field mills by clearing sites per the standard included in the requirement
- Conduct an Operational Acceptance Test of the LPLWS to be overseen by a Det 11 2 WS observer
- Establish configuration management and control of LPLWS. The configuration of the LPLWS was frozen for the Operational Acceptance test (OAT) in December 1987 although three sites were not certified
- Establish appropriate configuration identification for all aspects of the LPLWS
- Begin a LPLWS replacement project (Phase II) where NASA would design and provide improved field mill sensors and ESMC would provide improved processing and display capability

A scientist, Michael Maier, who was personally involved in the infrastructure upgrades, recalled that the goal was to certify LPLWS Phase I as operational prior to the return to flight of Space Shuttle in October 1988 (See Appendix VII). The ultimate goal was a "Phase II" project to replace the existing LPLWS with new mills of a better, more modern design. This was not completed until 1997.

For Phase II a panel of experts -- Krider, Charles B. Moore (NMIMT), and Willett -- was convened by John McBrearty (KSC/TM-PCO-3) to make recommendations for the design of new field mill sensors, the associated network data system, and the data-display system, and to review proposed new designs when they became available. The panel met at KSC on 21 September 1987 to develop design recommendations. The panel formally transmitted nine recommendations for Phase I and 23 recommendations for Phase II to NASA in a letter dated 28 September 1987 (unpublished). The Phase I recommendations dealt with improved calibration methods, site inspection and maintenance and concepts of operations. They concluded with a recommendation to commence Phase II as soon as possible. The Phase II recommendations dealt with many facets of the engineering design (both hardware and software) of the next generation of field mills. An important suggestion was to mount the mills in an inverted position with the sensor head pointing downward rather than upward as was customary at the time. This dramatically reduced the amount of cleaning and maintenance required by eliminating contamination by bird droppings and minimizing the amount of dirt and sea salt that accumulated on the sensor.

The field mill panel met again at the Eastern Space and Missile Center on 17-19 February 1988 and at the University of Arizona in Tucson on 4 March 1988 to review and comment on a proposed sensor design and on the performance of a prototype that had been assembled by Michael Stewart for NASA/MSFC. As a result, a set of eighteen additional recommendations for the electronic design was submitted to NASA in a letter dated 22 March 1988 (unpublished). The panel apparently reviewed the draft Request for Proposals for the new LPLWS in early 1989. Presumably contract award, fabrication, installation, and calibration proceeded over the next few years, but operational acceptance of the new system was not completed until 1997. See Appendix VIII, Section A8.1.1 for a description of the current, now seriously aging, LPLWS.

### 5.1.2 Lightning Detection and Ranging (LDAR)
As a direct consequence of AC 67, the LDAR system which was developed and tested as described in section 3.4.5.3 was made an operational system. LDAR provided additional lead time to issue warnings before cloud to ground lightning occurred which led to the establishment of a "two phase" lightning policy at KSC and CCAFS in the mid-90s. A "Phase I Lightning Watch" is issued when lightning is forecast within 5 nautical miles of a designated facility within 30 minutes, and a "Phase II Lightning Warning" is issued when lightning is "immanent or occurring" within 5 nmi. Ongoing sensitive operations may continue under a Phase I condition but must be shut down immediately when Phase II is issued. Previously, only lightning warnings were issued with the 30 minute lead time and sensitive operations had to stop immediately. There were frequent false alarms resulting in unnecessary work stoppages. Adoption of the two phase lightning policy saves KSC/CCAFS several million dollars in labor costs annually from reduced down time.

### 5.1.3 CGLSS Upgrade
After 1984, the system continued to be developed and evaluated and was eventually accepted into the ER inventory as a fully certified system on 24 July 1989. Between 1989 and 1994, the CGLSS was upgraded to a network of five LLP (Lightning Location and Prediction, Inc.) Model 141 Advanced Lightning Direction Finders (ALDFs), and during 1995-1998, the system was converted to a 6-station, short-baseline network of medium-gain IMPACT (IMproved Accuracy from Combined Technology) sensors (Cummins et al., 1998).

The present CGLSS system has an effective range of about 100 km and covers the KSC-ER launch and operations areas with good accuracy and high detection efficiency (Ward et al, 2008). The CGLSS is operated 24 hours a day, 7 days a week by the Range Technical Services (RTS) Contractor, and the data are sent in real-time to the 45 WS for use in operations.

### 5.1.4 Radar Display Upgrade
When the WSR-74C was acquired by the Eastern Range in 1984, it used display software developed by McGill University. In 1997, this was replaced with commercial software ("IRIS") marketed by Sigmet, Inc. The IRIS software permitted the generation of tailored launch support products such as radar cross sections that could not be generated by the McGill package.

### 5.1.5 Additional Eastern Range Systems
The Eastern Range network of boundary layer wind profilers (Appendix VIII section A8.4.2), rain gauges (Appendix VIII section A8.3.4) and offshore data buoys (Appendix VIII section A8.3.5) were all added to the ER infrastructure during the 1990s. The National Weather Service at Melbourne began its exploration of the

operational use of local, high-resolution numerical weather prediction systems (Appendix VIII, Section A8.6) during the same period.

The 50 MHz Doppler Radar Wind Profiler (Appendix VIII section A8.4.2) was built at KSC by Tycho, Inc. as an experimental proof of concept test bed in 1988 under the direction of the Shuttle Natural Environments Branch at Marshall Spaceflight Center. MSFC soon determined that the standard commercial "consensus" algorithm provided by Tycho was inadequate for either research or operational applications. As a result, they developed a new "median-filter first-guess" (MFFG) algorithm that provided more frequent, more accurate wind profiles than the consensus methodology (Schumann et al., 1999). The original MFFG required more computational resources than the wind profiler hardware could support so it could not be used in real time for operations, but provided excellent data for evaluating the potential value of the instrument. An evaluation by the AMU (section 5.2.3 below) in 1992 showed that the DRWP and MFFG combination produced wind profiles of high quality suitable for operations. As a result, the AMU was tasked to develop an operational version of the MFFG software. By reducing the size and increasing the efficiency of the code they were able to make it run on the operational computer system and added a real-time user interface for day of launch quality control.

### 5.1.6    *Kodiak Alaska Launch Complex*
To support a 2001 launch of an Athena vehicle carrying Air Force and NASA payloads (see Section 5.3 below) from the Kodiak, Alaska spaceport, an adequate weather infrastructure had to be created. Prior to the Athena mission, the site had no weather radar or lightning detectors and only two ten-meter wind towers. Working with the 45th Space Wing, the Alaskan Aerospace Development Corporation and Lockheed Martin Corporation, The KSC Weather Office coordinated procurement, testing and installation of a basic weather radar, a lightning detector, a surface field mill, upper air balloons for measuring profiles of wind and thermodynamic variables, and additional surface wind sensors. In addition, on day of launch an aircraft was contracted to provide real-time weather observations for LLCC evaluation.

## 5.2    Organization

### 5.2.0    *Space Shuttle Headquarters Support Office/Weather Office*
In the late 1980s, the Space Shuttle Program created a NASA headquarters office to oversee all weather support to the program including launch support at KSC and landing support at KSC, WSSH, Edwards AFB and the overseas emergency landing sites. The office was staffed by two senior civilians and a senior military liaison officer from the Air Force. Although the formal title was the Space Shuttle Headquarters Support Office/Weather Office, it was generally known as the Weather Support Office (WSO).

### 5.2.1    *Lightning Advisory Panel*
Following the AC 67 disaster, NASA and the Air Force were eager to identify an existing instrumented aircraft that could measure electric fields aloft in support of space-launch operations -- an operational ABFM. A joint workshop was held in early 1988 at the Cape Canaveral Air Force Station (CCAFS), during which many scientists familiar with airborne field measurements outlined an appropriate system design and a plan of flight operations (Barnes and Metcalf, 1988). A few individual scientists were also asked by NASA/WSO or Air Force Space Command to evaluate an ABFM that had been developed by Stanford Research Institute (SRI) -- at that time the only, appropriately instrumented, high-performance aircraft -- which was eventually determined not to be suitable for launch support.

A group of scientists and engineers led by Hugh Christian at the NASA Marshall Space Flight Center (MSFC) took their cue from the CCAFS workshop, in which they participated vigorously. In 1988 they began a major effort to instrument another Learjet as a demonstration project. The primary goal of the MSFC group, after achieving a credible calibration, was to obtain statistics on the field magnitudes aloft during conditions that violated the LLCC of the time. It was expected that such statistics could be used to make these LLCC less restrictive. MSFC conducted four field deployments during 1990, 1991, and 1992 to gather data for statistical analysis, as described in Section 5.4.1 on the ABFM I.

Because of the potential consequences to safety of any change in the LLCC, around 1990 the newly formed NASA WSO (see 5.2.0 above) and the Air Force Space Division asked an ad hoc group of experts to review MSFC's efforts to calibrate its ABFM. These scientists, who became known collectively as the Peer Review Committee (PRC), were drawn both from the old "Heritage Committee" (Heritage, 1988) and from the more recent individual efforts to evaluate the SRI aircraft. After the first year of the MSFC field program, the PRC's role expanded to reviewing the statistical significance of data collected by ABFM I for revision of the LLCC. Statisticians from The Aerospace Corporation were enlisted to help with this effort.

In 1992 MSFC formally proposed some LLCC changes that were based on the summer ABFM I dataset. In response, the NASA/WSO and the Air Force 45th Weather Squadron at CCAFS (45WS) formally agreed that outside experts in the fields of lightning and atmospheric electricity were needed to evaluate this proposal and any future changes to the LLCC. The formerly ad hoc PRC became officially recognized as the standing committee for that purpose, although its role remained informal and no official charter was written. The original membership of the standing committee consisted of E. Philip Krider (University of Arizona, Chair), James E. Dye (NCAR), Harry Koons (Aerospace Corp.), W. David Rust (NOAA), Richard L. Walterscheid (Aerospace Corp.), and John C. Willett (Air Force Geophysics Laboratory). In addition to having the needed expertise, the members represented a broad cross section of the relevant academic and government organizations. (The most appropriate NASA representative, Hugh Christian, was not on the PRC because it was his ABFM program that was under review.)

The PRC held its first face-to-face meeting at MSFC in August, 1992, to review the first MSFC-proposed rule changes. Over time the PRC's role gradually expanded from reviewing the proposals of others to include drafting its own proposed rule changes and even recommending additional experiments that could increase safety and launch availability. The most significant example of the latter was its December, 1999, recommendation to the KSC Weather Office to have an ABFM II experiment. A detailed outline of the PRC/LAP's activities is given in Appendix II, and a description of its operating methods can be found in Section 5.5.1.

James E. Dye left the PRC sometime after March, 1994, and was reappointed in December, 2003, while he was serving as the PI of the ABFM II campaign (see Section 6.4). In the meantime, the name, 'PRC,' was changed to ' Lightning Advisory Panel (LAP)' in February, 1997, when it was provided with a still-unofficial charter, and in February, 1999, Hugh Christian was added to bring a NASA perspective to the LAP. Sadly, Harry Koons passed away suddenly and unexpectedly on 11 May 2005, and his statistical expertise, penetrating logic, and clear expression were missed by all. Eventually, Dr. Koons was replaced by Paul O'Brien (also of the Aerospace Corporation) in October, 2007. The current membership of the LAP is Krider (chairman), Christian, Dye, O'Brien, Rust, Walterscheid, and Willett.

### 5.2.2 KSC Weather Office
In the late 1980's, Shuttle Program managers at KSC established a local office to oversee the weather infrastructure and weather research it was funding in whole or in part at KSC and the Eastern Range. The office was initially called the KSC Weather Projects Management Office, but later the name was shortened to the KSC Weather Office.

As part of a NASA cost-cutting reorganization in 1995, the agency-wide responsibility for weather support to the Shuttle program was transferred from the NASA Headquarters Weather Office (which was abolished) to the KSC Weather Office. Shortly thereafter, KSC was named the lead Center for launch and payload processing for NASA expendable launch vehicles (ELV) regardless of where launched. That resulted in the

KSC Weather Office also being responsible for weather support to NASA ELV launches from sites as diverse as the Eastern and Western Ranges, Kwajalein Atoll, Wallops Island, VA and Kodiak, Alaska.

### 5.2.3 Applied Meteorology Unit (AMU)

In 1991 NASA, the Air Force and the National Weather Service collaborated in the establishment of an Applied Meteorology Unit co-located with Range Weather Operations at CCAFS and managed by the KSC Weather Office, as recommended by the Theon and NRC reports (see Section 5.0.2). The AMU provides technology development, evaluation and transition services to the 45WS, SMG and the NWS office at Melbourne, FL as well as directly to the Shuttle program. Many of the highly specialized tools used for weather support to both manned and unmanned operations on the Eastern Range and at KSC were developed by or transitioned to operations by the AMU (Bauman et al., 2004).

## 5.3   Operations

The period covered by this chapter was free from major weather-related accidents or mishaps. On 29 September 2001, the first launch of a large rocket from the Kodiak Alaska launch complex took place successfully. An Athena booster placed in orbit the NASA Starshine satellite and three Air Force satellites: PICO, PC Sat and Sapphire.

## 5.4   Research

The period following AC 67 was one of vigorous research activity into atmospheric electricity, lightning and triggered lightning. Methodologies ranged from flying instrumented aircraft into thunderstorm-associated clouds to triggering lightning deliberately with small rockets. Several of the major experimental programs are described below after a general discussion of the methodologies for making airborne electric field measurements.

### 5.4.0 Methodologies for Airborne Electric Field Measurements

Before presenting a summary of the airborne studies conducted near KSC, we will first discuss the measurement of electric fields from an aircraft platform and the challenges encountered. The measurement of electric fields using rocket-borne sensors will encounter similar difficulties but neither rocket nor balloon measurements are discussed here because our focus is on aircraft measurements. Airborne measurements of the electric field inside clouds, especially thunderstorms, are difficult because of the distortion of the ambient field by the aircraft and the extra field that is created by charge on the aircraft due to the engine exhaust, the impaction of hydrometeors, and corona discharge. Although the three spatial components of the electric field vector and the charge (or potential) on the aircraft itself can be determined from four independent measurements, five or more sensors provide the ability to verify the calibration and permit a redundant determination of all four variables. Some redundancy is necessary, and knowledge of charge on the aircraft is helpful, when assessing the accuracy of the ambient-field measurement. A further complication is that space charge is always being emitted by an aircraft in the high-field environment of thunderstorms. If the space charge passes close to one or more of the sensors, it can contaminate the reading or make the measurement of one or more components of the ambient field impossible (Jones et al., 1993).

Most electric field sensors used on research aircraft in recent years have been of the rotating-vane type, i.e. field mills, such as the ones described by Bailey and Anderson (1987), Winn (1993), and more recently by Bateman et al., (2007). A grounded, rotating, segmented vane, or shutter, alternately exposes and shields fixed sensor plates to/from the electric field. Some aircraft flown in the studies near KSC and elsewhere have used cylindrical electric field mills of the type described by Kasemir (1964, 1972), wherein rotating sensor plates are directly exposed to the electric field. The use of one cylindrical mill permits the measurement of two spatial components of the field, but using two mills, for example, one mounted axially on the nose and one mounted transversely under the fuselage, permits the measurement of three components of the field. The in-cloud measurements from a transversely mounted cylindrical mill can be compromised by the impaction of hydrometeors on the sensing plates, however. For the NCAR sailplane, a small diameter, Kasemir-type, cylindrical field mill designed by William Winn was mounted on a boom that was centered ahead of the nose (Dye et al., 1989). Later a different type of cylindrical mill was used on the sailplane in CaPE. This mill had a cylindrical, rotating outer shutter that alternately exposed and covered fixed stator plates on the top and bottom and then on the left and right so that the charge induced on each plate could be measured as well as the charge on the aircraft (Dye et al., 1989).

A calibration of the full field-mill system (as opposed to a calibration of each individual field mill) is the most difficult requirement to achieve. This calibration is necessary to separate the effects of the aircraft itself from those of the ambient field. Roll and pitch maneuvers either in fair weather fields in clear air or in quasi-steady fields below anvils is the method that is most commonly used procedure to calibrate the aircraft measurements. A fair-weather calibration makes severe demands of the instrumentation, because the individual sensors must be capable of a wide dynamic-range to accurately measure fields ranging from a few tens of V/m to hundreds of kV/m. The fields below anvils are often not as steady as one would like. Another part of the procedure is to artificially charge the aircraft in fair weather with an on-board high voltage supply and corona "stinger" to determine the response of the individual mills to this charge. For example, Jones et al., (1991, Vol I) and Jones et al. (1993) used a stinger to calibrate the New Mexico Tech SPTVAR aircraft. Mach and Koshak (2007) have described a mathematical procedure that was used to calibrate four different aircraft, including the Citation that was used in the ABFM II campaign (2000-2001). An early version of this procedure was used for the calibration of the Lear jet 28/29 that was used in ABFM I (1990-1992). Other system calibration techniques include using a numerical solution of Laplace's equation on a mesh representation of the CV-580 aircraft (Laroche, 1986) and laboratory measurements on a scale model (Laroche et al., 1989 for a French Transall aircraft; and Kositsky et al., 1991 for the Aeromet Learjet 36A). An absolute calibration of a field mill system can also be accomplished by flying past a calibrated field mill on the ground (Mach and Koshak, 2007), on a tethered balloon (Winn, 1993)), or by in-flight comparisons with a previously calibrated aircraft.

Winn (1993) proposed a novel calibration method that addresses both the distortion of the external field by the aircraft and the problems caused by aircraft charging. This method does not require systematic placement, or even knowledge of the placement, of the individual field mills on the aircraft, nor does it require knowledge of the magnitude and direction of the ambient-field. The only requirements are that the ambient field be constant in magnitude and direction during aircraft maneuvers and that the field sensed by each mill be a linear combination of the three spatial components of the ambient field vector and the charge on the aircraft. One drawback of Winn's method is that it is not able to directly determine the aircraft charge (or potential). Mo et al. (1998) have taken advantage of Winn's calibration method when instrumenting the South Dakota School of Mines and Technology T-28 aircraft with field mills. The sensors are placed in distinctly unconventional locations, where they are well away from the streams of space charge that the aircraft emits while flying in high fields.

### 5.4.1   *Airborne Field Mill Program (ABFM I)*
NASA sponsored an airborne measurement campaign in 1990-1992 to make direct measurements of the electric field vector in thunderstorm-related clouds, with the hope of gaining sufficient knowledge to make further improvements in the LLCC. This program comprised two summer campaigns, 1990 and 1991, and two winter campaigns, 1991 and 1992.

A Learjet 28/29 operated by NASA Langley Research Center and instrumented by NASA Marshall Space Flight Center was flown from 6 July to 21 August 1990, 14 February to 18 March 1991, 8 July to 25 August 1991, and 19 January to 6 March 1992. The aircraft had turbofan engines and a ceiling of 45,000 ft , a flight duration of 1 to 1.5 hrs, a typical flight speed for penetrations of 200 m/s. It was instrumented with 5 field mills from MSFC, an early version of those described by Bateman et al., (2007). Calibration of the field mill system was an early version of the procedure described by Mach and Koshak (2007). The aircraft also flew an ice patch that sensed the impaction of ice particles and a King liquid water sensor. The combination of these two instruments allowed differentiation between all-ice and mixed-phase regions.

The main objectives of this project were to evaluate the Lightning Launch Commit Criteria and to modify them as necessary to maintain the same or better level of safety while making the rules less restrictive. The goals of the campaigns were to quantify the electric field strengths inside each type of cloud identified in the LLCC as a function of the cloud type, characteristics, and development. Missions were developed for each cloud type ranging from mature thunderstorms, to developing cumulus, to anvils (attached and detached), debris clouds, thick clouds and disturbed weather. Categorizing the relationships between the cloud type, electric field, and radar reflectivity quickly became an overarching objective of the mission.

The flight strategy varied depending on the mission and type of cloud. The overriding objective was to determine whether the cloud violated one or more of the LLCC, and if so which one(s), and to relate the measured electric field to the radar reflectivity of the cloud.

Mature thunderstorm: The main objective was to measure the decrease of the cloud electric field as a function of distance from the storm, and for this the aircraft was flown on repeated radials, toward and away from the cloud. There was an effort not to penetrate too deeply into the cloud, and the aircraft altitude was often varied between radials.

Developing cumulus: An effort was made to measure the electric field produced by cumulus clouds before the cloud top penetrated the zero degree isotherm. The aircraft made repeated cross-sections as close as possible to the cloud top and ascended with the cloud top. A particular interest was to quantify the altitude of the cloud top at first sign of electrification and the electric field strength when the cloud top reached the -10 and -20 °C temperature levels.

Anvil clouds (both attached and detached): Anvils were studied with the primary objectives of determining how the electric field decayed with time and distance from the core of the storm. These studies were hampered because of the limited flight duration of the aircraft. Often, a cloud would be investigated until the plane had to depart due to low fuel, and then it would then be refueled and take-off for a second mission with the intent to continue penetrations of the same cloud. The flight strategy was to combine repeated radials along the length of the anvil from the core of the storm (for attached) to the end of cloud (axial), mixed with traverse cross-sections. Altitudes were varied when allowed by the FAA air traffic controllers at the Miami and Jacksonville ARTCC. Spiral assents and descents were performed at the beginning and end of mission, if permitted.

Debris clouds: were investigated using the same flight strategies as for anvil clouds, except the flight legs were usually shorter and spirals were attempted more often.

Thick Clouds and Disturbed Weather: Similar flight strategies were used on both of these cloud types. Generally, disturbed weather passes rapidly over the Cape during the wintertime and is associated with frontal systems. The plane typically flew paths that were perpendicular to the cloud motion and repeated penetrations of the same region of the cloud were attempted to follow regions of embedded convection. For "thick clouds", the initial penetrations were made at cloud base and were often followed by cloud top penetrations. The bulk of the penetrations were made in the center of the cloud.

During the four aircraft deployments, penetrations were made through a variety of clouds at different stages of development. The winter studies were primarily of thick-layer clouds and clouds associated with disturbed weather, and the summer studies focused mainly on growing cumulus clouds, anvils, and debris clouds. Measurements from the WSR-74C weather radar at Patrick Air Force Base were used extensively in the data analysis, especially to calculate Vertically Integrated Reflectivity (VIR) above 1 km and above 4 km (approximately the 0 °C level). VIR-1 and VIR-4 (termed VII in the reports) are the sum of the reflectivity (in dBZ) in a column above the aircraft from either 1 or 4 km to 15 km. (ABFM Analysis Group, 1991, 1992a, 1992b, 1992c ). Both gridded Cartesian and radar-coordinate reflectivity values were used during real-time operations and for post flight analysis. However, only gridded reflectivity data were used when calculating VIR-1 and VIR-4. Because of calibration issues, the WSR-74C radar could not be used quantitatively to relate the measured electric fields to radar reflectivity in the ABFM-I campaign. Despite the calibration uncertainties, the measured reflectivities are believed to be close to the true values, and therefore, they can be used to characterize the clouds that were being investigated. (Merceret et al, 2008).

A summary of findings from the ABFM I program (MSFC Feb 92 and Oct 92 reports and Mach et al., 1992) follows:

ABFM I Winter studies
1. For thick layer clouds (Rule D in the reports) there is a strong correlation between the electric field inside the clouds and VIR-4 and also VIR-1. The correlation for VIR-1 was slightly better than for VIR-4. (ABFM Analysis Group, 1992a and 1992b). Significant electric fields might have been missed because of aircraft icing concerns while penetrating significant mixed phase clouds.
2. In layer clouds, the electric fields at the ground were not always correlated with electric fields aloft.
3. Fields $\geq$ 5 kV/m were not encountered in clouds at altitudes warmer than the 0 °C isotherm, and the highest altitudes with fields that were $\geq$ 5 kV/m were at temperatures near -10 °C, suggesting that the main electrical hazard was between 0 and -10 °C.
4. For disturbed weather clouds (Rule E in the reports) there was a strong correlation between the electric field aloft and VIR-4, and the correlation with VIR-1 was not as good as VIR-4.
5. For disturbed weather clouds all penetrations with fields > 5 kV/m were between 10 and 21 kft (approximately 0 to -10 °C).
6. Fields measured at the ground under disturbed weather clouds correlated well with fields aloft, but there were 3 exceptions where fields at the surface were < 1 kV/m and fields aloft were > 5 kV/m.
7. Penetrations that showed fields > 3kV/m were generally associated with regions of higher reflectivity (embedded convection).
8. Cloud thickness and the electric field strength were very poorly correlated.

ABFM I Summer studies
1. The fields outside storms with lightning decreased to < 5 kV/m at a distance of 5 nm (9 km).
2. For the 87 cumulus clouds that were sampled, the observed electric field depended strongly on the cloud top height (temperature) as defined by the (uncalibrated) 10 dBZ reflectivity. The fields in clouds with 10 dBZ tops lower than the 0 °C level did not exceed 3 kV m$^{-1}$. Fields > 3-5 kVm$^{-1}$ did not develop in cumulus clouds until the echo tops had grown above the -10 °C level ($\sim$ 6.4 km msl). Cumulus clouds did not produce natural lightning until the tops were above the -20 °C level. Fields at the edge of clouds with echo tops higher than the -20 °C level could be > 50 kV m$^{-1}$ (Merceret et al., 2008)
3. As long as cumulus clouds were within 5-6 nm of the LPLWS network, that system was able to indicate a hazard aloft as long as the fields aloft were > 5 kV/m. There were no cases when the fields aloft were >5 kV/m and fields at the surface were <1 kV/m.
4. Measurements of fields as a function of distance from the cloud edge showed that even with fields of tens of kV m$^{-1}$ inside electrically active convective clouds, the fields external to those clouds decayed to < 3 kV m$^{-1}$ within 15 km of cloud edge as shown in Figure 5.4.1-1 This was true even for clouds that were producing lightning and had tops above -20 °C.

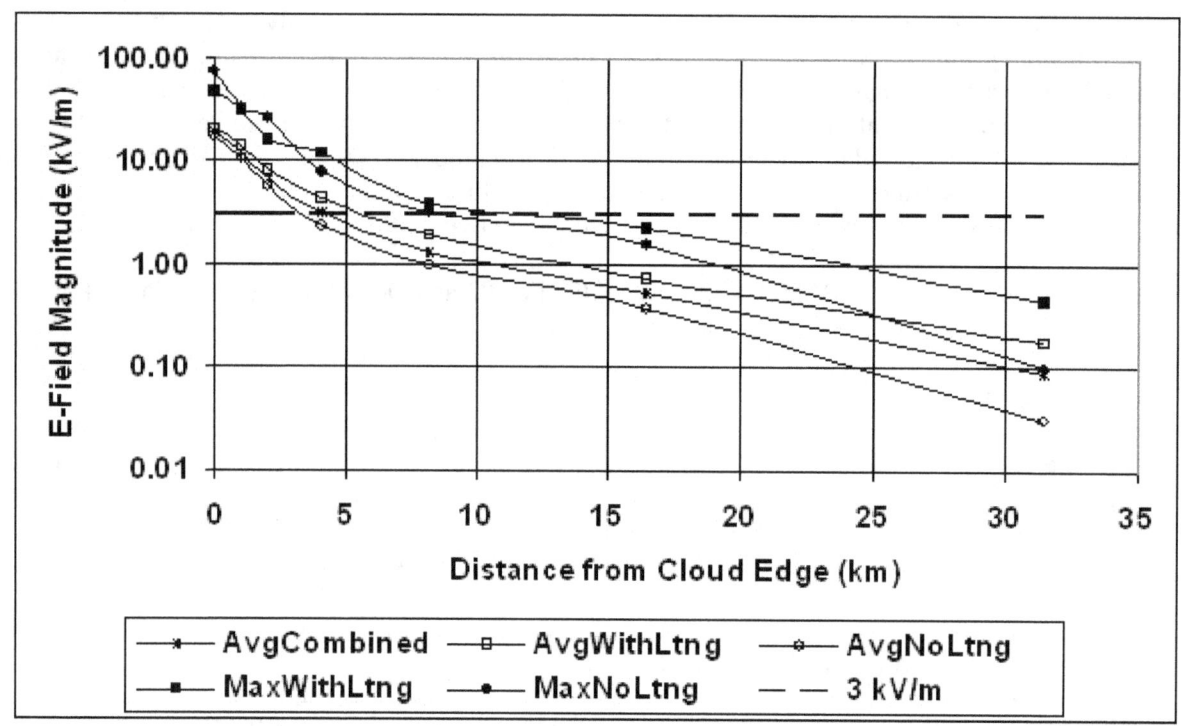

**Figure 5.4.1-1 Electric field magnitude as a function of distance from cloud edge**

Note: Reproduced from Merceret et al. (2008), Figure 1

### 5.4.2 Convection and Precipitation/Electrification Experiment (CaPE)

CaPE was a large observational program conducted during July and August of 1991 with multi-agency funding and investigators from many organizations and universities. One of the five main research themes was identification of the relationships of the co-evolving wind, water and electric fields within convective clouds. The scientific overview and operations plan for CaPE (Foote, 1991) describes the many facilities, radars and aircraft used in CaPE and gives detailed flight plans and how they would be utilized to achieve the goals embodied in the five themes. In this section we briefly describe the operational strategy and some findings for Theme 1 – the co-evolution of wind, water and electric fields in convective clouds. A key contribution of CaPE compared to earlier studies of electrification in Florida cumulus clouds was the close coordination of detailed in-situ electrical and microphysical measurements with multi-parameter radar measurements of the same clouds.

The aircraft primarily involved in the electrification studies were the South Dakota School of Mines and Technology T-28, the NCAR Explorer Sailplane, the NCAR King Air, and the NOAA P-3 aircraft. In addition to the electric field measurements, all four of the aircraft had a capability for cloud microphysical measurements. Plans were also made to have the NASA Langley Learjet (concurrently flying for ABFM I) participate in coordinated flights with the other aircraft, but obligations to the ABFM I campaign precluded that possibility. The NASA ER-2 aircraft sometimes flew over storms at high altitude and made electric field, radar, and radiometric measurements. The Univ. of Wyoming King Air was instrumented with an array of microphysical sensors and a wind measuring capability and participated in the CaPE studies.

The T-28, NCAR King Air, and NCAR sailplane were all instrumented with rotating-vane field mill sensors that were built at New Mexico Tech and were similar to those described by Winn (1993). Additionally, the sailplane also flew a new Winn-designed, rotating cylindrical mill on a nose boom that had the capability of measuring the induced charge on left, right, bottom, top and bottom plates (Dye et al., 1989). The field sensors on the NOAA P-3 were rotating-vane field mills designed by the Desert Research Institute. The calibration of the field mill system on each of these aircraft used pitch and bank maneuvers and artificial charging of the aircraft similar to what is described in Jones et al., 1993 and in section 5.4.1 of this document. An absolute calibration of the King Air system was not completed, but the values were thought to be approximately a factor of 2 high in comparison to flybys with other aircraft (Dye et al., 1992). Each of the 4 aircraft had a capability for measuring the cloud liquid water content and the concentrations of cloud and precipitation particles together with their sizes and types.

The two King Air aircraft made repeated, stacked penetrations of growing cumulus clouds in the early stage of development, with the Wyoming King Air flying slightly above cloud base and the NCAR King Air flying at mid-cloud levels near 0 °C and then ascending as the cloud top ascended until the electric field exceeded 10 kV/m. When conditions permitted, the sailplane entered growing cumulus clouds from the side near the freezing level while both King Airs were investigating the cloud, and after finding updraft, spiraled upward in the cloud. Prior to the sailplane entering the cloud, the NCAR King Air would change to an altitude either well above or well below the sailplane. If the cloud evolved into a larger stage of development, both King Air aircraft would exit that cloud and look for other targets. At this point the T-28 was used to penetrate regions of stronger convection sometimes in coordination with the sailplane and or the P-3. The P-3 flew the NOAA under-fuselage 3 cm Doppler radar, so it was used for coordinated Doppler radar studies in addition to the electrical studies. The aircraft were closely coordinated with each other and with multiple Doppler radars from the CaPE operations center, where aircraft track could be superimposed on radar plots. The primary radars involved were the NCAR CP-2 Doppler 10 cm multi-parameter, polarization diversity radar and the NCAR 5 cm Doppler radars CP-3 and CP-4, all 3 of which were dedicated to CaPE objectives. Additionally, the WSR-88D NEXRAD radar at Melbourne and the WSR-74C radar at Patrick Air Force Base also scanned the storms with their normal scan strategy; and NMIMT supplied a mobile 3 cm dual polarization coherent and incoherent unit operated in conjunction with the CP-2.

Previous airborne and radar studies in Florida had shown (e.g. Hallett et al., 1978) that large raindrops form in the updrafts well before the cloud top reaches the freezing level. When the cloud grows above the freezing level the raindrops ascend in the updraft until a temperature of about -8 to -10 °C is reached at which point some rain drops freeze and the cloud rapidly glaciates. The combined aircraft and multi-parameter radar studies from CaPE confirmed and strengthened this scenario. The polarization radar measurements from several investigators showed that a strong ZDR (differential reflectivity, a measure of mean particle oblateness) column indicative of large drops extended up to approximately -8 to -10 °C, at that point the ZDR signal would decrease and become capped by an enhanced linear depolarization ratio (LDR) indicative of ice formation (Willis et al., 1994; French et al., 1996; Ramachandran et al., 1996; Jameson et al., 1996; and Bringi et al., 1997). The aircraft measurements of cumulus clouds and the KSC field mill network both showed that the initial electrification of cells occurred just after the appearance of an enhanced LDR signature. For example, in a study of 3 clouds that examined measurements from the KSC field mill network, Jameson et al., (1996) showed that the onset of electrification coincided with the appearance of a significant volume of ZDR (indicating large rain drops) above -7 °C level and was accompanied by the nearly simultaneous appearance of radar depolarization indicative of the supercooled drops beginning to freeze. This is illustrated in figure 5.4.2-1 taken from Jameson et al.

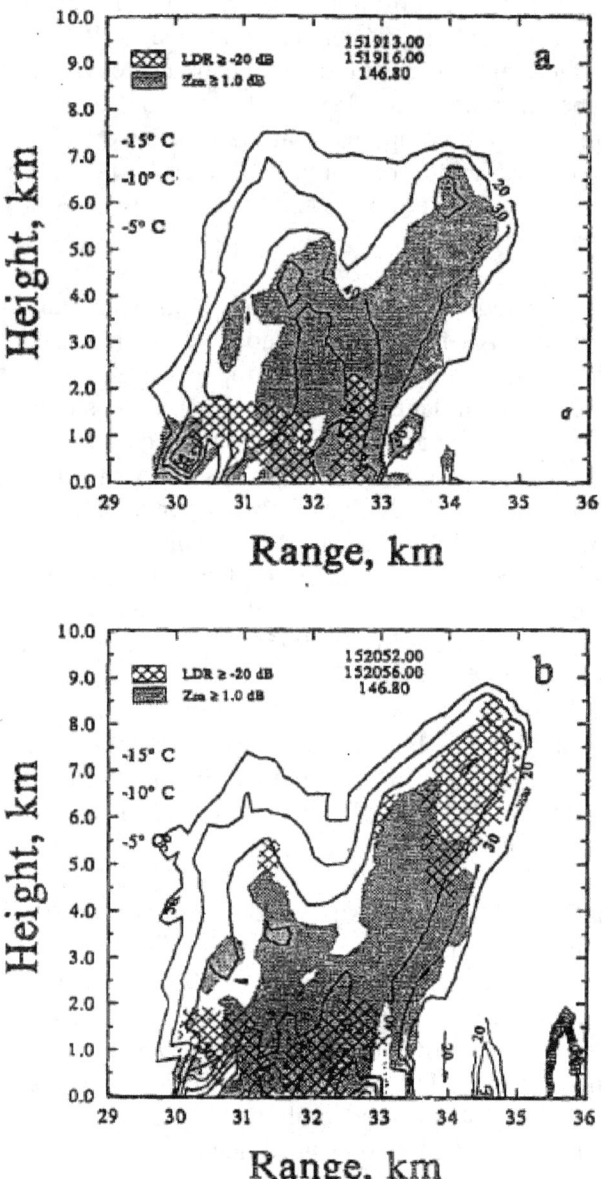

Fig. 2. Radar elevation scan at constant azimuth through a portion of storm B on 19 July 1991 showing (a) regions of supercooled raindrops where $Z_{DR} \geq 1$ dB well above the freezing level and 100 s later (b) where the $Z_{DR}$ has been replaced by LDR $\geq -20$ dB. Large LDR below 2 km are likely artifacts of ground clutter and are excluded from subsequent analyses.

**Figure 5.4.2-1  Radar elevation scan from Jameson et al., 1996**

Similarly, measurements from the NCAR King Air and sailplane showed that some clouds with tops at -8 °C had very limited development of ice and at fields that were only 2 kV m$^{-1}$ or less. Cases that had ice development but limited convective growth above the 7 km level (~-10 °C) produced fields of 5 to 10 kV m$^{-1}$ (Dye et al., 1992). An example of one growing cumulus that was penetrated by the King Air near or slightly above the 0 °C level eight separate times is shown in figure 5.4.2-1. Even though the radar cloud top had previously reached almost -20 °C, at 1420, the maximum observed field was ~3 kV m$^{-1}$. The field did not exceed 10 kV m$^{-1}$ until the pass at 1431 just after the development of new growth had started.

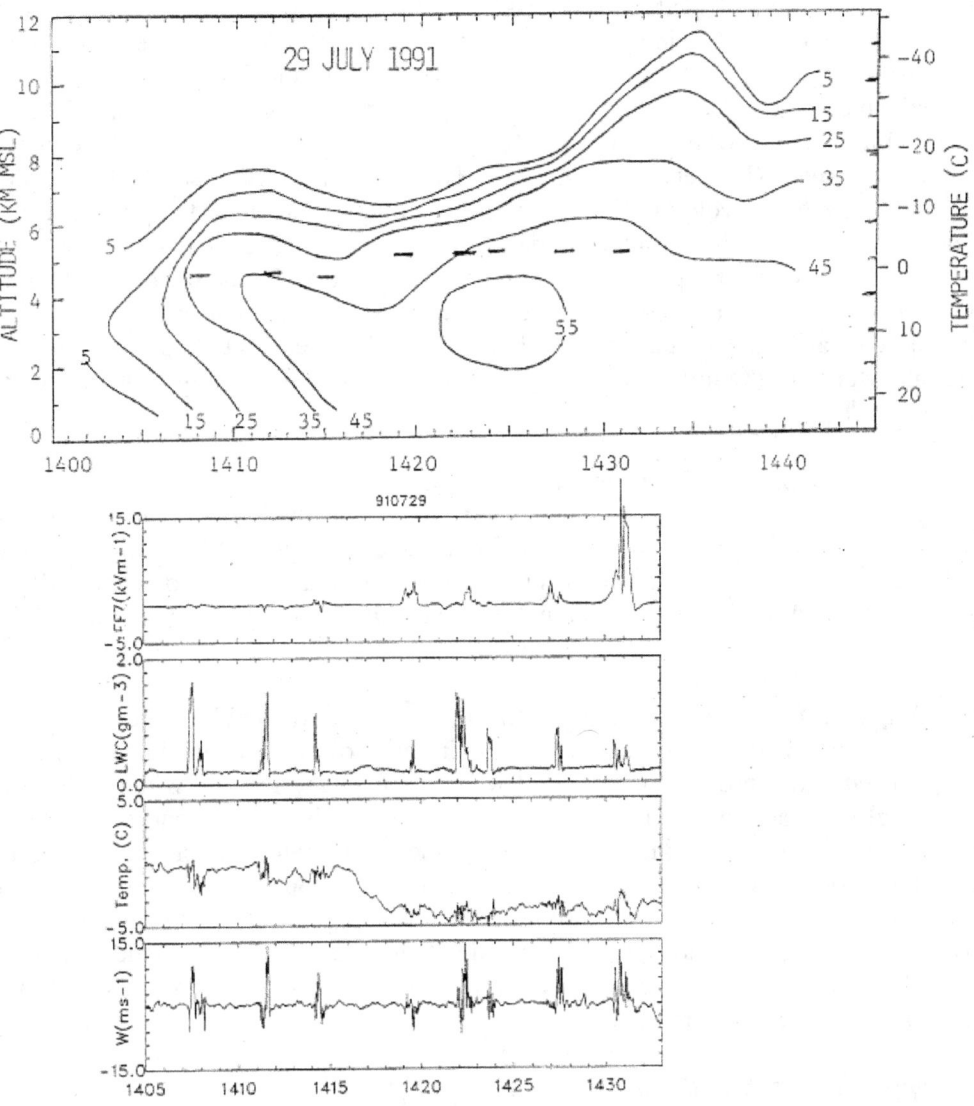

**Figure 5.4.2-2  Growing cumulus example adapted from Dye et al., 1992**

Note:  Figure represents local time. Top panel: Maximum reflectivity plotted as a function of altitude and time. The short solid dashes show the times of cloud penetrations by the NCAR King Air. Bottom panel:  Plots of the vertical component of the electric field (EFZ), liquid water content (LWC), temperature, and vertical velocity (W) plotted as a function of time.

It should be noted that in the high plains such as in New Mexico, Montana and Colorado where cloud bases are much higher and cloud base temperatures are colder (typically 5 to 10 °C compared to 15 to 20 °C in Florida), large drops do not form below the freezing level. The first precipitation is formed via an ice process: Small ice particles are nucleated in the cloud. These embryonic particles then grow by diffusional growth from transfer of water vapor from the air to the ice particles until they are large enough to fall relative to cloud droplets. As these ice crystals fall and grow sufficiently large they begin to collide with and accrete supercooled water droplets to become snow pellets (graupel). This ice phase process of precipitation development which dominates in the high plains (e.g. Dye et al., 1974) takes much longer than the development of precipitation through the warm rain process in Florida clouds.

Unlike New Mexico where initial electrification is associated with cloud vertical growth, the initial electrification to fields >1 kV/m in Florida sometimes occurs during the decay phase of the cloud (Breed et al., 1992) and can be seen in figure 5.4.2-2. The electrification of cumulus clouds in Florida is similar to that found for New Mexican clouds by Dye et al. (1989), but in Florida clouds often become electrified once the cloud top exceeds -7 km (-10 °C), whereas in New Mexico cloud tops exceed 8 km (-20 °C) before electric fields become greatly enhanced (Dye et al., 1989, 1992). Even though millimeter-sized raindrops and reflectivity of 50 to 60 dBZ can develop in Florida cumulus clouds when cloud tops were wholly warmer than 0 °C, the combined aircraft and multi-parameter radar measurements showed that significant electric fields do not develop until there is development of ice and mixed phased regions. Inside the clouds that did become significantly electrified, both millimetric graupel and numerous smaller ice particles were found in the presence of supercooled water (Dye et al., 1992; Willis et al., 1994; French et al., 1996; Bringi et al., 1997), thus showing the importance of the mixed phase region between -10 and -20 °C for the electrification of Florida convective clouds.

### 5.4.3 Rocket Triggered Lightning Program (RTLP) Results
"Rocket-triggered lightning" is initiated by a small rocket towing a grounded wire aloft under a thunderstorm (St. Privat D'Allier Group, 1985). The rocket-and-wire technique for triggering lightning was pioneered by Newman et al. (1958, 1967). The key to its success is likely an observation by Brook et al. (1961) that the sufficiently rapid introduction of a grounded conductor into a high-field region might actually initiate the discharge.

The KSC Rocket-Triggered-Lightning Program (RTLP) provided an invaluable source of data on, and a better understanding of, lightning processes during a key period in the evolution of such knowledge. The primary triggering technique used in this program was so-called "classical triggering," in which a small rocket lifts a grounded wire aloft below an active thunderstorm, and produces a lightning discharge that is usually guided by the wire to the launch platform and any instrumentation connected thereto. The primary categories of results from the RTLP that are relevant for space operations are (1) data in support of test standards, (2) exposure of hardware to real lightning strikes, (3) test and calibration of lightning-detection systems, (4) better understanding of the triggering conditions, (5) better understanding of the "attachment process," and (6) data to constrain theoretical models. Each of these areas is treated in turn below. Wherever possible, preference is given to references arising directly from the RTLP.

#### 5.4.3.1 Data in Support of Test Standards
The best and most comprehensive dataset on the peak currents, peak current derivatives, and current rise times produced by subsequent return strokes was compiled by the RTLP, primarily during 1985, 1987, and 1988 (Leteinturier et al., 1990, 1991). The largest peak currents produced by strikes to ground-based objects are generally found in first return strokes (e.g., Rakov and Uman, 2003, Table 4.4), which do not occur in rocket-triggered flashes, but the subsequent return strokes in triggered and natural lighting are believed to be similar (e.g., Le Vine et al., 1989). Further, the peak electric-field derivatives and the rise times of the fast-rising portions in natural first and subsequent stroke waveforms are very similar (Bailey et al., 1988; Willett et al., 1990; Krider et al., 1992; Willett and Krider, 2000). Finally, Willett et al. (1989a) found approximately linear relationships for peak derivative and rise time between field change and current in rocket-triggered return strokes. (These similarities themselves were first documented with data recorded during the RTLP.) Thus it is reasonable to conclude that the RTLP data on lightning current rise times and derivatives are directly relevant for lightning test standards, even for first strokes, although the peak currents, charge transfers, and action integrals are not.

Peak current derivatives and current rise times are especially relevant to the coupling of damaging signals into electromechanical and electronic systems. For example, the induced EMF in a conducting loop is directly proportional to the time derivative of the magnetic flux passing through that loop. The magnetic field derivative inside the loop, in turn, is directly proportional to the current derivative in the nearby conductor (or lightning channel) that is producing the magnetic field. Further, the peak current that is induced in an inductive circuit is proportional (at least over short times) to the rate of rise (d$i$/d$t$) of this nearby current. If the circuit of interest is protected by a Faraday shield, the penetration of the transient magnetic field through that shield, hence the induced current in the circuit, also increases with the duration of that transient, or the duration (or rise time) of the fast-rising source current. The knowledge that there are very large current transients with very short current rise times in lightning return strokes was relatively new to the community prior to the RTLP and was based primarily on recent remote measurements of the electromagnetic fields produced by strokes in natural lightning (e.g., Uman and Krider, 1982). As a result of the RTLP, however, the current derivatives and rise time in triggered subsequent return strokes were well known from direct measurements. Test standards given, for example, by Plumer (1992) fully reflected these observed peak current derivatives and current rise times.

### 5.4.3.2 *Exposure of Hardware to Real Lightning Strikes*

The kinds of hardware that have been tested by exposure to rocket-triggered lightning at the RTLP include protection systems for structures, radomes, power lines, and even nuclear devices. Here we list a few examples of such tests that were conducted at the RTLP. During 1985 Lawrence Livermore and Sandia National Laboratories tested a prototype protective canister for nuclear devices against rocket-triggered-lightning strikes (Melander et al., 1988)]. Rubinstein et al. (1991, 1994) describe measurements of induced voltages on an overhead power-distribution line by rocket-triggered discharges striking 20 m from one end of the line during 1986. Testing of electric power equipment was conducted during 1987 and 1988 by Power Technology Incorporated for the Electric Power Research Institute. In 1990 Fisher and Schnetzer (1991) measured actual damage to metal samples that were directly exposed to rocket-triggered lightning, as a calibration for laboratory testing of burn-through.

### 5.4.3.3 *Test and Calibration of Lightning-Detection Systems*

There are three characteristics of lightning-detection systems that might be tested or calibrated against rocket-triggered lightning at the RTLP. The most obvious is location accuracy, since the location of the triggered discharges is precisely known. The primary detection systems at the time (the National Lightning Detection System (e.g., Orville, 2008) -- called the NLDN -- and a medium-range version of it that was deployed around KSC (e.g., Krider, 1988) -- called the LLP system, later CGLSS) depended on radio direction finding (DF) and triangulation from multiple DFs to locate strikes. DFs are notorious for "site errors" -- errors in reported direction due to natural or man-made electromagnetic inhomogeneities of their surroundings. Statistical techniques have been developed for computing site-error corrections for DFs based on the self consistency of results from a multi-station network over many strikes distributed throughout its domain (e.g., Hiscox et al., 1984), but there remains a need for ground-truth verification. Examples of such verification using flashes triggered by the RTLP include Maier and Jafferis (1985) and Maier (1991).

In addition to location accuracy, there are the two related questions of amplitude accuracy and detection efficiency. In general, it was shown by Willett et al. (1988, 1989a) that there is an approximately linear relationship between peak electric-field change (proportional to the peak magnetic-field change that is measured by DFs in the "far field") and the peak current, as predicted by the simple transmission-line model. Stroke magnitude (in terms of inferred peak current) has been calibrated against peak currents directly measured in the RTLP (Orville, 1991; Idone et al., 1993). To our knowledge, detection efficiency has never been checked relative to RTLP data.

### 5.4.3.4 *Understanding of the Triggering Conditions*

Arguably the most important impact of the RTLP on spaceflight operations has come from a better understanding of the triggering process and the triggering conditions. This is a large subject that will only be outlined here, but the key contributions provided by the RTLP will be highlighted.

A basic understanding of the physics of air breakdown in long gaps had already been obtained from experiments on long laboratory sparks, especially during the 1970s (*e.g.*, Les Renardières Group, 1977, 1981). Unfortunately, the longer the spark, the lower the marginal ambient field -- that is, the increase in applied potential per unit increase in gap length (for "critical" time to voltage crest and 50% breakdown probability) -- required to sustain its propagation. The longest sparks then available in the laboratory implied a marginal field of about 55 kV/m for gap lengths of 27 m (Pigini et al., 1979). This was generally believed to be much larger than the average fields in which lightning discharges could propagate. For example, Pierce (1971) had suggested that lightning could be triggered in ambient fields of several kilovolts per meter if the ambient potential spanned by the triggering conductor (building, rocket, aircraft, etc.) reached a few megavolts. There seemed to be a considerable gap between lightning and sparks that could be produced in the laboratory, hence there was a clear need for the rocket-triggering experiments.

Figure 5.4.3-1 shows a histogram of all the successful and unsuccessful trigging attempts versus the electric field measured at the ground over the eight-year lifetime of the RTLP (including 1983, when the program was begun at a site south of Melbourne, FL) (Jafferis, 1995). It is evident that foul-weather (negative) surface fields of 3 kV/m or greater over land provide a high probability of success. The fact that lightning was triggered in fields that were greater than 1 kV/m certainly justifies a 1 kV/m surface-field criterion in the current LLCC.

Under a thunderstorm, the ambient field a few hundred meters aloft is considerably larger than that at the surface under conditions like those when triggering was successful (*e.g.*, Standler and Winn, 1979). This was demonstrated at the RTLP triggering site near the Mosquito Lagoon by Soula and Chauzy (1991), who used several electric-field sensors suspended by a captive balloon along an insulating tether at altitudes ranging from 80 to 800 m. Using the same instrumentation, Chauzy et al. (1991) and Soula and Chauzy (1991) showed that lightning could be triggered in fields aloft of 50 - 65 kV/m using grounded and ungrounded triggering wires of 200 - 300 m length. Since only four triggering attempts, all of which were successful, were made while the balloon sensors were operating, however, it must be assumed that triggering can occur in lower ambient fields. Indeed it was later shown that lightning could be triggered in ambient fields as low as 10 - 20 kV/m (Willett et al., 1999).

In the foul-weather fields generally encountered at KSC, triggered discharges were initiated by positive "leaders" (highly ionized, conducting, filamentary channels extending into virgin air) propagating upward from the tips of the triggering rockets (Rakov and Uman, 2003, Section 7.2.1). Important new information about such leaders was obtained from the RTLP. With optical and near-ultraviolet (UV) observations Idone (1992) showed that between the rocket and cloud base, they propagate at speeds of up to almost $10^6$ m/s. This propagation speed was often found to increase with altitude, sometimes abruptly in association with a marked decrease in channel tortuosity [Idone and Orville, 1988]. Even more interestingly, Idone [1992] showed that this positive-leader propagation is not smooth and steady, instead comprising a relatively regular sequence of optical "pulses" separated by roughly 20 µs of time and spanning several meters of height. The structure of individual pulses was revealed by the UV streak photographs to consist of "a bright, thin stem of typically 3-5 m in length which is surrounded in the direction of propagation by a diffuse, hemispherical corona brush that extends outward about 5-10 m."

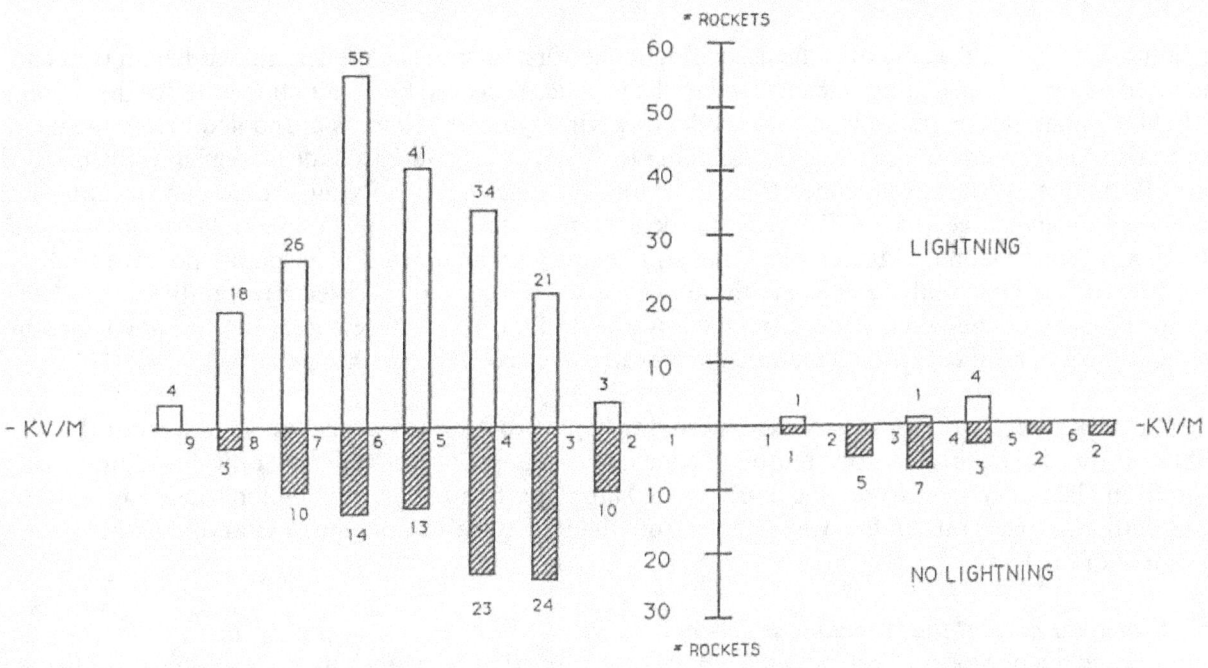

TRIGGERED LIGHTNING SUMMARY AT KSC FROM 1983 TO 1991

242 ROCKETS TRIGGERED LIGHTNING ( READINGS NOT AVAILABLE FOR 34)

167 ROCKETS DID NOT TRIGGER LIGHTNING (READINGS NOT AVAILABLE FOR 50)

**Figure 5.4.3-1  Triggering success vs. surface field during the RTLP**

Detailed electrical recordings were also made during the initiation and ascent of positive leaders in the RTLP (e.g., Laroche et al., 1989, Fig. 5). Much of this electrical phenomenology has been summarized in a review paper by Willett (1992). Synchronized records of channel-base current and close electric-field change show that brief current pulses with amplitudes of a few tens of Amperes and repetition intervals of several tens of microseconds are superimposed on a gradually increasing steady current of only a few Amperes. The current pulses coincide with steps in the electric field record, implying discrete pulses of charge deposition, presumably near the tip of the extending leader. Because of their similar time history, these electrical events are believed to correspond to the optical pulses mentioned above, although no synchronized optical and electrical records are available to date.

Positive-leader onset is typically preceded hundreds of milliseconds by a series of brief, irregularly spaced, current pulses, sometimes called "precursors," with amplitudes of a few Amperes and repetition intervals on the order of 30 ms (e.g., Barret, 1986). These precursors, in turn, have been shown by Laroche et al. (1988) to consist of individual current and electric-field pulses like those in the leader onset itself, or groups of a few such pulses with similar repetition intervals, but no optical records of these events are available. This observation, however, has led to the hypothesis that the precursors are essentially attempted leaders that are unable sustain their propagation. If this hypothesis is correct, it implies that air breakdown, and even leader initiation, generally begin long before the rocket has ascended to sufficient altitude to produce self-sustaining leader propagation. This tentative conclusion points to leader "viability" (the ability to propagate indefinitely in the ambient field) as the most important criterion for lightning triggering by long, thin conductors.

In the "altitude-triggering" technique, the ascending rocket first unspools a predetermined length (typically a few hundred meters) of insulating line attached to the ground, followed by conducting wire for the remainder of its flight. (Sometimes an initial length of conducting wire, typically 50 m, is unspooled before the insulating section of line to "encourage" the triggered lightning to attach to ground-based measurement instrumentation.) This new technique, which was intended to better simulate triggering by a flying aircraft or spacecraft, was first exploited to advantage at the RTLP (e.g., Laroche *et al.*, 1988, 1989, 1991). In a field of foul weather polarity it was found that the triggered discharge still begins with an upward-propagating positive leader from the tip of the rocket, apparently identical to those in classical triggering, followed a few milliseconds later by a downward-propagating negative leader from the lower end of the conducting wire. (To our knowledge this technique has yet to be successfully attempted in elevated fields of fair-weather polarity.)

Since the same sequence of events has been inferred to begin most strikes to instrumented aircraft (Boulay et al., 1988; Mazur, 1989), there is good reason to conclude that the positive-leader-viability conditions (as deduced from classical rocket-triggering experiments) are important in determining the triggering conditions to flying aircraft and spacecraft. In this way, results from the RTLP are still playing a vital role in the development of the LLCC.

### 5.4.3.5 *Understanding of the Attachment Process*
The term, "attachment process," refers to the details of the junction between a natural-lightning discharge and an object on the earth (e.g., Rakov and Uman, 2003, Section 4.5). For negative cloud-to-ground lightning this process comprises the initiation of an upward positive leader from the grounded object, its connection to the downward negative ("stepped") leader from the thundercloud that caused it, and the ultimate development of an upward-propagating return stroke. A better understanding of this process is obviously key to the lightning protection of grounded objects.

The attachment process can be simulated, albeit at relatively low intensity, by a variation on altitude triggering in which a short length of grounded, conducting wire is unspooled from the ascending rocket prior to the insulating segment. As described in Section 5.4.3.4 above, this leads to a downward, negative leader from the lower end of the triggering wire, which in turn causes the emission of a positive leader from the grounded segment. The first successful experiment of this type was conducted at the RTLP in 1989 (Laroche et al., 1991).

The attachment process for negative, subsequent return strokes was also investigated during the RTLP by Idone (1990) and by Willett et al. (1989a), who showed optically and electrically that there are probably non-negligible upward-connecting discharges that meet downward dart or dart-stepped leaders.

### 5.4.3.6 *Constraints on Theoretical Models*
The most important RTLP data for constraint of theoretical modeling have already been mentioned in Section 5.4.3.4 above. Having concluded that the positive leader is the controlling phenomenon, we can also conclude that those leaders that initiate rocket-triggered discharges are significantly different from those studied to date in the laboratory. Thus there is a need for further theoretical and laboratory-experimental work to develop adequate models of these leaders. Such models could then be used (in concert with other, yet to be developed, models of the effects of rocket-exhaust plumes) for quantitative prediction of the ambient fields necessary for

spacecraft triggering of lightning. To date one detailed, self-consistent, physical model of the positive leader has been developed (Bondiou et al., 1994), and an unrelated, semi-empirical model of the triggering conditions has been proposed (Bazelyan and Raizer, 2000, Section 4.1.1). Nevertheless, the theoretical problem is far from solved, and more sophisticated rocket-triggering experiments will undoubtedly be needed to complete this process.

Thottappillil and Uman (1993) have demonstrated the value of RTLP measurements for testing of various return-stroke models. Other observations made as part of the RTLP that have already been, or will be, useful in lightning theory include direct measurements of the time evolution of return stroke (luminous) diameter by Idone (1992); the first wide-band measurements by Willett et al. (1989b) of the "narrow bipolar pulses" that had been discovered by Le Vine (1980); deductions about the behavior of return-stroke currents above ground by Willett et al. (2008); and the first fairly comprehensive measurements of close leader/return-stroke fields by Hubert and Hubert (1986) and by Rubinstein et al. (1992, 1995).

### 5.4.4   Stanford Research Institute (SRI)

In the late 1980s the Stanford Research Institute instrumented the Aeromet Inc. Learjet 36C with 8 field mills. Both calibration and research flights were flown from 12 August 1989 to 6 September 1989 to investigate electric fields of growing and towering cumulus clouds. The Learjet 36C has a ceiling of 45,000 ft (13.7 km), flight duration of ~ 5 hrs, a flight speed of 175 - 200 m s$^{-1}$, is certified for flight in known icing conditions, and lightning hardened. In addition to the field mills and instruments for determining aircraft position, attitude, and state parameters, the aircraft carried an array of microphysical instruments capable of determining particle concentration, size and types and liquid water content. The field mills are those described by Kositsky et al. (1991a). Calibration of the airborne field mill system was the scale-model charge-transfer technique of Kositsky et al., (1991) and Kositsky and Nanevicz (1991). A conductive scale-model of the aircraft was placed in a uniform electric field and the charge density was measured at the locations of the field mills. In-flight artificial charging of the aircraft and flight maneuvers in clear air were used to verify the scale model results.

Penetrations were made into cumulus and towering cumulus congestus clouds. Additionally passes were made outside cloud, but near the edge of clouds too large to penetrate. Flights into clouds containing ice or only water hydrometeors showed significant charging of the aircraft. All three components of the ambient electric field appeared to be contaminated and proportional to those of the aircraft potential (Kositsky et al., 1991). However, field measurements outside of the cloud were found to be reliable. The results presented in Giori and Harris-Hobbs (1991) and Harris-Hobbs and Giori (1994) include only clear air measurements. Measurements from the WSR-74C radar at Patrick Air Force Base were used in the analysis to help determine cloud edge.

Results of maximum electric fields observed at cloud edge from the penetration of growing cumulus clouds at different stages of development are shown in Table 2 of Giori and Harris-Hobbs (1991) and reproduced below in Figure 5.4.4-1. B-1 refers to cumulus clouds with cloud top temperatures at or warmer than +5 °C; B-2 to cloud top temperatures at or warmer than -10 °C; B-2 to cloud top temperatures at or warmer than -20 °C; and B-4 to cumulonimbus or thunderstorm clouds. The results show that the cumulus clouds they studied did not develop significant electric fields until cloud top temperatures were colder than -10 °C.

Other findings reported in Giori and Harris-Hobbs (1991) and Harris-Hobbs and Giori (1994) are:
1. For seven clouds (41 penetrations; 21 passes near cloud edge) they examined the relationship between the maximum field strength at cloud edge (Emax) and the distance from the cloud at which the field drops to slightly above fair weather fields. In all categories from B-1 through B-4, they found that the standoff distances in the LLCC were much larger than the distances from cloud edge for the fields to drop to fair weather values.
2. A study of the cloud volume containing radar reflectivity >30 dBZ above 0°C showed a correlation with Emax at cloud boundary.

3. They suggested that radar measured parameters might prove useful in determining safe standoff distances.
4. Anvil debris clouds studied seemed to become rather weakly electrified after they detached from the parent storm.
5. Three passes in a layered cloud (debris cloud in current terminology), showed that the electric field remained strong for at least 30 min.

### Table 2
### SUMMARY OF MAXIMUM ELECTRIC FIELDS
### MEASURED JUST OUTSIDE CUMULUS CLOUDS

| LCC | Total Samples | Range of $|E|_{max}$ (kV/m) | | Average of $|E|_{max}$ (kV/m) | Standard Deviation (kV/m) |
|-----|------|---------|---------|------|------|
|     |      | Minimum | Maximum |      |      |
| B-1 | 19 | 0.0 | 0.53 | 0.14 | 0.15 |
| B-2 | 4  | 0.1 | 1.0  | 0.52 | 0.47 |
| B-3 | 33 | 0.4 | 45   | 10.1 | 9.7  |
| B-4 | 16 | 0.0 | 11   | 4.5  | 3.1  |

**Figure 5.4.4-1  Table 2 reproduced from Gori and Harris-Hobbs (1991)**

### 5.4.5    *New Mexico Institute of Mining and Technology Studies*

The New Mexico Institute of Mining and Technology/Office of Naval Research Special Purpose Test Vehicle for Atmospheric Research (SPTVAR) aircraft was flown from 14 Sept. to 4 Nov. 1988 and again from 21 July to 1 Sept. 1989. The SPTVAR is a modified Schweizer 845 aircraft airframe powered by a 150 kW Lycoming engine with capability of flying through clouds up to 30,000 ft (9 km). It has a flight endurance of 3 to 3.5 hours and flies at 50 m s$^{-1}$. For the studies near Kennedy Space Center it was instrumented with 5 rotating-vane electric field mills of the type described by Winn (1993). It had the capability of measuring all three components of the electric field and the potential due to charge on the aircraft. Additionally, the aircraft was instrumented to measure ambient temperature, cloud liquid water content, pressure, air speed and aircraft position (VOR/DME and Loran C). Gyroscopes provided measurements of heading, roll and pitch. Some measurements were telemetered to the ground. The capability of displaying electric fields at the ground during flight was developed.

The SPTVAR was directed to a cloud system to determine if fields were present and to explore the intensity, duration and extent of that field when it appeared that a cloud system would preclude launch of a space vehicle because of a potential electrical hazard, For cumulus clouds with tops below the freezing level, repeated penetrations were made midway between cloud base and cloud top. For taller clouds repeated penetrations were made near the 0°C level through the thickest part the cloud. This flight level was flown to minimize icing but also to be near the larger fields within the cloud. Most of the flights were into cumulus clouds but some flights were also made into debris clouds and adjacent to cumulonimbus clouds too large to penetrate. A few layer clouds and disturbed weather clouds were also investigated.

The main goals of the project were: 1) Develop and demonstrate techniques for measuring electric field aloft and locating regions of charge during flight. 2) Characterize the conditions within and near clouds at KSC that are presently identified as a threat to space launch vehicles. 3) Study the correlation between the electric fields aloft and those from the KSC ground-based field mill network for a variety of electrified clouds.

The studies conducted by SPTVAR are documented in 3 reports. Part I (Jones et al., April 27, 1990) describes the aircraft and instrumentation; project goals; flight plans for studying cloud over KSC; summary of operations flights for 1988; conclusions and recommendations. It also contains appendices on field mill calibration; determination of the vector field, a track plotting program "SPT" and individual flight summaries for 1988. Part II (Jones et al., April 27, 1990) presents a case study of the 4 Nov 1988 flight in which the SPTVAR found fields aloft less than 1.6 kV/m while the KSC field mill network showed fields >3 kV/m at the surface. Part III (Jones et al., Aug 21, 1990) describes flights made during the summer 1989 deployment and the overall findings and conclusions based on all data gathered.

General finding from these studies are:
1.  An aircraft instrumented to measure the electric field vector can be an effective method for assessing electrical conditions within selected clouds over KSC.
2.  For cumulus clouds with temperatures everywhere warmer than 0 °C, no significant electrification was encountered. Three marginally electrified clouds had 5 dBZ tops to about the -20 °C level. It appeared that dynamic growth above the -20 °C level was necessary for the clouds to become significantly electrified.
3.  The contour plots of the surface electric field gave misleading representations of charge locations outside of the network due to unjustified closing of the field contours near the edge of the network.
4.  On one occasion with a widespread cloud layer, the electric field disturbances coincided with cells of embedded convection.
5.  Screening layers on the sides of clouds, if they existed, were insufficient to mask the presence of charge within the clouds.
6.  Usually the electric field aloft is much stronger than at the ground, but one exception was documented in Part II.

# 5.5    Lightning LCC

### 5.5.1    *The LAP Process*
The success that the LAP has had in reviewing and improving the LLCC and the associated Definitions is based primarily on three factors:  Expertise, experience, and collaboration.

*Expertise and Scientific Experience.* The constitution and evolution of the LAP (formerly the PRC) has been described in Section 5.2.1. The expertise and experience of the individual LAP members are briefly summarized in Appendix III. Collectively, their experience with lightning science and lightning's effects on space vehicles spans the entire period from the Apollo XII lightning strike in 1969 to the present. Their scientific expertise encompasses not only lightning physics and measurement, but cloud physics and measurement, statistics and advanced mathematics, and other aspects of atmospheric physics and even astrophysics. The current LAP members represent a broad cross section of the relevant knowledge, scientific experience, and institutional perspectives. Therefore, the LAP is uniquely qualified to advise NASA and the Air Force on matters relating to the LLCC.

*LAP Experience Promotes Collaboration.* The initial role of the PRC was fairly narrow -- essentially to review some proposed changes to the LLCC generated by MSFC personnel after the ABFM I campaigns (see Section 5.4.1). Over the past two decades, however, a dual purpose has evolved. On the one hand, the LAP has re-written sections of the LLCC in response to new scientific knowledge and instrument capabilities (*e.g.*, new understanding of the electrical environment that can trigger lightning), and it has also proposed new experiments to advance that knowledge in areas that are important for the LLCC (*e.g.*, the ABFM II experiment and the use of radar to evaluate anvil and debris clouds -- see Section 6.4). On the other hand, the operational launch-weather community (particularly the KSC Weather Office and the 45WS, but also the SMG, the 30WS at Vandenberg AFB, and even meteorologists at the KLC) have brought questions and concerns to the LAP for various reasons. The dialog between the LAP and the launch-weather community has become a close collaboration on important tasks, such as clarifying the text of the rules, and has increased both launch availability and safety.

Rather than risk future lightning incidents, the LAP has always chosen to start with general rules that are known to be conservative, and it has recommended additional instrumentation (*e.g.*, field mills) to cover unexpected conditions that may be hazardous. Close collaboration between the LAP, the 45WS, and the KSC Weather Office has proved to be an efficient way to identify individual launch restrictions that have a significant impact on launch availability or that may have an unduly high false-alarm rate. Attention has then focused on these situations, with the operational community gathering relevant data for the LAP to review, and solutions have been found that increase launch availability without compromising safety. By identifying the sources of false alarms, and by recommending new ways of detecting and eliminating them, the LAP has been able to incrementally reduce the initial conservatism of the general rules and to safely increase launch availability. For example, the LPLWS has thus been converted from a launch-availability negative into a means of relaxing certain rules through exceptions (*e.g.*, the current field-mill exception to the long-duration, 3 NM standoff requirements in the absence of VAHIRR by both the Detached-Anvil and the Debris-Cloud Rules).

Unfortunately increased launch availability (without compromising safety) comes at the price of greater complexity in the language of the LLCC because of the addition of numerous exceptions, and many of these exceptions require specialized measurements and analysis to verify them. Here again, the close working relationship that has developed between the LAP and the 45WS has been an invaluable tool for clarifying the meaning of the LLCC and has led to more efficient application of them during launch operations.

*The Future of Collaboration.* Logical truth-tables (for examples, see the figures in Appendix II) were originally introduced by Dr. Harry Koons as a technique for identifying and eliminating any inconsistencies between different sections of a given complex rule. Recently the LAP has been using truth tables not only to eliminate contradictions, but also to remove redundancies between sections of such a rule. Any inconsistency or redundancy can lead to ambiguity, which makes the rules more difficult to understand and apply. The truth tables can also make the structure of the more complex rules easier to grasp. The LAP is currently exploring whether graphical truth-tables might be beneficial as training materials for launch weather personnel.

### 5.5.2   LLCC Revisions – 1991
As mentioned in Section 5.0.3, Heritage (1988) published a set recommended LLCC together with a detailed scientific rationale (their Chapter 7, further justified by their Chapters 1 - 3). These recommendations are quoted in Appendix I, Section A1.8. After considerable discussion among the interested parties, a new set of launch rules that was uniform across the Shuttle and ELVs was finally agreed upon. The first operational post-AC 67 rules are well represented by the "Space Shuttle LLCC (1991)" discussed in this section.

In each of the following sections, major changes from the preceding set of rules are mentioned (for a more complete discussion of the differences, see the corresponding entry in Appendix II), and only rules that are significantly changed are given in paraphrase to save space and reduce complexity, as was done with the current rules in Section 1.1. Note that these paraphrases are invariably given in *italics* to distinguish them from the actual LLCC, which are quoted verbatim in the corresponding sections of Appendix I.

*Space Shuttle LLCC (1991) [LCN 00048, 01/28/91].* These rules may be found in Appendix I, Section A1.9.1. Many of them were similar or identical to the recommendations of Heritage Committee. The primary difference was that these LLCC include no provision for the use of electric fields measured aloft by one or more ABFMs, as had been done briefly during the Apollo-Soyuz and Viking missions (see Sections 3.3.3 and 3.3.4) and was also recommended by the National Research Council (1988) and the PRC (see Section 5.2.1). The "Triboelectrification Rule" recommended by the Heritage Committee was also omitted here, presumably because these rules were for Shuttle only, and the Shuttle had been hardened against triboelectrification. The 1991 Shuttle LLCC are simple enough to quote in full here as a starting point for the following sub-sections:

Even when constraints are not violated, if any other hazardous conditions exist, the Launch Weather Officer will report the threat to the Launch Director. The Launch Director may HOLD at any time based on the instability of the weather.

The Launch Weather Officer must have clear and convincing evidence the following constraints are not violated.

A. Do not launch if any type of lightning is detected within 10 nautical miles of the launch site or planned flight path within 30 minutes prior to launch, unless the meteorological condition that produced the lightning has moved more than 10 nautical miles away from the launch site or planned flight path.

B. Do not launch if the planned flight path will carry the vehicle:
   1. Through cumulus clouds with tops higher than the 5 °C level; or
   2. Through or within 5 nautical miles of cumulus clouds with tops higher than the -10 °C level; or
   3. Through or within 10 nautical miles of cumulus clouds with tops higher than the -20 °C level; or
   4. Through or within 10 nautical miles of the nearest edge of any cumulonimbus or thunderstorm cloud including its associated anvil.

C. Do not launch if, for Ranges equipped with a surface electric field mill network, at any time during the 15 minutes prior to launch time the one minute average of absolute electric field intensity at the ground exceeds 1 kilovolt per meter (1kV/m) within 5 nautical miles of the launch site unless:
   1. There are no clouds within 10 nautical miles of the launch site; and
   2. Smoke or ground fog is clearly causing abnormal readings.
   NOTE: For confirmed instrumentation failure, continue countdown.

D. Do not launch if the planned flight path is through a vertically continuous layer of clouds with an overall depth of 4,500 feet or greater where any part of the clouds are located between the 0 °C and the -20 °C temperature levels.

E. Do not launch if the planned flight path is through any cloud types that extend to altitudes at or above the 0 °C temperature level and that are associated with disturbed weather within 5 nautical miles of the flight path.

F. Do not launch through thunderstorm debris clouds, or within 5 nautical miles of thunderstorm debris clouds not monitored by a field mill network or producing radar returns greater than or equal to 10 dBz.

G. Definitions
   1. Debris Cloud: Any cloud layer, other than a thin fibrous layer, that has become detached from the parent cumulonimbus within 3 hours before launch.
   2. Disturbed Weather: Any meteorological phenomenon that is producing moderate or greater precipitation.
   3. Cumulonimbus Cloud: Any convective cloud which exceeds the -20 °C temperature level.
   4. Cloud Layer: Any cloud broken or overcast layer or layers connected by cloud elements, e.g. turrets from one cloud layer to another.
   5. Planned Flight Path: The trajectory of the flight vehicle from the launch pad through its flight profile until it reaches the altitude of 100,000 feet. The flight path may vary plus or minus 0.5 nautical miles horizontally up to an altitude of 25,000 feet.
   6. Anvil: Stratiform or fibrous cloud produced by the upper level outflow from thunderstorms or convective clouds. Anvil debris does not meet the definition if it is optically transparent.

### 5.5.3    LLCC Revisions – 1992

*Space Shuttle LLCC (1992) [LCN 00230R04, 03/20/92]*. These rules may be found in Appendix I, Section A1.9.2. They are nearly identical to the above except for relatively minor changes to the Field-Mill Rule (C), which may be paraphrased as follows:

Surface Electric Fields. *Do not launch if the electric field within 5 nm of the launch site has exceeded ±1 kV/m in the past 15 minutes. Exceptions are allowed if (1) all clouds within 10 nm of the launch site have not been associated with convective clouds with tops colder than -10 °C for at least 3 hours and are (a) thin fibrous (optically transparent) clouds or (b) less than 2/8 coverage with tops warmer than +5 °C, and if (2) smoke, ground fog, or a maritime inversion with fields between +1000 and +1500 V/m is clearly causing the elevated readings.*

### 5.5.4    LLCC Revisions – 1995

*Space Shuttle LLCC (1995) [LCN 00520R03, 04/25/95]* (see Appendix I, Section A1.9.4). This major rule change was based a new set of LLCC recommended the year before by the PRC (see Appendix I, Section 1.9.3) and had more to do with reorganization and clarification than with substantive changes. Nevertheless, the following significant differences between the PRC's recommendations and the published rules drew a formal "letter of concern" from the PRC: 1) "Rationales" had been added to all of the rules without PRC knowledge or review. 2) The PRC-recommended Triboelectrification Rule had been dropped (only for Shuttle, it appears.) 3) An Electric-Fields-Aloft Rule granting exceptions to the Thick-Cloud, Disturbed-Weather, and Debris-Cloud Rules (the PRC's version of the Heritage Committee's recommendation) had also been dropped.

Substantive changes from the previous operational version of the LLCC included insertion of the word, "transparent," to limit the application of some LLCC, some relaxation of the Cumulus Clouds Rule, more explicit treatment of Anvil Clouds, and application of the Surface Electric Fields Rule all along the flight path. The PRC-recommended Electric-Fields-Aloft Rule is also paraphrased below, although it was *never* part of the Shuttle rules:

Cumulus Clouds. *Do not fly within 10 nm of any cumulonimbus or thunderstorm cloud, including nontransparent parts of its attached anvil, nor within 10 nm of a nontransparent detached anvil for the first hour after detachment, nor within 5 nm of any cumulus cloud with a top colder that -10 °C. Do not fly through any cumulus cloud with its top colder +5 °C. An exception is allowed for a cumulus cloud with its top between ±5 °C if that cloud is not producing precipitation and if a field mill within 3 nm (or the altitude of the -5 °C level, if smaller than 3 nm) of the farthest edge of that top, and any other mills within 5 nm of the flight path, have shown greater than –100 V/m, but less than +1000 V/m, for at least 15 minutes.*

Surface Electric Fields. *Do not fly within 5 nm of any electric field mill that has shown readings in excess of ±1 kV/m in the past 15 minutes. Exceptions are allowed if (1) all clouds within 10 nm of the flight path are (a) transparent or (b) have tops warmer than +5 °C and have not been associated with convective clouds with tops colder than -10 °C for at least 3 hours and if (2) a known benign source near the mill is clearly causing the elevated readings.*

Electric Fields Aloft. *The "Thick Cloud Layers," "Disturbed Weather," and "Debris Clouds" rules need not be applied if, during the 15 min prior to launch time, the absolute value of the instantaneous vector electric field sampled within the cloud volume expected to be along the flight path is less than a designated function of altitude.*

### 5.5.5    LLCC Revisions – 1998

Space Shuttle LLCC (1998) [LCN 00824R03, 06/04/98]. This major change was adopted by the LAP on 5/20/1998, was documented by Krider et al. (1999), and is quoted in Appendix I, Section A1.9.5. It included a complete re-write of the rules to clarify and improve their logic. All Rationales were eliminated. Although all of the rules were re-ordered and changed enough to deserve paraphrases below, key changes were the removal of anvil clouds from the Cumulus Cloud Rule into a separate Anvil Clouds Rule and the addition of new Smoke Plume Rule. In the process, several of the rules were somewhat relaxed.

Lightning. *Do not fly within 10 nm of any type of lightning, or any convective cloud that has produced it, within the past 30 min. An exception is allowed if the cloud has moved beyond 10 nm and if an electric field mill within 5 nm of the lightning, and any other mills within 5 nm of the flight path, have shown less than ±1000 V/m for at least 15 minutes.*

Cumulus Clouds. *Do not fly within 10 nm of any cumulus cloud with a top colder than -20 °C, nor within 5 nm of any cumulus cloud with a top colder that -10 °C. Do not fly through any cumulus cloud with its top colder +5 °C. An exception is allowed for a cumulus cloud with its top between ±5 °C if that cloud is not producing precipitation and if a field mill within 2 nm of that top, and any other mills within 5 nm of the flight path, have shown greater than −100 V/m, but less than +500 V/m, for at least 15 minutes.*

Anvil Clouds:
Attached Anvil Clouds. *Do not fly within 10 nm of any non-transparent, attached anvil for at least 30 minutes after the last lightning discharge occurs in the parent cloud or anvil, nor within 5 nm for 3 hours after the last lightning. Never fly through such a cloud.*

Detached Anvil Clouds. *Do not fly within 10 nm of a non-transparent, detached anvil for 30 minutes after the last lightning discharge occurs in the detached anvil (or in its parent cloud before detachment), nor within 5 nm for 3 hours after the last such lightning. Do not penetrate such an anvil for 3 hours after detachment, nor for 4 hours after it has produced lightning. Exception: Flight is allowed up to the edge of a detached anvil between 30 minutes and 3 hours after the last lightning if the radar reflectivity of all parts of that cloud within 5 nm of the flight path has been less than 10 dBZ, and if a field mill within 5 nm of the cloud (and any other mills within 5 nm of the flight path) have shown less than ±1000 V/m, for at least 15 minutes.*

Debris Clouds. *Do not fly within 5 nm of a non-transparent debris cloud for 3 hours after it detaches or decays from its parent cloud and for 3 hours after it has produced lightning. Exception: Flight is allowed up to the edge of a debris cloud at any time if the radar reflectivity of any part of that cloud within 5 nm of the flight path has been less than 10 dBZ, and if a field mill within 5 nm of the cloud (and any other mills within 5 nm of the flight path) have shown less than ±1000 V/m, for at least 15 minutes.*

Disturbed Weather. *Do not fly through any non-transparent cloud associated with disturbed weather that has cloud tops colder than 0 °C and that, within 5 nm of the flight path, either contains moderate or greater precipitation or shows evidence of melting precipitation.*

Thick Cloud Layers. *Do not fly through a non-transparent cloud layer that is thicker than 4500 feet and contains temperatures between 0 °C and -20 °C, nor through any non-transparent cloud layer that is connected to such a thick cloud layer within 5 nm of the flight path. An exception is allowed if the thick cloud layer is cirriform, is entirely colder than -15 °C, contains no liquid water and has never been associated with a convective cloud.*

Smoke Plumes. *Do not fly through any cumulus cloud that develops from a smoke plume for 60 minutes after it has detached from that plume.*

Surface Electric Fields. *Do not fly within 5 nm of any electric field mill that has shown readings in excess of ±1 kV/m in the past 15 minutes. The field threshold can be raised to ±1.5 kV/m if all clouds within 10 nm of the flight path are transparent or if they (a) have tops warmer than +5 °C and (b) have not been part of convective clouds with tops colder than -10 °C for at least 3 hours.*

### 5.5.6    *Proposed Field Mill Aloft Rule*

As we have seen in Section 3.3, fields aloft were used briefly during the ASTP/Viking era as a means of relaxing several of the contemporary cloud-related rules. Based on his report on ASTP, Arabian [1976] clearly believed that a successful operations plan involving several ABFM platforms had been worked out that would allow the successful use of such an exception.

After the AC-67 disaster, the Hosler Committee [National Research Council, 1988] revived this idea, recommending "implementation of the following actions as rapidly as possible:  instrumenting an aircraft to measure electric fields aloft [and other parameters] along the launch/landing paths..." The Heritage Committee [Heritage, 1988] went further in their 1988 review of the A/C-67 disaster by proposing specific exceptions of the following form for several of their proposed, cloud-related rules:

A. The electric field intensity at the ground (for ranges that have a ground field mill system) has remained below 1 kV/m within 5 nmi of the launch pad [There are no clouds within 10 nmi of the launch pad at the planned time of launch]; and

B. The absolute value of the instantaneous critical electric field intensity [within the clouds] sampled along the flight path [during the 15 minutes prior to launch] is less than the critical value, $E_c$, shown as a function of altitude in Figure 7-1.

(The referenced figure gave an altitude profile of $E_c$ that decreased from 5 kV/m at the surface through 1 kV/m at about 45,000 ft in proportion to atmospheric pressure.)

A similar exception was advocated by the PRC in their first formal LLCC report [Koons and Walterscheid, 1996], and was continued by the LAP in their second LLCC report [Krider et al., 1999]. It was dropped from their proposed LLCC sometime between 1998 and 2003, after Hugh Christian joined the LAP in February 1999 and had begun making a strong case against it. Dr. Christian argued successfully (1) that there was no scientific consensus on the critical field profile that had been proposed by the Heritage Committee (although such a consensus may have evolved over the last few years -- see the Rational Document) and (2) that, contrary to Mr. Arabian's earlier position, no viable operations plan had been developed for use of an ABFM in support of launches. The latter was a position that had been espoused by the weather-support community all along and which was the reason that such an ABFM exception had never been implemented in the operational LLCC.

# Chapter 6   ABFM II and Beyond (2000 – 2010)

## 6.0    Background

On 1 Dec 1999, the LAP recommended that NASA sponsor a second airborne field mill program (ABFM II) to "validate physics-based relationships between the decay of cloud electric fields and cloud properties and to verify that these cloud properties, and also the presence of electric generators aloft, can be inferred from ground-based measurements."

Initial ABFM II funding for limited planning, instrumentation development and testing was obtained from the Shuttle Launch Integration Manager, Don McMonagle, in mid-1999. Funding for the field programs in 2000 and 2001 was obtained from a variety of sources but the largest contributor by far was the Shuttle program. The total cost of ABFM II by the time data analysis was largely completed in 2005 was about $3M. Details of the program are provided in section 6.4 of this chapter. Application of the data to LLCC development and some limited additional analysis continued through 2009.

The ABFM II field program and subsequent extensive efforts to properly calibrate and synchronize the multiple *in-situ* and remote sensing data it produced resulted in the largest, most comprehensive and reliable data set of its kind ever collected. The size and quality of the data set provided the LAP sufficient information to recommend major revisions to the LLCC in 2005 and again in 2008. The data set is accessible to the public for further research at <http://abfm.ksc.nasa.gov/>.

## 6.1    Infrastructure

Except as noted below in this section, the weather infrastructure at KSC/CCAFS remained the same as that presented in Section 5.1 during the period covered by this chapter.

During the first decade of the 21st century, the consolidation of KSC and CCAFS weather infrastructure under the management of the Eastern Range continued. The NASA-developed Lightning Detection and Ranging (LDAR) system had proven its value, but it was aging and needed replacement. A commercial version of LDAR had been developed by Global Atmospherics, Inc. (subsequently acquired by Vaisala, Inc.) under a Space Act Agreement with KSC. The 45th Space Wing purchased the commercial version of LDAR from Vaisala. The new system, called LDAR II, is now the operational system and the KSC LDAR was decommissioned in November 2008.

As this is being written, a new, 5 cm, Doppler, dual-polarization weather radar is undergoing acceptance testing at the Eastern Range. The new radar is scheduled to replace the antique WSR-74C as soon as it is accepted by the Range, currently expected by late summer 2010.

During this same period, Vandenberg Air Force Base saw significant changes in range weather equipment. The Range Standardization and Automation system, now known as the Western Range Weather System (WRWS), and Space Launch Range System Contract range systems upgrade projects affected just about every aspect of how the Western Range conducts business. There were changes to equipment, communications, procedures and instructions that govern weather for launch and flight operations. The 30th Space Wing operationally accepted the WRWS in June 2008.

## 6.2    Organization

There were no significant changes in the organization of weather support for America's space program during the period presented in this chapter.

## 6.3    Operations

On 1 February 2003, the Space Shuttle Columbia disintegrated during its reentry into the atmosphere while attempting to land at KSC. All seven crew members died. Although this accident did not involve lightning or the LLCC, a brief description is provided here because some of the lessons learned have affected the processes and procedures used by NASA and the Air Force in assessing issues (like the LLCC) that may affect flight safety. The material in this section is based on the Columbia Accident Investigation Board Report (Columbia Accident Investigation Board, 2003, hereafter CAIB).

The final mission of Space Shuttle Columbia, designated as STS-107, was launched from complex 39A on 16 January 2003 for an intensive science mission. At 81.7 seconds after launch, when the Shuttle was at about 65,820 feet and traveling at Mach 2.46 (1,650 mph), a large piece of hand-crafted insulating foam came off an area where the Orbiter attaches to the External Tank. At 81.9 seconds, it struck the leading edge of Columbia's left wing. This event was not detected by the crew on board or seen by ground support teams until the next day, during detailed reviews of all launch camera photography and videos. This foam strike had no apparent effect on the daily conduct of the 16-day mission, which met all its objectives. (CAIB, p11.)

On 1 February, with an extremely successful mission behind them, the STS-107 crew prepared for their return to a planned morning landing at KSC. The de-orbit burn to slow Columbia down for re-entry into Earth's atmosphere was normal, and the flight profile throughout re-entry was standard. Time during entry is determined from "entry interface", an altitude of 400,000 feet where the Orbiter begins to experience the effects of Earth's atmosphere. Entry Interface for STS-107 occurred at 8:44:09 a.m. on February 1. Unknown to the crew or ground personnel, because the data is recorded and stored in the Orbiter instead of being transmitted to Mission Control at Johnson Space Center, the first abnormal indication occurred 270 seconds after Entry Interface. (CAIB, p.12). By the time Columbia passed over the coast of California in the pre-dawn hours of February 1, at Entry Interface plus 555 seconds, amateur videos show that pieces of the Orbiter were shedding. Analysis indicated that the Orbiter continued to fly its pre-planned flight profile, although, still unknown to anyone on the ground or aboard Columbia, her control systems were working furiously to maintain that flight profile. Finally, over Texas, just southwest of Dallas-Fort Worth, the increasing aerodynamic forces the Orbiter experienced in the denser levels of the atmosphere overwhelmed the catastrophically damaged left wing, causing the Orbiter to fall out of control at speeds in excess of 10,000 mph (ibid).

The Board concluded that "Columbia re-entered Earth's atmosphere with a pre-existing breach in the leading edge of its left wing in the vicinity of Reinforced Carbon-Carbon (RCC) panel 8. This breach, caused by the foam strike on ascent, was of sufficient size to allow superheated air (probably exceeding 5,000 degrees Fahrenheit) to penetrate the cavity behind the RCC panel. The breach widened, destroying the insulation protecting the wing's leading edge support structure, and the superheated air eventually melted the thin aluminum wing spar. Once in the interior, the superheated air began to destroy the left wing." (ibid).

The Board identified a number of specific causes of the accident, including schedule pressure (CAIB, section 6.2) and management's failure to realize the danger and address it (CAIB, section 6.3). Of more direct relevance to this History and its potential value, the Board found the general cause of the accident was a "broken safety culture (CAIB, section 7.4). An entire chapter (CAIB, chapter 8) was entitled "History as a Cause". The applicable lesson is that attitudes, processes and procedures matter as much as technology. That is a lesson fully applicable to the LLCC.

## 6.4    ABFM II Research

This section describes ABFM II in detail.

### 6.4.1    Overview

The ABFM II project was lead jointly by NCAR and MSFC and had other participants from the Univ. of Arizona, NASA KSC, NOAA, the Univ. of North Dakota and other organizations. The University of North Dakota Citation 2 aircraft was deployed from 2 June 2000 through 29 June 2000 and from 22 May 2001 through 29 June 2001. It was also deployed during the first half of February 2001 to investigate thick clouds and long anvils originating from the Gulf of Mexico, but because of severe drought in central Florida the deployment was terminated early. The UND Citation 2 is a twin-engine fanjet with an operational ceiling up to 43,000 ft (13 km), an endurance of 4 1/2 hours, normal flight speeds of 100 to 120 m sec$^{-1}$, and certification for known icing conditions.

### 6.4.2    Experimental Design

The main general goal of the project was to obtain measurements within anvils, debris clouds and other cloud systems to identify possible relationships between electric field, particle concentration, size and types, and the radar reflectivity and structure of the clouds. A specific goal was to try to determine the decay rate of electric field in anvils near KSC to see if the observed decay rates were consistent with decay rates theoretically predicted in a simple model developed by John Willett for this project.

Initial penetrations into the anvils were normally made near the core of the storm. Then subsequent, repeated passes were made at different distances downstream or at different altitudes. When Air Traffic Control would permit (which was not very often) spiral descents/ascents were made to obtain vertical profiles of the anvil or sometimes in debris or trailing stratiform clouds. Vertical sampling in individual anvils was often incomplete but over the period of the two summer campaigns penetrations were made over a variety of altitudes at different locations relative to the anvil top and bottom.

The flights were coordinated with the WSR-74C 5 cm radar at Patrick Air Force Base and WSR-88D 10 cm NEXRAD radar at Melbourne Florida. ABFM scientists in the Range Operations Control Center viewing radar cross-sections helped to direct the aircraft. Both cloud-to-ground (CG) and intra-cloud (IC) lightning were detected and recorded using the KSC Cloud to Ground Lightning Surveillance System (CGLSS) and the Lightning Detection and Ranging System (LDAR), as well as the ground-based field mill network.

### 6.4.3    Instrumentation

The UND Citation was instrumented with 6 rotating-vane field mills described by Bateman et al. (2007) (See Figure 6.4.3-1). The calibration of the field mill system is described in Mach and Koshak (2007).

**Figure 6.4.3-1  A field mill on the side of the Citation fuselage**

Note:  Conductive paint surrounds the Mill. The Cloud Particle Imager appears below it.

The Citation had an extensive array of microphysical probes and sensors including the Particle Measuring Systems (PMS) FSSP, 1-DC, and 2-DC, and the Stratton Park Engineering Company (SPEC) Cloud Imaging Probe (CPI) and High Volume Particle System (HVPS). Thus particles from a few microns to more than 10 mm in size could be counted, sized and imaged. Additionally there was a King liquid water probe and a Rosemount Icing detector for measuring and detecting the presence of supercooled liquid water (See Figure 6.4.3-2). The aircraft, instrumentation, radars, and lightning detection systems are described in the Appendices of Dye et al. (2003).

**Figure 6.4.3-2 The UND Citation Wing-mounted cloud physics sensors**

Note: From left to right, the High Volume Particle Sampler, King liquid water sensor, the Rosemount temperature sensor and the Forward Scattering Spectrometer Probe.

### 6.4.4    The 2000 field campaign

In 2000, all of the research flights took place in June. There were twelve days on which research flights were conducted, and on three of those days there was sufficient time for two flights, providing a total of fifteen research flights for the deployment. In addition, calibration flights over the Shuttle landing facility (See Figure 6.4.4-1) and over the ocean were conducted as needed.

Initial penetrations were often made near to, but at a safe distance from, the convective cores of storms. Then subsequent passes were made in the anvil at different distances downstream to examine the decay of the electric field with both time and distance. When Air Traffic Control (ATC) would allow, spiral ascents or descents were made through the anvils, but these were not nearly as frequent as was desirable for studying the anvil vertical structure. As a result, sampling of the vertical structure of the anvil of individual storms was often incomplete.

**Figure 6.4.4-1  UND Citation calibration flight at the SLF**

Note:  A field mill from the KSC/CCAFS network appears on a tripod at the lower right and the U. Arizona mobile field mill appears at the lower left.

The Citation was flown with 2 pilots and three scientific observers: a flight scientist, a scientist to operate and monitor the UND data system and microphysical instruments, and a scientist or engineer to operate and monitor the field mill measurements. Decisions on where to fly were interactive between crew in the aircraft and aircraft coordinators at the KSC Range Operations Control Center (ROCC).

There were normally 2 or more ABFM team members on the ground in the ROCC. Aircraft position was telemetered to the ROCC and overlaid on the WSR-74C near real-time vertical and horizontal displays generated using Sigmet software. In the ROCC one ABFM team member communicated with the aircraft for both safety and scientific purposes while one person operated the 74C radar display system to produce desired cross sections. Often there was an additional person in the ROCC to guide the University of Arizona mobile field mill vehicle to the near vicinity of storms of interest and especially those being investigated by the Citation. Unfortunately, storms were often too distant from KSC for the mobile field mill to be deployed directly to storms studied by the aircraft.

### 6.4.5   The 2001 Field campaign
There were actually two field campaigns in 2001, a winter campaign in February and a summer campaign in May and June. With the hope of improving the "Thick Cloud Rule", the month of February was selected for a program targeted at thick layered clouds because they are climatologically most likely to occur in central

Florida in the late winter. In addition to the instrumented ABFM II aircraft described above, a ground-based 35 GHz scanning Doppler, dual-polarized cloud radar (Martner et al, 2002) was provided by NOAA's Environmental Technology Laboratory. Unfortunately, during the entire three week February deployment, no clouds of the type covered by the Thick Cloud Rule occurred over central Florida.

The summer program ran from 20 May to 30 June 2001 with the primary objective being to increase the sample sizes in anvil and debris clouds using similar strategies as those used in 2000. Based on lessons learned in 2000, some adjustments were made in the coordination with the FAA Miami and Jacksonville Air Route Control Centers in an attempt to generate more opportunities for vertical profiles, but those opportunities remained rare. The weather cooperated for the summer program and research flights took place on eighteen days.

Over the period of the two summer campaigns anvils were sampled at a wide variety of altitudes in different locations relative to anvil top and bottom. Thus, the observations in aggregate are felt to be representative of conditions in anvils of Florida thunderstorms. On some occasions, stair step horizontal passes were made through the anvil or at other times passes were made along the downwind axis. In other cases the aircraft arrived during the decay stage of the anvil, but these cases are also important because we know the lightning and reflectivity history of that storm relative to the aircraft flights.

### 6.4.6   *Quality Control and Synchronization of Data*
Quality control (QC) and synchronization of the ABFM II data took the science team nearly two years because it had to be absolutely right if the results were to be reliable enough for the development of launch commit criteria. QC and synchronization efforts were divided among various science team institutions with MSFC taking the lead on calibration of the airborne field mills and NCAR taking the lead on the remaining QC and synchronizations with support from the entire team.

Time synchronization was necessary because of the many separate timing sources in the various data streams. The cloud physics and electric field measurements on the aircraft were recorded separately. The KSC/CCAFS instrumentation including the critical lightning instrument had its own source of timing. The weather radars, WRS-74C and WSR-88D each had its own separate timing source. A timing error on the order of a second or more could easily result in attempting to correlate data from one system measured while the aircraft was in cloud with another taken when the aircraft was actually outside of a cloud since the aircraft was flying at a speed greater than 100 m/s.

Spatial synchronization was also necessary. The aircraft position was well known as a function of time because of the high quality of the on-board navigation system. That position had to be connected to specific features displayed by two different weather radars, each at its own location with its own scan strategy. This required reconciling a set of locally Cartesian aircraft coordinates (latitude, longitude and altitude) with two different sets of spherical radar coordinates (altitude angle, azimuth angle and range) and placing all of the data into a common coordinate system.

Quality control of the data could not be automated because of the types (including images) and variety, and was extremely labor intensive and, hence, lengthy. The output of each instrument was examined by someone familiar with its characteristics and weaknesses. Data that seemed questionable at first glance was examined in detail and either validated or flagged as bad. In addition to screening the data from each instrument separately, the data from multiple sensors were checked for mutual consistency with each other and with the general meteorological environment.

After QC and synchronization had been accomplished, an integrated data set was placed on an NCAR website where the entire science team could access it to conduct their analysis. As noted in Section 6.0, the data set is now available to the public for further research.

### 6.4.7 Major Results – VAHIRR

While there were many findings of interest from ABFM II, there were two that resulted in major revisions to the LLCC. The first, the development of the VAHIRR quantity, is discussed in this section. The second, the possibility of safely reducing the standoff distances in the rules, is presented in the next section.

One motivation for ABFM II was the hope that there existed a radar-derived quantity that could give a reliable indication of when electric fields aloft did not exceed 3 kV/m. After exploring a dozen or so potential candidates for such a quantity, the team was able to establish that the Volume Averaged Height Integrated Radar Reflectivity (VAHIRR) met the requirement. The computation of VAHIRR is described in Dye et al. (2006), Krider et al (2006) and Merceret et al (2006) and is too complex to discuss in detail here. Simply, it is the product of the average radar reflectivity in dBZ multiplied by the cloud thickness in a specified volume. The specified volume is centered on the point at which VAHIRR is being calculated, bounded laterally by a square box 11 km on a side, with its bottom at the 0C isotherm and its top at the top of the highest clouds in the box. If VAHIRR was less than 10 dBZ-km at a point aloft, then the probability that the electric field was greater than 3 kV/m at that point was less than one in ten thousand. Major revisions to the LLCC based on VAHIRR were proposed in 2005 and 2008. These revisions permitted the launch constraints for anvil and debris clouds to be significantly relaxed with no reduction in safety.

### 6.4.8 Major Results – Stand-off Distances

The other finding that permitted safe relaxation of the anvil and debris cloud constraints was that electric fields outside of anvil and debris clouds remained below 3 kV/m even when fields as high as 70 kV/m or more were present in the clouds (Merceret et al, 2008). As a result, the "stand-off distance" from the edge of anvil and debris clouds that the LLCC require be maintained for several hours after the last lightning in the cloud or its parent thunderstorm could be reduced from 5 to 3 NM. Compare the 2005 version of the LLCC (Appendix I Section A1.10) with the current version in Appendix I, Section A1.11, to see the full impact of the changes which are actually somewhat more complex than a simple reduction in the standoff distance.

### 6.4.9 Other Results

Some general findings (Dye et al., 2003, 2006a) from the ABFM II investigation of anvils and debris clouds are:

1.  When electric fields were strong there was a surprisingly degree of consistency of particle types and concentrations in different size ranges; the smaller particles in anvils appear to be frozen cloud droplets, particles >100 to 500 μm were primarily irregularly shaped with few pristine crystals, and most particles >500 μm were aggregates; there was no evidence of supercooled water within the anvils. The largest aggregates were frequently as large as 5 to 6 mm in the dense part of anvils and sometimes larger. Surprisingly even near the edges of anvils 1- 2 mm aggregates were often found.
2.  As expected, the best correlation between radar reflectivity and particle concentration was for particles >3mm in size. Interestingly, the measurements also showed that as the aircraft flew from regions of weak reflectivity into regions with higher reflectivity the particle concentrations in all size ranges -- small, medium and large – increased together. That is, particles over the entire particle size spectrum increased to reach maximum concentration in the regions of greatest reflectivity. The high reflectivity was not only a result of a few large particles, but also a very broad spectrum of particles.
3.  Strong electric fields (>10 kV m-1) in anvils were usually associated with regions of higher reflectivity. The spatial transition to strong fields was usually quite abrupt as shown in Figure 6.4.9-1 below. In the figure, the top panel shows the time history of particle concentrations measured by the following instruments: PMS FSSP (1 to 48 μm), light, solid line = total conc. on right scale; PMS 2D-C (30 μm to ~3 mm), bold line = total conc., dashed line = conc. >1 mm on left scale; PMS 1D-C (15 to 960 μm), dotted line = total conc. on left scale. The second panel shows the radar reflectivity and temperature at the aircraft position as well as the aircraft bank angle. The solid bold line is the radar,

the thin solid line the temperature, and the dashed line the bank angle. The third panel shows the radar reflectivity curtain above and below the aircraft track from the NEXRAD radar at Melbourne, FL (KMLB). The bold line is the aircraft altitude. The bottom panel shows the vertical component of the electric field, Ez as a light line on a linear scale (left) and the resultant vector field magnitude, Emag as a bold line on a logarithmic scale (right). Entry into the strong fields occurred at about 21:07:20 UTC and exit occurred about three minutes later.

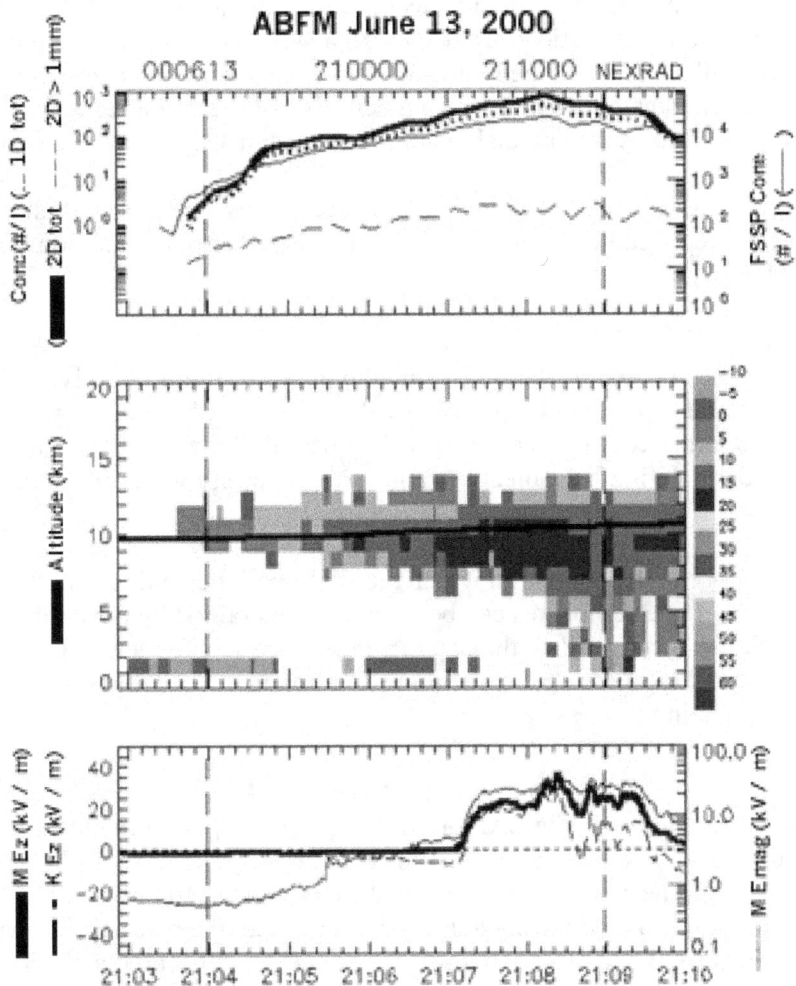

**Figure 6.4.9-1  Example of the abrupt electric field transition in an anvil**

4.  There was a threshold behavior between electric field and reflectivity such that below about 5 to 10 dBZ the electric field was less than 3 kV m-1. This threshold behavior between electric field and reflectivity was observed in debris clouds as well as anvils (Dye et al., 2008) and lead to the development of the radar-based parameter Volume Averaged Height Integrated Radar Reflectivity (VAHIRR) (Dye et al., 2006b).

5.  The abrupt transition in anvils from fields ~1 kV m-1 to fields >3 kV m-1 occurred well within the cloud so that outside of the anvils and debris clouds the electric fields dropped rapidly to values less than those considered a threat for triggering lightning (Merceret et al., 2008).

6.  Based on the measured particle size distributions across anvils and a simplified model of the decay of electric field in a quiescent anvil, it was estimated that in the dense parts of anvils it could take as long

as 1 1/2 hours for the electric field to decay from thunderstorm strength fields of 50 kV/m to fields near zero, whereas on the edge of the anvil the decay from an initial field of 50 kV/m to near zero could occur in only a few minutes (Willett and Dye, 2003). This large variation is primarily a result of the large variation in observed total particle cross-sectional area (i.e. concentration and size) across the anvils.

7. Comparison between the observed decay times of electric field in anvils were generally consistent with those calculated from the simple model of field decay using observed ice particle size distributions, but only one case permitted a meaningful comparison (Willett and Dye, 2003). Further carefully designed field investigations are needed to adequately validate this model.

8. A few long-lived anvil cases developed into stratiform-like cloud layers in which both electric fields and reflectivity increased and persisted for long periods of time. Both reflectivity and electric fields became relatively uniform over appreciable horizontal distances, in one case over 70 km (Dye and Willett, 2006).

9. The WSR-88D (NEXRAD) and WSR-74C radar reflectivities agreed to within typical uncertainties found in well-calibrated weather radars over the range of reflectivity commonly found in these clouds (Dye et al., 2004, p 61.)

### 6.4.10 Lessons Learned

One thing that quickly became apparent is that following the time evolution of individual cloud volumes with an aircraft is extremely difficult. More careful planning will be required in future experiments of this kind if it is desired to look at the temporal decay of electric fields in a parcel. Pre-flight planning sessions before every campaign and every mission to define key objectives and approaches are wise no matter what the focus of the campaign or mission is.

The ability to conduct an airborne experiment in which the flight path depends on the location of rapidly developing and moving convective complexes can be severely constrained by the ability and willingness of the air traffic control authorities to accommodate the experimenter's request for access to airspace. Airspace flexibility should be considered in selecting the operating location for field programs of this kind, and the FAA should be involved in the planning process early on.

## 6.5    Lightning LCC

This section briefly summarizes not only the LAP discussions directly pertaining to revisions of the LLCC, but also their planning and discussions relating to the ABFM II field program and subsequent data analysis. The two cannot really be separated. The research was designed from the start to inform revision of the LLCC, and the LLCC changes flow directly from new knowledge gained through the research. As with section 5.5, the details are provided in Appendix II and only a summary of the key events and conclusions is given here.

### 6.5.1    ABFM II Planning Discussions

Early in 1999, the KSC Weather Office obtained a small amount of funding from the Space Shuttle program to install field mills on a research aircraft that would be taking part that summer in a research project in the Pacific. This "target of opportunity" would permit the mills to be installed and flight tested without the program having to pay for the flight hours or travel expenses since these were already funded by the research project. Also in 1999, the KSCWO was able to assemble combined funding commitments from the Shuttle program and other NASA sources to support the dedicated field program in 2000 that became ABFM-II described in Section 6.4 above.

The LAP had begun preliminary discussions for a follow-on program to ABFM-I in February 1999. In January 2000, after funding had been identified, the LAP held a formal ABFM-II planning meeting at KSC to design the overall strategy for the 2000 summer field program, including scientific goals and required instrumentation. In addition, logistics, schedule and budget issues were addressed. The members of the LAP, representatives from the 45th Weather Squadron, the Spaceflight Meteorology Group, the Applied Meteorology Unit, the University of North Dakota, the Air Force Research Laboratories the KSC Weather Office, and several KSC administrative offices attended.

At the beginning of the field program in June, final science and operational planning meetings were held. Coordination meetings took place between the University of North Dakota pilots and the FAA to facilitate access to the air space necessary for the planned observational strategy. Additional planning, frequently by telephone, preceded each individual mission during the summer.

Planning for the winter and summer 2001 campaigns followed the same pattern, but took advantage of the experience gained during the 2000 campaign, especially with regard to improved coordination with the FAA.

Planning did not stop after the end of the field campaigns. In November 2001 an ABFM II workshop was held in Melbourne, Florida at the offices of the AMU contractor, ENSCO, Inc. The LAP and the ABFM science team extensively discussed issues relating to the calibration and archiving of the field data. Specific areas of research were assigned to specific investigators in order to avoid duplication of effort and provide a coordinated, unified approach to maximizing the utility of the data for potential revisions to the LLCC. On a smaller scale, such planning continued during teleconferences until the final submission of ABFM-II-based recommended LLCC changes by the LAP in 2009.

### 6.5.2  LLCC Revisions – 2005

The analysis of the ABFM II data led to a determination that the VAHIRR quantity (section 6.4.7 above) could serve as the basis for changes in the Anvil rules (both attached and detached) that would significantly reduce the number of unnecessary scrubs and delays caused by these rules. The resulting recommendation was adopted by NASA and the Air Force. The Shuttle version is presented in detail in Appendix I, section A1.10. These rules were very similar to the current rules except that they did not provide for the use of VAHIRR for debris clouds, and the stand-off distances for both anvil and debris clouds were longer. Another revision (2009) was required to make those changes. A short summary of the changes as implemented in 2005 by the Space Shuttle program follows.

*Space Shuttle LLCC (2005) [LCN 01166R01, 06/03/05].* MAJOR Changes:  This major change was the first application of both Harry Koons's truth-table analysis (see the figures in Appendix II), and his statistical analysis of the new ABFM II dataset, to both clarify and relax the anvil rules. It is quoted in full in Appendix I, Section A1.10. In spite of numerous wording changes throughout, however, there were no substantive changes to any of the other rules, except indirectly through changes in the definitions. The key definition change was in the radar definition of the cloud boundary, which was lowered from 10 dBZ to 0 dBZ, significantly tightening the rules and closing a potentially serious loophole.

A major new exception was provided for both Attached and Detached Anvil Clouds, in terms of a new radar parameter called the Volume-Averaged, Height-Integrated Radar Reflectivity (VAHIRR). The new versions of the two anvil rules are paraphrased below:

Anvil Clouds:
Attached Anvil Clouds.
*Do not fly within 10 nm of any non-transparent, attached anvil for at least 30 minutes after the last lightning discharge occurs in the parent cloud or anvil, nor within 5 nm for 3 hours after the last lightning. Never fly through such a cloud.*

*Exception: Flight is allowed through or within any distance of an attached anvil after 30 minutes if that anvil is colder than 0 °C and if its VAHIRR is less than 10 dBZ-km everywhere along the flight path.*

Detached Anvil Clouds.
*Do not fly within 10 nm of a non-transparent, detached anvil for 30 minutes after the last lightning discharge occurs in the detached anvil (or in its parent cloud before detachment), nor within 5 nm for 3 hours after the last such lightning. Do not penetrate such an anvil for 3 hours after detachment, nor for 4 hours after it has produced lightning.*

*Two kinds of exceptions are allowed: 1) Flight is allowed up to the edge of a detached anvil between 30 minutes and 3 hours after the last lightning if the radar reflectivity of all parts of that cloud within 5 nm of the flight path has been less than 10 dBZ, and if a field mill within 5 nm of the cloud (and any other mills within 5 nm of the flight path) have shown less than ±1000 V/m, for at least 15 minutes. 2) Flight is allowed through or within any distance of such an anvil after 30 minutes if the cloud is colder than 0 °C and if its VAHIRR is less than 10 dBZ-km everywhere along the flight path.*

After considerable negotiation with the LAP, the FAA adopted essentially this version of the LLCC, translated into their own vocabulary and format to become what they call the LFCC, which are quoted in Appendix I, Section A1.12 "Current (through 25 March 2010) FAA LLCC (14 CFR 417, Appendix G)."

### 6.5.3    LLCC Revisions – 2009

The LAP continued to analyze and discuss the ABFM II data following the development of the 2005 recommendations discussed above. Two additional major improvements to the rules were found to be justified by the analysis: application of the VAHIRR concept to debris clouds in a manner similar to their application to anvils, and reduction of the stand-off distances for both anvil and debris clouds (section 6.4.8 above). The final product was the recommendation for adoption of the currently effective rules summarized in Chapter 1, section 1.1 and presented in detail (Space Shuttle Version) in Appendix I, section A1.11. The LAP's recommendations were presented in 2008 in a format requested by the Federal Aviation Administration (FAA) for application to the civilian spaceports that were under development throughout the United States. The Space Shuttle program adopted these recommendations in substance in 2009 but did not keep the FAA format. A summary of the LAP recommendation follows.

This version, quoted in Appendix I, Section A1.13, includes major changes in both language and substance. In view of the apparent necessity of working with the FAA in the future, the LAP chose to attempt writing these LLCC using FAA language from the beginning.

This is the second application of Dr. Harry Koons's logical analysis and of the ABFM II dataset as further analyzed statistically by Paul O'Brien and Frank Merceret. The standoff distances from Attached-Anvil, Detached-Anvil and Debris Clouds were reduced from 5 to 3 NM under many conditions, and a VAHIRR exception was added to the Debris Cloud Rule. Other significant changes included placing the definitions first (to emphasize that they are an important part of the rules), making Surface Electric Fields the first rule (to emphasize its importance as the key physical parameter for triggering lightning), adding an explicit requirement that all LFCC must be individually satisfied, narrowing the definition of "Anvil," and explicitly including a Triboelectrification Rule. Since this rule set has already been paraphrased in Section 1.1, only the significantly changed sections will be repeated here.

All of the LLCC must be satisfied.

Attached Anvil Clouds.
*Do not fly within 10 nm of any non-transparent, attached anvil for at least 30 minutes after the last lightning discharge occurs in the parent cloud or anvil, nor within 5 nm for 3 hours after the last lightning. Never fly within 3 nm of such a cloud.*

*Two kinds of exceptions are allowed: 1) Flight is allowed up to 3 nm from an attached anvil at any time if it is colder than 0 °C everywhere within the prescribed distances of the flight path. 2) Flight is allowed through or within any distance of an attached anvil at any time if that anvil is colder than 0 °C and if its VAHIRR is less than 10 dBZ-km everywhere within 1 nm of the flight path.*

Detached Anvil Clouds.
*Do not fly within 10 nm of a non-transparent, detached anvil for 30 minutes after the last lightning discharge occurs in the detached anvil (or in its parent cloud before detachment), nor within 3 nm for 3 hours after the last such lightning. Do not penetrate such an anvil for 3 hours after detachment, nor for 4 hours after it has produced lightning.*

*Three kinds of exceptions are allowed: 1) Flight is allowed up to 3 nm from such an anvil during the first 30 minutes if the detached anvil is colder than 0 °C everywhere within a 10 nm of the flight path. 2) Flight is allowed up to the edge of a detached anvil between 30 minutes and 3 hours after the last lightning if the radar reflectivity of all parts of that cloud within 5 nm of the flight path has been less than 10 dBZ, and if a field mill within 5 nm of the cloud (and any other mills within 5 nm of the flight path) have shown less than ±1000 V/m, for at least 15 minutes. 3) Flight is allowed through or within any distance of such an anvil at any time if the cloud is colder than 0 °C and if its VAHIRR is less than 10 dBZ-km everywhere within the prescribed distances of the flight path.*

Debris Clouds.
*Do not fly within 3 nm of a non-transparent debris cloud for 3 hours after it detaches or decays from its parent cloud and for 3 hours after it has produced lightning.*

*Two kinds of exceptions are allowed to the 3 nm standoff requirement: 1) Flight is allowed up to the edge of a debris cloud at any time if the radar reflectivity of any part of that cloud within 5 nm of the flight path has been less than 10 dBZ, and if a field mill within 5 nm of the cloud (and any other mills within 5 nm of the flight path) have shown less than ±1000 V/m, for at least 15 minutes. 2) Flight is allowed through or within any distance of such a debris cloud at any time if the cloud is colder than 0 °C everywhere within 5 nm of the flight path and if VAHIRR is less than 10 dBZ-km everywhere within prescribed distances of the flight path.*

Triboelectrification.
*Do not fly through any cloud (transparent or not) that is colder than -10 °C at vehicle velocities less than 3000 ft/s unless the vehicle has been treated or hardened against surface discharges.*

# Chapter 7   Conclusions and Lessons Learned

The development of the LLCC and their associated Definitions in the post-AC 67 era has been an ongoing process of collaboration between the LAP (originally the PRC), the NASA/KSC Weather Office (originally the NASA-HQ Weather Support Office), and the Air Force 45WS. As we have seen, the first set of lightning launch commit criteria was developed in the wake of the Apollo XII incident, after it was recognized that a launch vehicle can trigger lightning. After this incident there was considerable involvement of the scientific community at large, and the resulting LLCC were detailed by the Apollo Program in its official accident investigation (NASA, 1970, p.50). Thereafter, the LLCC evolved gradually and without formal coordination over a succession of manned and unmanned vehicles until the AC 67 accident. This event demonstrated that the triggered-lightning hazard was not being taken seriously by launch weather personnel and that the rules were not being applied uniformly.

In reaction to the AC 67 accident, a relatively small, blue-ribbon panel of experts, the "Heritage Committee," was empowered to undertake an independent study of the issues and draft a much more comprehensive and scientifically justified set of LLCC [Heritage, 1988, Chapter 7]. Shortly thereafter, a subset of the Heritage Committee was asked to serve on the first standing committee, the PRC, to evaluate and improve the LLCC (see Section 5.2.1).

Since the original formation of the PRC/LAP, a dual-track process has evolved. The LAP initiates, re-writes, or re-phrases LLCC in response to new scientific knowledge, and the LAP proposes experiments to advance knowledge in areas that are important for launch support. The LAP also receives questions and concerns about the LLCC and their implementation from launch-weather personnel (particularly the KSC Weather Office and the 45WS, and more recently the SMG, the 30WS, and even people at the Kodiak Launch Complex). Out of what might easily have become an adversarial relationship due to an imperfect science and the need to apply the LLCC in the real world, certain mutual goals have emerged, especially having rules that both improve safety and increase launch availability. This process has resulted in growing confidence in, and attention to, the LLCC and the associated Definitions. In this chapter we try to summarize the most significant conclusions and recommendations that have emerged from this process.

Summarized in a single sentence, the LLCC must be safe, physically valid, complete, not unnecessarily restrictive, self-consistent, clear, and faithfully implemented. Each of these requirements and its importance is briefly elaborated below:

Safety. First and foremost, the LLCC must be safe. No lightning strikes to vehicles during launch, triggered or natural, will occur when the rules are followed. Safety is assured by making the LLCC both physically valid and complete, as outlined in the following two sections. The LAP process for achieving these and the other goals outlined below has been discussed in Section 5.5.1 above. It bears repeating, however, that launch safety and launch availability are competing needs that must be carefully balanced.

Physical Validity. A sound physical basis for the LLCC is an obvious prerequisite for their safe and successful application. Furthermore, if individual launch rules could easily be challenged on technical grounds, they would quickly be ignored, and the whole LLCC process would lose credibility. The LAP has used the scientific literature and the outcomes of recommended experiments to ensure that the LLCC cover the meteorological conditions that will produce hazardous electric fields aloft. The science that is behind the current rules will be discussed in detail in the Rationale document that is planned to accompany this History.

Completeness. Safety requires that the LLCC identify all meteorological conditions that will create high electric fields aloft. Without this completeness, accidents will occur that will call the whole process into question. Rather than risk any such accidents, the LAP has started with LLCC that are conservative, and additional instrumentation (*e.g.*, field mills) has been recommended to cover unexpected conditions that might be hazardous. By identifying the sources of false alarms, and by recommending new methods of detecting and eliminating them, the LAP has incrementally reduced the initial conservatism and increased launch availability without sacrificing safety. Nevertheless, there will always be a tradeoff between completeness and launch availability in questionable weather situations.

Launch Availability. Close collaboration between the LAP, the 45WS, and the KSC Weather Office has emerged as an efficient way to identify individual launch restrictions that have a significant impact on launch availability or that appear to have an unduly high false-alarm rate. When attention is focused on these situations, solutions can often be found that increase launch availability without compromising safety.

Consistency. There are two important aspects of self-consistency in the LLCC. 1) Within each individual rule, the sub-sections should be mutually consistent to prevent confusion and incorrect evaluation by the Launch Weather Team (LWT). This issue has been addressed with logical truth-table analysis, which can not only eliminate inconsistencies between sections of a given rule but also remove redundancy that can lead to ambiguity. 2) Because all rules in the LLCC must be evaluated simultaneously, it is helpful, though not always possible, for each rule to be consistent with the others. There will occasionally be ambiguity over which rule applies to a given weather situation, because of the inherent subjectivity in evaluation, and in such situations all potentially applicable rules must be satisfied.. In these situations, an apparent inconsistency may worry the LWT, but if the most conservative criterion is applied, that will insure safety.

Clarity. The LLCC must be clearly written to ensure proper understanding by both customer personnel, who must accept them prior to any countdown, and the LWT, which must apply them during the launch. Because it is critical for the LWT to evaluate the LLCC quickly and accurately under pressure, the rules themselves should be as clear and easy to understand as possible. The close working relationship that has developed between the LAP and the 45WS has been invaluable for clarifying the rules and has led to more efficient operational use. Nevertheless, increased launch availability without compromising safety comes at the price of greater complexity. The desire for lower false-alarm rates has led to the addition of numerous exceptions to the basic LLCC that require additional conditions to be satisfied. The LAP has introduced truth tables to facilitate better understanding of the rules and make them self-consistent (see figures in Appendix II). These truth-tables are also beneficial as training materials.

Faithful Implementation. Even a perfect set of LLCC and definitions cannot assure safety unless they are fully and correctly applied during launch operations. The following are requirements for safe implementation: 1) The rules must be coordinated with, and certified by, all customers prior to use to ensure that they understand and accept the assumptions and associated risks. 2) The LWT must be rigorously and correctly trained, certified, tested, and routinely retested during the conditions of a launch countdown. 3) The LLCC must continue to be evaluated on the basis of data derived from a reliable, accurate, and robust weather and atmospheric-electricity infrastructure. 4) Each member of the LWT must be clearly convinced that each of the LLCC has been completely satisfied. 5) The LWT must be managed by a Senior Weather Officer who ensures that each LWT team member is aggressively and strictly assessing the LLCC and is freely and constructively communicating data and assessments with other LWT members. 6) The Senior Weather Officer in charge must ensure that external factors like "launch fever" or visitors, especially VIPs, do not impede or bias the LWT evaluations; and that manager must handle all queries, comments, and concerns from higher headquarters and other senior management and not on the formal customer Launch Decision Team. 7) The LAP notes that the civilian Launch Weather Officers (LWOs) have made a significant contribution to launch safety through rigorous training, careful observation, and conscientious application of the LLCC. It is essential to preserve this 'corporate memory' by maintaining a sufficient number of civilian LWO positions.

# Appendix I    The Lightning Launch Commit Criteria (LLCC) and Associated Definitions

*A1.1    Post Apollo XII Lightning LCC (LLCC) 1970*
Space vehicle will not be launched if nominal flight path will carry vehicle:
- Within 5 sm of a cumulonimbus (thunderstorm) cloud;
- Within 3 sm of anvil associated with a thunderstorm;
- Through cold front or squall line clouds which extend above 10K feet;
- Through middle cloud layers 6000 feet or greater in depth where the freeze level is in the clouds;
- Through cumulus clouds with tops at 10,000 feet or higher.

*A1.2    Skylab (1973) LLCC*
Space vehicle will not be launched if nominal flight path will carry vehicle
- Within 5 sm of a cumulonimbus (thunderstorm) cloud or within 3 sm of an associated anvil;
- Through cold front or squall line clouds which extend above 10K feet;
- Through middle cloud layers 6000 feet or greater in depth where the freeze level is in the clouds;
- Through cumulus clouds with tops at 10,000 feet or higher

*A1.3    Apollo-Soyuz Lightning LLCC 1975*
(Heritage, 1988, p. 4-4)
Space vehicle will not be launched if the nominal flight path will carry vehicle:
A. Through a cumulonimbus (thunderstorm) cloud;
B. Within 5 sm of a cumulonimbus (thunderstorm) cloud or within 3 sm of an associated anvil. This rule may be relaxed at the discretion of the Launch Director if the electric field at the launch pad is less than 1 kV//m with a very narrow launch window
C. Through cold front or squall line clouds which extend above 10K feet;
D. Through middle cloud layers 6000 feet or greater in depth where the freeze level is in the clouds;
E. Through cumulus clouds with tops at 10,000 feet or higher;
F. Rules C, D, and E above may be relaxed at the discretion of the Launch Director when electric field measurements in the launch pad area are stable and measure less than 1 kV/m.
G. Rules C, D, and E above may be further relaxed provided that airborne and ground electric field measurements are less than or equal to 3 kV/m as measured by the ground based field mill network and less than or equal to EC(H) where EC(H) varies linearly from 3 kV/m at the surface to 15 kV/m at 25K feet and remains constant above that altitude. This rule may be applied only if vertical field measurements along the flight path in a 3-mile area are within the described envelope and there are no rapid fluctuations of about 3 kV/m at about 1 minute intervals within the 3 mile area measured by the ground mills.

*A1.4    AC 38 (1976) LLCC*
Space vehicle will not be launched if nominal flight path will carry vehicle:
- Within 5 sm of a cumulonimbus (thunderstorm) cloud;
- Within 3 sm of anvil associated with a thunderstorm;
- Through cold front or squall line clouds which extend above 10K feet;
- Through middle cloud layers 6000 feet or greater in depth where the freeze level is in the clouds;
- Through cumulus clouds with tops at 10,000 feet or higher.

*A1.5    Original Space Shuttle LLCC 1979*
(Heritage, p. 4-7, Table 4-3)
Space vehicle will not be launched if nominal flight path will carry vehicle:
- Through a cumulonimbus (thunderstorm) cloud;
- Within 5 sm (7.8 km) of a cumulonimbus (thunderstorm) cloud or the edge of an associated anvil that is within 5 sm of its radar cell. At the discretion of the Launch Director the anvil may be penetrated if the static electric field is less than 1kV/m and the 5 sm rule is maintained.
- Through or within 5 sm of any other clouds where radar shows virga or precipitation and tops extend (or will extend at the time of launch) above the -10 °C temperature altitude and four or more field mills within the launch site have changed more than 500 V/m.
- Through single layer clouds in the dissipation stage and where these clouds have activated the delta electric field contours within a period of 10 minutes before launch. The maximum electric fields associated with subject cloud as measured on the ground must not exceed 2 kV/m. The electric field meter readings for double layer clouds cannot exceed 500 V/m.

*A1.6    1986 Shuttle Lightning Launch Commit Criteria (LLCC)*
(Heritage, p. 4-8, Table 4-4)
Do not launch if vehicle path is:
- Within 5 nm of a cumulonimbus cloud or the edge of associated anvil cloud;
- Within 5 nm of any convective cloud whose top extends to -20 °C isotherm with virga/precipitation;
- Through any cloud where precipitation is observed;
- Through dissipating clouds in which the electric field network has detected lightning within 15 minutes prior to launch;
- Through any cloud if ground level electric field at launch site is greater than + or - 1000 V/m.

*A1.7    Atlas/Centaur 67 Lightning Launch Commit Criteria (1987)*
(Heritage, 1988, p. 4-9, Table 4-5)
Flight path of the vehicle should not be:
A. Through a thunderstorm/cumulonimbus cloud.
B. Within 5 miles of thunderstorm/cumulonimbus cloud or 3 miles of associated anvil top.
C. Through cold front/squall line associated clouds with tops 10,000 ft or higher.
D. Through middle-level cloud layers, 6,000 ft or greater in depth, when the freezing level is in the clouds.
E. Through cumulus clouds with the freezing level in the clouds.

*A1.8    Heritage LRC Recommended LLCC (1988)*
(Heritage, 1988, pp. 7-1 through 7-6)
LAUNCH CONSTRAINTS FOR THE AVOIDANCE OF VEHICLE-TRIGGERED LIGHTNING,
NATURAL LIGHTNING, AND VEHICLE ELECTRIFICATION
The following launch constraints must be observed. They are designed to minimize the hazards to launch vehicles after launch from vehicle-triggered lightning, natural lightning, and vehicle electrification resulting from interactions with the environment.

Good Sense Rule: If hazardous conditions exist that approach the launch constraint limits or if hazardous conditions are believed to exist for any other reason, an assessment of the nature and severity of the threat shall be made and reported to the Test Director and Mission Director.

Constraint I. The vehicle will not be launched if any type of lightning is detected within 10 nmi of the launch site and within 30 minutes prior to launch time. The 30-minute time period in Constraint I may be terminated at the discretion of the Test Director and Mission Director if the meteorological condition that produced the lightning has moved more than 10 nmi away from the launch complex and planned flight path.

Constraint II. The vehicle will not be launched if the planned flight path is:
- a1.   Through cumulus clouds with tops that extend to an altitude above the +5 °C temperature level; or
- a2.   Through or within 5 nmi of cumulus clouds with tops that extend to an altitude higher than the -10 °C temperature level; or
- a3.   Through or within 10 nmi of cumulus clouds with tops that extend to an altitude higher than the -20 °C temperature level; or
- a4.   Through or within 10 nmi of the nearest edge of either a cumulonimbus cloud or a thunderstorm including the associated anvil; or
- b.   Through or within 5 nmi of a cloud layer, other than a thin transparent fibrous layer, that has become detached from a parent cumulonimbus cloud; or
- c.   Through any cloud types that extend to altitudes at or above the 0 °C temperature level and that are associated with disturbed weather within 5 nmi of the flight path; or
- d.   Through a vertically continuous layer of clouds with an overall depth of 4,500 ft, or greater, where any part of the clouds is located between the 0 °C and the -20 °C temperature levels.

The 10-nmi distance in Constraint IIa may be reduced to 5 nmi at the discretion of the Test Director and Mission Director if, in the 15 minutes prior to launch time:
- a.   The electric field intensity at the ground (for ranges that have a ground field mill system) has remained below 1 kV/m within 5 nmi of the launch pad; and
- b.   The absolute value of the instantaneous critical electric field intensity sampled along the flight path is less than the critical value, Ec, shown as a function of altitude in Figure 7-1.

Constraint IIa,, IIa2, IIb, IIc, and IId may be dispensed with at the discretion of the Test Director and Mission Director if, in the 15 minutes prior to launch time:
- a.   The electric field intensity at the ground (for ranges that have a ground field mill system) has remained below 1 kV/m within 5 nmi of the launch pad; and
- b.   The absolute value of the instantaneous critical electric field intensity within the clouds sampled along the flight path is less than the critical value, Ec, shown as a function of altitude in Figure 7-1. (Figure A1.8-1)

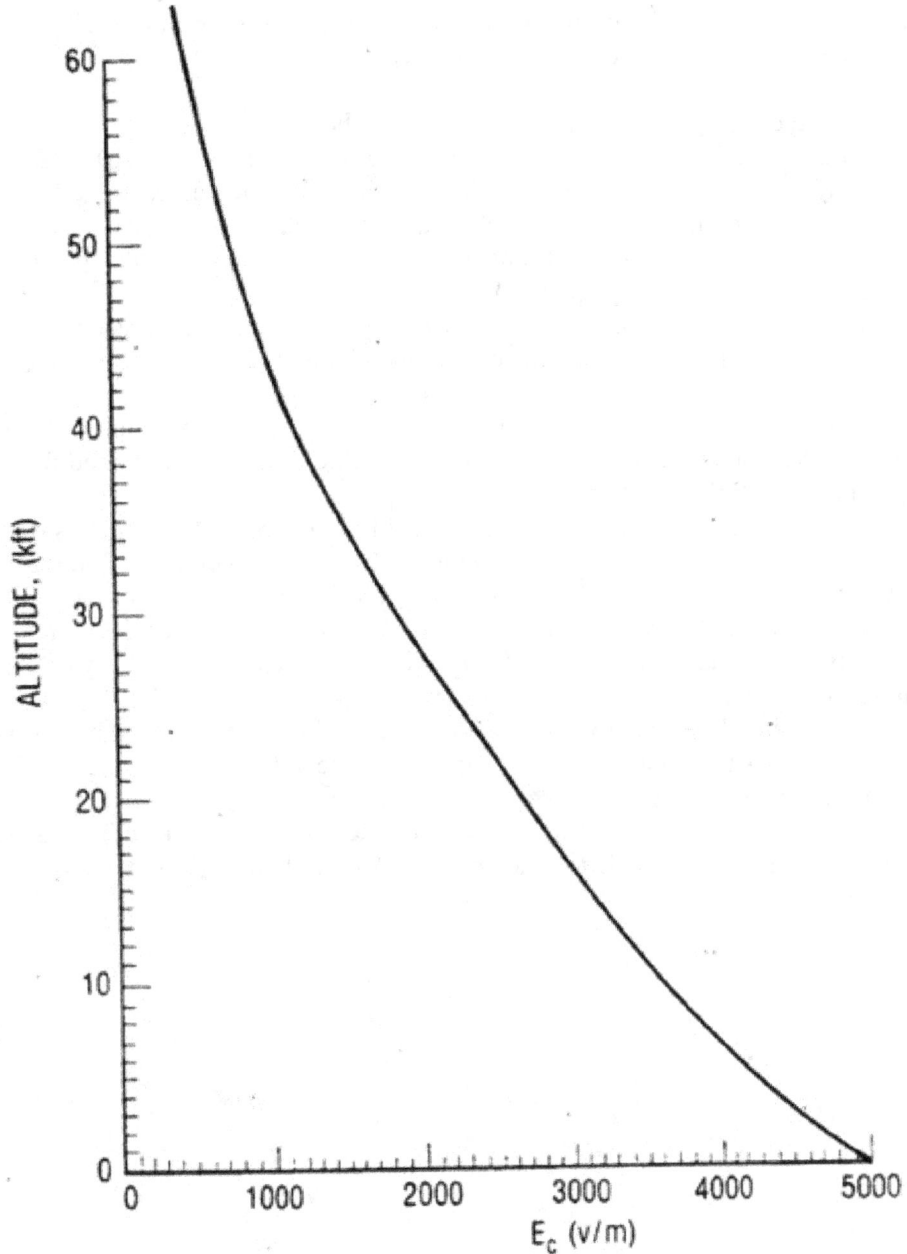

**Figure A1.8-1  Heritage (1988) Fig. 7-1, Instantaneous Critical Electric Field Intensity vs Altitude**

Constraint III. The vehicle will not be launched if, at any time during the 15 minutes prior to launch time, the electric field intensity at the ground exceeds 1 kV/m anywhere within a 5-nmi radius of the launch pad. Constraint III may be dispensed with at the discretion of the Test Director and Mission Director if:
  a.  There are no clouds within 10 nmi of the launch pad at the planned time of launch; and
  b.  The absolute value of the instantaneous critical electric field intensity sampled along the flight path during the 15 minutes prior to launch is less than the critical value, Ec shown as a function of altitude in Figure 7-1.

Constraint IV. A vehicle that has not been treated for surface electrification will not be launched if the planned flight path will go through clouds in the altitude regime extending upward from the -10 °C temperature level to the altitude at which the vehicle velocity exceeds 3000 ft/sec.

Definitions. The following definitions apply to the constraints above:
a. The term "nearest edge" of a cloud, as used in Constraint IIa, refers to the 10-dBZ level as measured by a weather radar or the visible edge. NOTE: The distance to the nearest edge is omnidirectional and includes the top of the cloud when the vehicle is above a cloud.
b. The term "disturbed weather", in Constraint IIc, refers to any meteorological phenomenon that is producing moderate or greater precipitation.
c. "Vertically continuous cloud layer", as used in Constraint IId, refers to any array of clouds, not necessarily all of the same type, in which some cloud elements connect the entire thickness of the array.
d. "Electric field intensity", as used in Constraints II and III, means the absolute value of the 1-minute time average of the instantaneous, ground-level electric field intensity obtained from an analysis of ground-based electric field mill data.
e. The term "critical value", as pertains to the electrical field intensity and which is used in Constraints II and III, refers to the absolute value of the instantaneous minimum electric field that will just cause electrical breakdown. This value will be a function of altitude as shown in Figure 7-1. (Other factors, not considered herein, that may contribute to the electrical breakdown of the atmosphere include vehicle velocity and spatial extent of the electric field. The presence of nearby or distant lightning or cloud conditions also may be contributing factors.)
f. A vehicle is considered "treated" for surface electrification in Constraint IV if all surfaces of the vehicle susceptible to precipitation particle impact have been treated to assure:
    1. That the surface resistivity is less than $10^9$ ohms/square, and
    2. That all conductors on surfaces (including dielectric surfaces that have been treated with conductive coatings) are bonded to the vehicle structure by a resistance that is less than $10^5$ ohms.

*A1.9    Space Shuttle LLCC (1992 – 1998)*

*A1.9.1  Space Shuttle LLCC (1992)[LCN# 00230R04]*
4.5  Natural and Triggered Lightning Constraints.

NOTICE:  ANY CHANGES TO THIS SECTION WILL REQUIRE COORDINATION WITH EASTERN
     SPACE AND MISSILE CENTER RANGE SAFETY OFFICE.

Even when constraints are not violated, if any other hazardous conditions exist, the Launch Weather Officer will report the threat to the Launch Director. The Launch Director may HOLD at any time based on the instability of the weather. The Launch Weather Officer must have clear and convincing evidence the following constraints are not violated.

A.  Do not launch if any type of lightning is detected within 10 nautical miles of the launch site or planned flight path within 30 minutes prior to launch, unless the meteorological condition that produced the lightning has moved more than 10 nautical miles away from the launch site or planned flight path.

B.  Do not launch if the planned flight path will carry the vehicle:

(1)  Through cumulus clouds with tops higher than the 5 degrees Celsius level; or

(2)  Through or within 5 nautical miles of cumulus clouds with tops higher than the -10 degrees Celsius level; or

(3)  Through or within 10 nautical miles of cumulus clouds with tops higher than the -20 degrees Celsius level; or

(4)  Through or within 10 nautical miles of the nearest edge of any cumulonimbus or thunderstorm cloud including its associated anvil.

C.  Do not launch if, for Ranges equipped with a surface electric field mill network, at any time during the 15 minutes prior to launch time the one minute average of absolute electric field intensity at the ground exceeds 1 kilovolt per meter (1kV/m) within 5 nautical miles of the launch site unless:

(1)  There are no clouds within 10 nautical miles of the launch site; and

(2)  Smoke or ground fog is clearly causing abnormal readings.

NOTE:  For confirmed instrumentation failure, continue countdown.

Relative to (1): the following clouds are acceptable if they have not been previously associated with a thunderstorm, or convective clouds with tops greater than -10 degree Celsius temperature level within the last three hours:

(a) Thin fibrous (optically transparent) clouds;

(b) Less than or equal to 2/8 cumulus/strato-cumulus/ stratus (CU/SC/ST) clouds with tops below or equal to the +5 degree Celsius temperature level. For example, 3/8 CU/SC/ST is not acceptable regardless of cloud top level, nor is any CU/SC/ST cloud amount above the +5 degree Celsius temperature level.

Relative to (2): This also includes a maritime inversion with an onshore/alongshore wind present over the electric field mills, causing those mills located near the ocean to be elevated with a positive polarity between 1kV/m and 1.5kV/m inclusive.

D.  Do not launch if the planned flight path is through a vertically continuous layer of clouds with an overall depth of 4,500 feet or greater where any part of the clouds are located between the zero (0) degree and the minus 20 (-20) degree Celsius temperature levels.

E.  Do not launch if the planned flight path is through any cloud types that extend to altitudes at or above the zero degree Celsius temperature level and that are associated with disturbed weather within 5 nautical miles of the flight path.

F.  Do not launch through thunderstorm debris clouds, or within 5 nautical miles of thunderstorm debris clouds not monitored by a field mill network or producing radar returns greater than or equal to 10 dBz.

G.  Definitions

(1)  Debris Cloud:  Any cloud layer, other than a thin fibrous layer, that has become detached from the parent cumulonimbus within 3 hours before launch.

(2)  Disturbed Weather:  Any meteorological phenomenon that is producing moderate or greater precipitation.

(3)  Cumulonimbus Cloud:  Any convective cloud which exceeds the -20 degree Celsius temperature level.

(4)  Cloud Layer:  Any cloud broken or overcast layer or layers connected by cloud elements, e.g. turrets from one cloud layer to another.

(5)  Planned Flight Path:  The trajectory of the flight vehicle from the launch pad through its flight profile until it reaches the altitude of 100,000 feet. The flight path may vary plus or minus 0.5 nautical miles horizontally up to an altitude of 25,000 feet.

(6)  Anvil:  Stratiform or fibrous cloud produced by the upper level outflow from thunderstorms or convective clouds. Anvil debris does not meet the definition if it is optically transparent.

ELECTRICAL CHARGE REGIONS CAN OCCUR IN CLOUDS WITH ALTITUDES AT OR ABOVE THE ZERO DEGREE CELSIUS ISOTHERM. THESE CHARGE REGIONS CAN PRODUCE NATURAL LIGHTNING DISCHARGES OR DISCHARGES THAT ARE TRIGGERED BY THE PROXIMITY OF LONG ELECTRICAL CONDUCTORS (LAUNCH VEHICLE PLUS CONDUCTIVE PORTION OF PLUME).

THE ABOVE CONSTRAINTS ARE FOR THE AVOIDANCE OF NATURAL OR TRIGGERED LIGHTNING AND ARE BASED ON THE KNOWN CLOUD TYPES WHICH CAN PRODUCE DISCHARGES AND THE DISTANCES TO THE CHARGE REGIONS AT WHICH DISCHARGES ARE KNOWN TO OCCUR.

4.5 Natural and Triggered Lightning Constraints.

NOTICE: ANY CHANGES TO THIS SECTION WILL REQUIRE COORDINATION WITH THE 30TH AND 45TH SPACE WING RANGE SAFETY OFFICES.

Even when constraints are not violated, if any other hazardous conditions exist, the Launch Weather Officer will report the threat to the Launch Director. The Launch Director may HOLD at any time based on the instability of the weather.

The Launch Weather Officer must have clear and convincing evidence the following constraints are not violated:

A.  Do not launch if any type of lightning is detected within 10 nautical miles (NM) of the flight path within 30 minutes prior to launch, unless the meteorological condition that produced the lightning has moved more than 10 NM away from the flight path.

NATURAL LIGHTNING IS AN OBVIOUS HAZARD (COMPARED TO VEHICLE TRIGGERED LIGHTNING) AND IS ALSO THE MOST DIRECT EVIDENCE ELECTRIC FIELDS ARE PRESENT WITH SUFFICIENT INTENSITY TO CAUSE TRIGGERED LIGHTNING. THE MEASURED FREQUENCY DISTRIBUTION OF THE DISTANCE BETWEEN SUCCESSIVE FLASHES TO GROUND APPROACHES ZERO NEAR 6 NM; THUS 10 NM PROVIDES A SAFETY FACTOR. THE 30 MINUTE TIME PERIOD WITHOUT LIGHTNING INDICATES THE STORM HAS DISSIPATED OR MOVED AWAY. HOWEVER, FORECASTERS MUST STILL REMAIN ALERT FOR REDEVELOPMENT OR FORMATION OF A NEW CELL.

B.  Do not launch if the flight path will carry the vehicle:

(1)  Through a cumulus cloud with its top between the +5.0 degrees Celcius (C) and -5.0 deg. C levels unless:

(a) The cloud is not producing precipitation;

-AND-

(b) The horizontal distance from the furthest edge of the cloud top to at least one working field mill is less than the altitude of the -5.0 deg. C level or 3 NM whichever is smaller.

-AND-

(c) All field mill readings within 5 NM of the flight path are between -100 V/m and +1000 V/m for the preceeding 15 minutes.

(2)  Through cumulus clouds with tops higher than the -5.0 deg. C level.

(3)  Through or within 5 NM (horizontal or vertical) of the nearest edge of cumulus clouds with tops higher than the -10.0 deg. C level.

(4)  Through or within 10 NM (horizontal or vertical) of the nearest edge of any cumulonimbus or thunderstorm cloud, including nontransparent parts of its anvil.

(5) Through or within 10 NM (horizontal or vertical) of the nearest edge of a nontransparent detached anvil for the first hour after detachment from the parent thunderstorm or cumulonimbus cloud.

NOTE: 'Cumulus' does not include Altocumulus or Stratocumulus.

B(1): CUMULUS (CONVECTIVE) CLOUDS CAN DEVELOP AND PRODUCE ELECTRIC CHARGE VERY RAPIDLY. NORMALLY THIS OCCURS WELL ABOVE THE FREEZING LEVEL. HOWEVER, THE +5.0 DEG. C LEVEL WAS SPECIFIED BECAUSE ELECTRIC CHARGE HAS BEEN SPECULATED TO OCCUR IN SOME 'WARM' CLOUDS IN THE TROPICS AND BECAUSE CUMULUS CLOUDS CAN BUILD VERY RAPIDLY. CHARGE SUFFICIENT TO TRIGGER LIGHTNING IS NORMALLY NOT DETECTED UNTIL THE CLOUD REACHES WELL ABOVE THE -10.0 DEG. C LEVEL. THUS, IF THE CLOUD IS NOT PRODUCING PRECIPITATION (PRECIPITATION IS BELIEVED TO BE A NECESSARY CONDITION FOR CLOUD ELECTRIFICATION) AND THE GROUND BASED FIELD MILLS ARE NOT MEASURING ELEVATED ELECTRIC FIELDS ASSOCIATED WITH THE ONSET OF CUMULUS CLOUD ELECTRIFICATION, THEN LAUNCH MAY BE PERMITTED THROUGH CLOUDS BELOW THE -5.0 DEG. C LEVEL.

B(2): AS CLOUD TOPS APPROACH -10.0 DEG. C, CHARGE MAY DEVELOP VERY RAPIDLY. THUS, NO LAUNCH IS PERMITTED THROUGH CLOUDS WITH TOPS AT OR ABOVE -5.0 DEG. C.

B(3): CLOUDS WITH TOPS ABOVE -10.0 DEG. C CAN CREATE ELECTRIC FIELDS OUTSIDE THE CLOUD WITH SUFFICIENT STRENGTH TO TRIGGER LIGHTNING. THUS, NO FLIGHT PATH IS PERMITTED WITHIN 5 NM, HORIZONTALLY OR VERTICALLY, OF CLOUDS FROM -10.0 DEG. C TO -20.0 DEG. C.

B(4): THE MOST DANGEROUS CLOUD IS THE CUMULONIMBUS WHICH TYPICALLY PRODUCES NATURAL LIGHTNING. THUS, THE DISTANCE CRITERION (10 NM) IS THE SAME AS FOR NATURAL LIGHTNING IN RULE A. DANGEROUS ELECTRIC CHARGE CAN ALSO BE ADVECTED INTO THE HIGH LEVEL ANVIL PRODUCED BY A CUMULONIMBUS CLOUD, THUS THE FLIGHT PATH MUST REMAIN 10 NM, HORIZONTALLY AND VERTICALLY, FROM THE ANVIL.

B(5): SIGNIFICANT CHARGE CAN REMAIN IN AN ANVIL FOR UP TO ONE HOUR AFTER IT DETACHES FROM ITS PARENT SOURCE CLOUD. THUS, THE SAME 10 NM DISTANCE CRITERION APPLIES AS IN RULE B(4). AFTER 1 HOUR THE DETACHED CLOUD IS TREATED PER RULE F.

C. Do not launch if, for Ranges equipped with a working surface electric field mill network, at any time during the 15 minutes prior to launch time the absolute value of any electric field intensity measurement at the ground is greater than 1000 V/m within 5 NM of the flight path unless:

(1) There are no clouds within 10 NM miles of the flight path except:

(a) transparent clouds,

-OR-

(b) clouds with tops below the +5.0 deg. C level that have not been associated with convective clouds with tops above the -10.0 deg. C level within the last 3 hours.

-AND-

(2)  A known source of electric field (such as ground fog) that is occurring near the sensor, and that has been previously determined and documented to be benign, is clearly causing the elevated readings.

NOTE:  For confirmed failure of the surface field mill system, the countdown and launch may continue, since the other lightning LCC completely describe unsafe meteorological conditions.

GROUND BASED FIELD MILLS (GBFM) JUST MEASURE THE ELECTRIC POTENTIAL AT THE EARTH'S SURFACE. THEY CAN ONLY INFER THE ELECTRIC FIELD IN AND NEAR CLOUDS ALOFT (WHICH PERHAPS DECREASE WITH THE HORIZONTAL DISTANCE CUBED AND THE CHARGE HEIGHT SQUARED); AND THEIR READINGS CAN BE MASKED BY INTERVENING LAYERS OF SPACE CHARGE BETWEEN THE MILLS AND THE CLOUD. MEASUREMENTS OF CHARGE ALOFT USING AIRCRAFT EQUIPPED WITH FIELD MILLS, AND EXPERIMENTS WITH ROCKETS FIRED INTO CLOUDS TO TRIGGER LIGHTNING, INDICATE GBFM MEASUREMENTS GREATER THAN 1000 V/M CAN BE INDICATIVE OF ELECTRIC FIELDS ALOFT SUFFICIENTLY HIGH TO TRIGGER LIGHTNING.

THERE ARE OTHER NEAR-SURFACE SOURCES OF ELECTRIC CHARGE NOT RELATED TO CHARGE ALOFT; FOR INSTANCE GROUND FOG, SMOKE, POWER LINES, SEA SPRAY,  ETC. IF THERE IS NO POSSIBLE SOURCE OF CHARGE ALOFT AND THERE IS A CONFIRMED, DOCUMENTED SOURCE OF NEAR-SURFACE CHARGE, THEN THIS RULE IS NOT VIOLATED.

NOTE: 'DOCUMENTED' MEANS SUFFICIENT DATA HAS BEEN GATHERED ON THE PHENOMENA TO PROPERLY STUDY IT, AND CIRCUMSTANCES CAUSING IT TO BE PRESENT ARE UNDERSTOOD AND WRITTEN IN A TECHNICAL REPORT.

D.  Do not launch if the flight path is through a vertically continuous layer of clouds with an overall depth of 4,500 feet or greater where any part of the clouds is located between the 0.0 deg. C and the -20.0 deg C levels.

THIS RULE COVERS STRATIFORM CLOUDS. ELECTRIFICATION PROCESSES PRIMARILY OCCUR WITHIN CLOUDS BETWEEN -0.0 AND -20.0 DEG. C, AND INCREASE WITH INCREASING CLOUD THICKNESS. IF CLOUD LAYERS ARE CONNECTED BY TURRETS, THE CLOUD LAYER DEPTH IS DETERMINED BY MEASURING FROM THE BASE OF THE LOWER CLOUD LAYER TO THE TOP OF THE HIGHER CLOUD LAYER. THE TURRETS INDICATE POSSIBLE CHARGE PRODUCING CONVECTION, AND THE CHARGE MAY BE ADVECTED INTO THE INDIVIDUAL LAYERS. THUS, AN INDIVIDUAL CLOUD LAYER MAY NOT BE 4500 FEET THICK BUT THE RULE MAY BE VIOLATED IF TWO LAYERS ARE CONNECTED AND THEIR LAYERS SUM TO 4500 FEET OR GREATER.

E.  Do not launch if the flight path is through any clouds that:

(1) Extend to altitudes at or above the 0.0 deg.C level, and

(2) Are associated with disturbed weather that is producing moderate (29 dBz) or greater precipitation within 5 NM of the flight path.

CLOUDS EXTENDING ABOVE THE FREEZING LEVEL AND SUFFICIENTLY ACTIVE TO PRODUCE MODERATE OR GREATER PRECIPITATION CAN ALSO CREATE ELECTRIC CHARGE. THIS CHARGE CAN BE ADVECTED INTO ASSOCIATED CLOUDS AND CAN CREATE ELECTRIC

FIELDS AT A DISTANCE. THUS, IF THE PRECIPITATION IS WITHIN 5 NM, THIS RULE IS VIOLATED.

F.  Do not launch if the flight path will carry the vehicle:

(1) Through any nontransparent thunderstorm or cumulonimbus debris cloud during the first 3 hours after the debris cloud formed from the parent cloud.

(2) Within 5 NM (horizontal or vertical) of the nearest edge of a  nontransparent thunderstorm or cumulonimbus debris cloud during the first 3 hours after the debris cloud formed from a parent cloud, UNLESS:

(a) There is at least one working field mill within 5 NM of the debris cloud;

-AND-

(b) All electric field intensity measurements at the ground are between +1000 V/m and -1000V/m within 5 NM of the flight path during the 15 minutes preceding the launch time;

-AND-

(c) The maximum radar return from the entire debris cloud is less than 10 dBz during the 15 minutes preceding launch time.

(3) The start of the 3-hour period is reckoned as follows:

(a) DETACHMENT. If the cloud detaches from the parent cloud:
The 3-hour period begins at the time when cloud detachment is observed or at the time of the last detected lightning discharge (if any) from the detached debris cloud, whichever is later.

(b) DECAY or DETACHMENT UNCERTAIN. If it is not known whether the cloud is detached or the debris cloud forms from the decay of the parent cloud:  The 3-hour period begins at the time when the parent cloud top decays to below the altitude of the -10.0 deg. C level, or at the time of the last detected lightning discharge (if any) from the parent cloud or debris cloud, whichever is later.

CLOUDS PRODUCED BY CUMULONIMBUS AND THUNDERSTORMS CAN RETAIN THEIR CHARGE EVEN AFTER THE PARENT STORM HAS DECAYED AND STOPPED PRODUCING CHARGE, OR AFTER THE CLOUD HAS BROKEN AWAY FROM THE CHARGE SOURCE. AIRBORNE FIELD MILL MEASUREMENTS SUGGEST MOST OF THE CHARGE HAS DECAYED WITHIN 90 MINUTES. HOWEVER, THE CRITERION WAS SET AT THREE HOURS DUE TO THE UNCERTAINTY REGARDING THE START OF THE CLOUD DECAY OR CLOUD BREAKOFF TIME, AND THE POSSIBILITY OF BRIEF INTERNAL CHARGE GENERATION WITHIN THE CLOUD EVEN AFTER BREAKING OFF FROM THE PARENT CLOUD.

SINCE NEGLIGIBLE RADAR RETURNS AND BENIGN GBFM READINGS INDICATE THE CLOUD IS NOT PRODUCING CHARGE, AND ANY RESIDUAL CHARGE HAS DECAYED SUBSTANTIALLY, THEN THE 5 NM STANDOFF CAN BE REDUCED TO JUST A 'FLIGHT THROUGH' PROHIBITION. NOTE:  BENIGN GBFM VALUES CAN NOT BE USED BY THEMSELVES TO CONCLUDE THE CLOUD CAN BE SAFELY PENETRATED, SINCE SHIELDING LAYERS CAN MASK SIGNIFICANT FIELDS (AND CHARGE) INSIDE THE CLOUDS FROM THE GBFM NETWORK.

TRANSPARENT DEBRIS CLOUDS ARE NOT CONSIDERED CAPABLE OF CARRYING SUFFICIENT CHARGE TO BE DANGEROUS.

G. Definitions/Explanations

(1) Anvil: Stratiform or fibrous cloud produced by the upper level outflow or blow-off from thunderstorms or convective clouds.

(2) Cloud Edge: The visible cloud edge is preferred. If this is not possible, then the 10 dBZ radar cloud edge is acceptable.

(3) Cloud Layer: An array of clouds, not necessarily all of the same type, whose bases are approximately at the same level. Also, multiple arrays of clouds at different altitudes that are connected vertically by cloud elements, e.g., turrets from one cloud array to another. Convective clouds (e.g., clouds falling under Rule B) are excluded from this definition unless they are imbedded with other cloud types.

(4) Cloud Top: The visible cloud top is preferred. If this is not possible, then the 13 dBZ radar cloud top is acceptable.

(5) Cumulonimbus Cloud: Any convective cloud with any part above the -20.0 deg. C temperature level.

(6) Debris Cloud: Any nontransparent cloud, that has become detached from a parent cumulonimbus cloud or thunderstorm, or results from the decay of a parent cumulonimbus cloud or thunderstorm.

(7) Documented: With respect to Rule C(2), 'documented' means sufficient data has been gathered on the benign phenomena to both understand it and to develop procedures to evaluate it, and the supporting data and evaluation have been reported in a technical report, journal article, or equivalent publication. For launches at the Eastern Range, copies of the documentation shall be maintained by the 45th Weather Squadron and the KSC Weather Projects Office. The procedures used to assess the phenomena during launch countdowns shall be documented and implemented by the 45th Weather Squadron.

(8) Electric Field (for surface based electric field mill measurements): The one-minute arithmetic average of the vertical electric field (Ez) at the ground, such as measured by a ground based field mill. The polarity of the electric field is the same as that of the potential gradient; that is, the polarity of the field at the ground is the same as that of the charge overhead.

(9) Flight Path: The planned flight trajectory including its uncertainties ('error bounds').

(10) Precipitating Cloud: Any cloud containing precipitation, producing virga, or having radar reflectivity greater than 13 dBZ.

(11) Thunderstorm: Any cloud that produces lightning.

(12) Transparent: Synonymous with visually transparent. Sky cover through which higher clouds, blue sky, stars, etc., may be clearly observed from below. Also, sky cover through which terrain, buildings, etc., may be clearly observed from above. Sky cover through which forms are blurred, indistinct, or obscured is not transparent.

4.4 NATURAL AND TRIGGERED LIGHTNING CONSTRAINTS.

NOTICE: ANY CHANGES TO THIS SECTION WILL REQUIRE COORDINATION WITH THE 30TH AND 45TH SPACE WING RANGE SAFETY OFFICES.

Natural and Triggered Lightning Launch Commit Criteria (LCC)

The Launch Weather Team must have clear and convincing evidence that the following hazard avoidance criteria are not violated.

Even when these criteria are not violated, if any other hazardous condition exists, the Launch Weather Team will report the threat to the Launch Director. The Launch Director may HOLD at any time based on the instability of the weather.

1. Lightning

a) Do not launch for 30 minutes after any type of lightning occurs in a thunderstorm if the flight path will carry the vehicle within 10 NM of that thunderstorm.

b) Do not launch for 30 minutes after any type of lightning occurs within 10 NM of the flight path unless:

(1) The cloud that produced the lightning is not within 10 NM of the flight path;

-and-

(2) There is at least one working field mill within 5 NM of each such lightning flash;

-and-

(3) The absolute values of all electric field measurements at the surface within 5 NM of the flight path and at the mill(s) specified in (2) above have been less than 1000 V/m for 15 minutes.

Notes:

i) Anvils are covered in Criterion 3.

ii) If a cumulus cloud remains 30 minutes after the last lightning occurs in a thunderstorm then Criterion 2 applies.

Definitions: Anvil, Electric Field Measurement at the Surface, Flight Path, Thunderstorm, Within

2. Cumulus Clouds

a) Do not launch if the flight path will carry the vehicle within 10 NM of any cumulus cloud with its cloud top higher than the -20 deg C level.

b) Do not launch if the flight path will carry the vehicle within 5 NM of any cumulus cloud with its cloud top higher than the -10 deg C level.

c) Do not launch if the flight path will carry the vehicle through any cumulus cloud with its cloud top higher than the -5 deg C level.

d) Do not launch if the flight path will carry the vehicle through any cumulus cloud with its cloud top between the +5 deg C and -5 deg C levels unless:

(1) The cloud is not producing precipitation;

-and-

(2) The horizontal distance from the center of the cloud top to at least one working field mill is less than 2 NM;

-and-

(3) All electric field measurements at the surface within 5 NM of the flight path and at the mill(s) specified in (2) above have been between -100 V/m and +500 V/m for 15 minutes.

Note: Cumulus clouds in Criterion 2 do not include altocumulus, cirrocumulus or stratocumulus.

Definitions: Cloud Top, Electric Field Measurement at the Surface, Flight Path, Precipitation, Within

3. Anvil Clouds

a) Attached anvils:

(1) Do not launch if the flight path will carry the vehicle through nontransparent parts of attached anvil clouds.

(2) Do not launch if the flight path will carry the vehicle within 5 NM of nontransparent parts of attached anvil clouds for the first 3 hours after the time of the last lightning discharge that occurs in the parent cloud or anvil cloud.

(3) Do not launch if the flight path will carry the vehicle within 10 NM of nontransparent parts of attached anvil clouds for the first 30 minutes after the time of the last lightning discharge that occurs in the parent cloud or anvil cloud.

b) Detached Anvils:

(1) Do not launch if the flight path will carry the vehicle through nontransparent parts of a detached anvil cloud for the first 3 hours after the time that the anvil cloud is observed to have detached from the parent cloud.

(2) Do not launch if the flight path will carry the vehicle through nontransparent parts of a detached anvil cloud for the first 4 hours after the time of the last lightning discharge that occurs in the detached anvil cloud.

(3) Do not launch if the flight path will carry the vehicle within 5 NM of nontransparent parts of a detached anvil cloud for the first 3 hours after the time of the last lightning discharge that occurs in the parent cloud or anvil cloud before detachment or in the detached anvil cloud after detachment unless:

(a) There is at least one working field mill within 5 NM of the detached anvil cloud;

-and-

(b) The absolute values of all electric field measurements at the surface within 5 NM of the flight path and at the mill(s) specified in (a) above have been less than 1000 V/m for 15 minutes;

-and-

(c) The maximum radar return from any part of the detached anvil cloud within 5 NM of the flight path has been less than 10 dBZ for 15 minutes.

(4) Do not launch if the flight path will carry the vehicle within 10 NM of nontransparent parts of a detached anvil cloud for the first 30 minutes after the time of the last lightning discharge that occurs in the parent cloud or anvil cloud before detachment or in the detached anvil cloud after detachment.

Note: Detached anvil clouds are never considered DEBRIS CLOUDS, nor are they covered by Criterion 4.

Definitions: Anvil, Debris Cloud, Flight Path, Nontransparent, Thunderstorm, Within

4. Debris Clouds

a) Do not launch if the flight path will carry the vehicle through any nontransparent parts of a debris cloud during the 3-hour period defined below.

b) Do not launch if the flight path will carry the vehicle within 5 NM of any nontransparent parts of a debris cloud during the 3-hour period defined below, unless:

(1) There is at least one working field mill within 5 NM of the debris cloud;

-and-

(2) The absolute values of all electric field measurements at the surface within 5 NM of the flight path and at the mill(s) specified in (1) above have been less than 1000 V/m for 15 minutes;

-and-

(3) The maximum radar return from any part of the debris cloud within 5 NM of the flight path has been less than 10 dBZ for 15 minutes.

The 3-hour period in a) and b) above begins at the time when the debris cloud is observed to have detached from the parent cloud or when the debris cloud is observed to have formed from the decay of the parent cloud top below the altitude of the -10 deg C level. The 3-hour period begins anew at the time of any lightning discharge that occurs in the debris cloud.

Definitions: Cloud Top, Debris Cloud, Electric Field Measurement at the Surface, Flight Path, Nontransparent, Within

5. Disturbed Weather

Do not launch if the flight path will carry the vehicle through any nontransparent clouds that are associated with a weather disturbance having clouds that extend to altitudes at or above the 0 deg C level and contain moderate or greater precipitation or a radar bright band or other evidence of melting precipitation within 5 NM of the flight path.

Definitions: Associated, Flight Path, Moderate Precipitation, Nontransparent, Weather Disturbance, Within

6. Thick Cloud Layers

Do not launch if the flight path will carry the vehicle through nontransparent parts of a cloud layer that is:

(1) Greater than 4,500 ft thick and any part of the cloud layer along flight path is located between the 0 deg C and the -20 deg C levels;

-or-

(2) Connected to a cloud layer that, within 5 NM of the flight path, is greater than 4,500 ft thick and has any part located between the 0 deg C and the -20 deg C levels;

unless the cloud layer is a cirriform cloud that has never been associated with convective clouds, is located entirely at temperatures of -15 deg C or colder, and shows no evidence of containing liquid water (e.g., aircraft icing).

Definitions: Associated, Cloud Layer, Flight Path, Nontransparent

7. Smoke Plumes

Do not launch if the flight path will carry the vehicle through any cumulus cloud that has developed from a smoke plume while the cloud is attached to the smoke plume, or for the first 60 minutes after the cumulus cloud is observed to have detached from the smoke plume.

Note: Cumulus clouds that have formed above a fire but have been detached from the smoke plume for more than 60 minutes are considered CUMULUS CLOUDS and are covered in Criterion 2.

Definitions: Flight Path

8. Surface Electric Fields

a) Do not launch for 15 minutes after the absolute value of any electric field measurement at the surface within 5 NM of the flight path has been greater than 1500 V/m.

b) Do not launch for 15 minutes after the absolute value of any electric field measurement at the surface within 5 NM of the flight path has been greater than 1000 V/m unless:

(1) All clouds within 10 NM of the flight path are transparent;

-or-

(2) All nontransparent clouds within 10 NM of the flight path have cloud tops below the +5 deg C level and have not been part of convective clouds with cloud tops above the -10 deg C level within the last 3 hours.

Notes:

i) Electric field measurements at the surface are used to increase safety by detecting electric fields due to unforeseen or unrecognized hazards.

ii) For confirmed failure of one or more field mill sensors, the countdown and launch may continue.

Definitions: Cloud Top, Electric Field Measurement at the Surface, Flight Path, Nontransparent, Transparent, Within

9. Definitions:

a) Anvil: Stratiform or fibrous cloud produced by the upper level outflow or blow-off from thunderstorms or convective clouds.

b) Associated: Used to denote that two or more clouds are causally related to the same weather disturbance or are physically connected. ASSOCIATED is not synonymous with occurring at the same time. An example of clouds that are NOT associated is air mass clouds formed by surface heating in the absence of organized lifting. Also, a cumulus cloud formed locally and a physically separated cirrus layer generated by a distant source are not associated, even if they occur over or near the launch site at the same time.

Subsidiary Definition: Weather Disturbance

c) Bright Band: An enhancement of radar reflectivity caused by frozen hydrometeors falling through the 0 deg C level and beginning to melt.

d) Cloud Edge: The visible cloud edge is preferred. If this is not possible, then the 10 dBZ radar reflectivity cloud edge is acceptable.

e) Cloud Layer: A vertically continuous array of clouds, not necessarily of the same type, whose bases are approximately at the same level.

f) Cloud Top: The visible cloud top is preferred. If this is not possible, then the 10 dBZ radar reflectivity cloud top is acceptable.

g) Cumulonimbus Cloud: Any convective cloud with any part above the -20 deg C temperature level.

h) Debris Cloud: Any cloud, except an anvil cloud, that has become detached from a parent cumulonimbus cloud or thunderstorm, or that results from the decay of a parent cumulonimbus cloud or thunderstorm.

Subsidiary Definition: Cumulonimbus Cloud

i) Electric Field Measurement at the Surface: The one-minute arithmetic average of the vertical electric field (Ez) at the ground measured by a ground based field mill. The polarity of the electric field is the same as that of the potential gradient; that is, the polarity of the field at the ground is the same as the dominant charge overhead.

Note: Electric field contours shall not be used for the electric field measurement at the surface.

j) Flight Path: The planned flight path including its uncertainties ("error bounds").

k) Moderate Precipitation: A precipitation rate of 0.1 inches/hr or a radar reflectivity factor of 30 dBZ.

l) Nontransparent: Opposite of Transparent. Sky cover through which forms are blurred, indistinct, or obscured is nontransparent.

Note: Nontransparency must be assessed for launch time. Sky cover through which forms are seen distinctly ONLY through breaks in the cloud cover is considered nontransparent. Clouds with a radar reflectivity of 10 dBZ or greater are also considered nontransparent.

Subsidiary Definition: Transparent

m) Optically Thin: Having a vertical optical thickness of unity or less at visible wavelengths.

n) Precipitation: Detectable rain, snow, sleet, etc., at the ground, or virga, or a radar reflectivity greater than 18 dBZ.

o) Transparent: Synonymous with optically thin. Sky cover is transparent if higher clouds, blue sky, stars, the disk of the sun, etc., can be distinctly seen from below, or if the sun casts distinct shadows of objects on the ground, or if terrain, buildings, lights on the ground, etc., can be distinctly seen from above.

Note: Visible transparency is required. Transparency must be assessed for launch time. Sky cover through which forms are seen distinctly ONLY through breaks in the cloud cover is considered NONtransparent.

Subsidiary Definitions: Nontransparent, Optically Thin

p) Thunderstorm: Any convective cloud that produces lightning.

q) Weather Disturbance: A weather system where dynamical processes destabilize the air on a scale larger than the individual clouds or cells. Examples of disturbances are fronts, troughs and squall lines.

r) Within: Used as a function word to specify a margin in all directions (horizontal, vertical, and slant separation) between the cloud edge or top and the flight path. For example, "WITHIN 10 NM of a thunderstorm cloud" means that there must be a 10 NM margin between every part of a thunderstorm cloud and the flight path.

Subsidiary Definitions: Cloud Edge, Cloud Top, Flight Path

1.0 Weather Rules.
This section contains the weather conditions that are a constraint to launching the Space Shuttle Vehicle (SSV). The weather constraints stated in this section shall not be waived.

1.4 Natural and Triggered Lightning Constraints.
Notice: Any changes to this section will require coordination with the 45th Space Wing Range Safety Office (RSO).

Natural and Triggered Lightning LCC

The Launch Weather Team (LWT) must have clear and convincing evidence that the following hazard avoidance criteria are not violated.

Even when these criteria are not violated, if any other hazardous condition exists, the LWT will report the threat to the Launch Director (LD). The LD may Hold at any time based on the instability of the weather.

A. Lightning

> 1. Do not launch for 30 minutes after any type of lightning occurs in a thunderstorm if the flight path will carry the SSV within 10 Nautical Miles (NM) of that thunderstorm.
>
> 2. Do not launch for 30 minutes after any type of lightning occurs within 10 NM of the flight path unless:
>> a. The cloud that produced the lightning is not within 10 NM of flight path;
>>
>> -and-
>>
>> b. There is at least one working field mill within 5 NM of each such lightning flash;
>>
>> -and-
>>
>> c. The absolute values of all electric field measurements at the surface, within 5 NM of the flight path and at the mill(s) specified in subparagraph 1.4.A.2.b, have been less than 1000 Volts/meter (V/m) for 15 minutes.

Notes:
i) Anvils are covered in section 1.4.C.
ii) If a cumulus cloud remains 30 minutes after the last lightning occurs in a thunderstorm, then section 1.4.B. applies.

Definitions: Anvil, Electric Field Measurement at the Surface, Flight Path, Thunderstorm, Within

B. Cumulus Clouds

> 1. Do not launch if the flight path will carry the SSV within 10 NM of any cumulus cloud with its cloud top higher than the -20 degrees Celsius (C) level.
>
> 2. Do not launch if the flight path will carry the SSV within 5 NM of any cumulus cloud with its cloud top higher than the -10 degrees C level.

3. Do not launch if the flight path will carry the SSV through any cumulus cloud with its cloud top higher than the -5 degrees C level.

4. Do not launch if the flight path will carry the SSV through any cumulus cloud with its cloud top between the +5 degrees C and -5 degrees C levels unless:

a. The cloud is not producing precipitation;

- and-

b. The horizontal distance from the center of the cloud top to at least one working field mill is less than 2 NM;

-and-

c. All electric field measurements at the surface within 5 NM of the flight path and at the mill(s) specified in subparagraph 1.4.B.4.b. above have been between -100 V/m and +500 V/m for 15 minutes.

Note: Cumulus clouds in section 1.4.B. do not include altocumulus, cirrocumulus or stratocumulus.

Definitions: Cloud Top, Electric Field Measurement at the Surface, Flight Path, Precipitation, Within

C. Anvil Clouds

1. Attached Anvil Clouds

a. Do not launch if the flight path will carry the SSV through or within 10 NM of a nontransparent part of any attached anvil cloud for the first 30 minutes after the last lightning discharge in or from the parent cloud or anvil cloud.

b. Do not launch if the flight path will carry the SSV through or within 5 NM of a nontransparent part of any attached anvil cloud between 30 minutes and three hours after the last lightning discharge in or from the parent cloud or anvil cloud unless both of the following conditions are satisfied:

1) The portion of the attached anvil cloud within 5 NM of the flight path is located entirely at altitudes where the temperature is colder than 0 degrees C;

-and-

2) The VAHIRR is less than +33 Range Corrected Reflectivity (dBZ)-thousand feet(kft) (+10 dBZ-kilometers (km)) everywhere along the portion of the flight path where any part of the attached anvil cloud is within the specified volume.

c. Do not launch if the flight path will carry the SSV through a nontransparent part of any attached anvil cloud more than 3 hours after the last lightning discharge in or from the parent cloud or anvil cloud unless both of the following conditions are satisfied:

1) The portion of the attached anvil cloud within 5 NM miles of the flight path is located entirely at altitudes where the temperature is colder than 0 degrees C;

-and-

2) The VAHIRR is less than +33 dBZ-kft (+10 dBZ-km) everywhere along the portion of the flight path where any part of the attached anvil cloud is within the specified volume.

2. Detached Anvil Clouds.

For the purposes of this section, detached anvil clouds are never considered debris clouds.

a. Do not launch if the flight path will carry the SSV through or within 10 NM of a nontransparent part of a detached anvil cloud for the first 30 minutes after the last lightning discharge in or from the parent cloud or anvil cloud before detachment or after the last lightning discharge in or from the detached anvil cloud after detachment.

b. Do not launch if the flight path will carry the SSV between 0 (zero) and 5 NM from a nontransparent part of a detached anvil cloud between 30 minutes and 3 hours after the time of the last lightning discharge in or from the parent cloud or anvil cloud before detachment or after the last lightning discharge in or from the detached anvil cloud after detachment unless subparagraph 1.4.C.2.b.1) or subparagraph 1.4.C.2.b.2) is satisfied:

1) This section is satisfied if all three of the following conditions are met:

a) There is at least one working field mill within 5 NM of the detached anvil cloud;

-and-

b) The absolute values of all electric field measurements at the surface within 5 NM of the flight path and at each field mill specified in subparagraph 1.4.C.2.b.1)a) have been less than 1000 V/m for 15 minutes;

-and-

c) The maximum radar return from any part of the detached anvil cloud within 5 NM of the flight path has been less than 10 dBZ for 15 minutes.

2) This section is satisfied if both of the following conditions are met:

a) The portion of the detached anvil cloud within 5 NM of the flight path is located entirely at altitudes where the temperature is colder than 0 degrees C;

-and-

b) The VAHIRR is less than +33 dBZ-kft (+10 dBZ-km) everywhere along the portion of the flight path where any part of the detached anvil cloud is within the specified volume.

c. Do not launch if the flight path will carry the SSV through a nontransparent part of a detached anvil cloud unless subparagraph 1.4.C.2.c.1) or subparagraph 1.4.C.2.c.2) is satisfied

    1) This section is satisfied if both of the following conditions are met.

        a) At least 4 hours have passed since the last lightning discharge in or from the detached anvil cloud;

        -and-

        b) At least 3 hours have passed since the time that the anvil cloud is observed to be detached from the parent cloud.

    2) This section is satisfied if both of the following conditions are met.

        a) The portion of the detached anvil cloud within 5 nautical miles of the flight path is located entirely at altitudes where the temperature is colder than 0 degrees C;

        -and-

        b) The Volume-Averaged, Height-Integrated Radar Reflectivity (VAHIRR) is less than +33 dBZ-kft (+10 dBZ-km) everywhere along the portion of the flight path where any part of the detached anvil cloud is within the specified volume.

Note: Detached anvil clouds are never considered debris clouds, nor are they covered by section 1.4.D.

Definitions: Anvil; Cloud Base; Cloud Edge; Cloud Top; Debris Cloud; Electric Field Measurements at the Surface; Field Mill; Flight Path; Volume-Averaged, Height-Integrated Radar Reflectivity (VAHIRR); Nontransparent; Specified Volume; Transparent; Within

D. Debris Clouds

    1. Do not launch if the flight path will carry the SSV through any nontransparent parts of a debris cloud during the 3 hour period defined below.

    2. Do not launch if the flight path will carry the vehicle within 5 NM of any nontransparent parts of a debris cloud during the 3 hour period defined below, unless:

    a. There is at least one working field mill within 5 NM of the debris cloud;

    -and-

    b. The absolute values of all electric field measurements at the surface within 5 NM of the flight path and at the mill(s) specified in section 1.4.D.2.a. have been less than 1000 V/m for 15 minutes;

    -and-

c. The maximum radar return from any part of the debris cloud within 5 NM of the flight path has been less than 10 dBZ for 15 minutes.

The 3 hour period in subparagraphs 1.4.D.2.a. and 1.4.D.2.b. begins at the time when the debris cloud is observed to have detached from the parent cloud or when the debris cloud is observed to have formed from the decay of the parent cloud top below the altitude of the -10 degrees C level. The 3 hour period begins anew at the time of any lightning discharge that occurs in the debris cloud.

Definitions: Cloud Top, Debris Cloud, Electric Field Measurement at the Surface, Flight Path, Nontransparent, Within

E. Disturbed Weather

Do not launch if the flight path will carry the vehicle through any nontransparent clouds that are associated with a weather disturbance having clouds that extend to altitudes at or above the 0 degree C level and contain moderate or greater precipitation or a radar bright band or other evidence of melting precipitation within 5 NM of the flight path.

Definitions: Associated, Flight Path, Moderate Precipitation, Nontransparent, Weather Disturbance, Within

F. Thick Cloud Layers

Do not launch if the flight path will carry the SSV through nontransparent parts of a cloud layer that is:
1. Greater than 4,500 foot thick and any part of the cloud layer along flight path is located between the 0 degrees C and the -20 degrees C levels;

-or-

2. Connected to a cloud layer that, within 5 NM of the flight path, is greater than 4,500 foot thick and has any part located between the 0 degrees C and the -20 degrees C levels unless the cloud layer is a cirriform cloud that has never been associated with convective clouds, is located entirely at temperatures of -15 degrees C or colder, and shows no evidence of containing liquid water (e.g., aircraft icing).

Definitions: Associated, Cloud Layer, Flight Path, Nontransparent

G. Smoke Plumes
Do not launch if the flight path will carry the SSV through any cumulus cloud that has developed from a smoke plume while the cloud is attached to the smoke plume, or for the first 60 minutes after the cumulus cloud is observed to have detached from the smoke plume.

Note: Cumulus clouds that have formed above a fire but have been detached from the smoke plume for more than 60 minutes are considered cumulus clouds and are covered in section 1.4.B.

Definitions: Flight Path

H. Surface Electric Fields
1. Do not launch for 15 minutes after the absolute value of any electric field measurement at the surface within 5 NM of the flight path has been greater than 1500 V/m.

2. Do not launch for 15 minutes after the absolute value of any electric field measurement at the surface within 5 NM of the flight path has been greater than 1000 V/m unless:

    a. All clouds within 10 NM of the flight path are transparent;

-or-

    All nontransparent clouds within 10 NM of the flight path have cloud tops below the +5 degrees C level and have not been part of convective clouds with cloud tops above the -10 degrees C level within the last 3 hours.

Notes:
i) Electric field measurements at the surface are used to increase safety by detecting electric fields due to unforeseen or unrecognized hazards.
ii) For confirmed failure of one or more field mill sensors, the countdown and launch may continue.

Definitions: Cloud Top, Electric Field Measurement at the Surface, Flight Path, Nontransparent, Transparent, Within

I. Definitions:

1. Anvil: Stratiform or fibrous cloud produced by the upper level outflow or blow-off from thunderstorms or convective clouds.

2. Associated: Used to denote that two or more clouds are causally related to the same weather disturbance or are physically connected. Associated is not synonymous with occurring at the same time. An example of clouds that are not associated is air mass clouds formed by surface heating in the absence of organized lifting. Also, a cumulus cloud formed locally and a physically separated cirrus layer generated by a distant source, are not associated, even if they occur over or near the launch site at the same time.
Subsidiary Definition: Weather Disturbance

3. Bright Band: An enhancement of radar reflectivity caused by frozen hydrometeors falling through the 0 degrees C level and beginning to melt.

4. Cloud Base: The visible cloud base is preferred. If this is not possible, then the 0 (zero) dBZ radar-reflectivity cloud base is acceptable.

5. Cloud Edge: The visible cloud edge is preferred. If this is not possible, then the 0 (zero) dBZ radar reflectivity cloud edge is acceptable.

6. Cloud Layer: A vertically continuous array of clouds, not necessarily of the same type, whose bases are approximately at the same level.

7. Cloud Top: The visible cloud top is preferred. If this is not possible, then the 0 (zero) dBZ radar reflectivity cloud top is acceptable.

8. Cumulonimbus Cloud: Any convective cloud with any part above the –20 degrees C temperature level.

9. Debris Cloud: Any cloud, except an anvil cloud, that has become detached from a parent cumulonimbus cloud or thunderstorm, or that results from the decay of a parent cumulonimbus cloud or thunderstorm.
Subsidiary Definition: Cumulonimbus Cloud

10. Electric Field Measurement at the Surface: The one-minute arithmetic average of the vertical electric field (Ez) at the ground measured by a ground based field mill. The polarity of the electric field is the same as that of the potential gradient; that is, the polarity of the field at the ground is the same as the dominant charge overhead. Note: Electric field contours shall not be used for the electric field measurement at the surface.
Subsidiary Definition: Field Mill

11. Field Mill: A specific class of electric-field sensor that uses a moving, grounded conductor to induce a time-varying electric charge on one or more sensing elements in proportion to the ambient electrostatic field

12. Flight Path: The planned flight path including its uncertainties ("error bounds").

13. Moderate Precipitation: A precipitation rate of 0.1 inches/hour or a radar reflectivity factor of 30 dBZ.

14. Nontransparent: Sky cover through which forms are blurred, indistinct, or obscured is nontransparent. Note: Nontransparency must be assessed for launch time. Sky cover through which forms are seen distinctly only through breaks in the cloud cover is considered nontransparent. Clouds with a radar reflectivity of 0 (zero) dBZ or greater, are also considered nontransparent.
Subsidiary Definition: Transparent

15. Precipitation: Detectable rain, snow, sleet, hail or graupel at the ground, or virga, or a radar reflectivity greater than 18 dBZ.

16. Specified Volume: is the volume bounded in the horizontal by vertical, plane, perpendicular sides located 5.5 km (3 NM) north, east, south, and west of the point on the flight track, on the bottom by the 0 degree C level, and on the top by the upper extent of all cloud.

17. Transparent: Sky cover is transparent if higher clouds, blue sky, stars, etc., can be distinctly seen from below, or if terrain, buildings, lights on the ground, etc., can be distinctly seen from above.
Note: Visible transparency is required. Transparency must be assessed for launch time. Sky cover through which forms are seen distinctly only through breaks in the cloud cover is considered nontransparent.
Subsidiary Definitions: Nontransparent

18. Thunderstorm: Any convective cloud that produces lightning.

19. Volume-Averaged, Height–Integrated Radar Reflectivity (VAHIRR): Product of the Volume-Averaged Radar Reflectivity (VARR) and the average cloud thickness within a specified volume relative to a point along the flight track (units of dBZ-km) where The VARR is the arithmetic average (in dBZ) of the cloud radar reflectivity within the Specified Volume. Normally, a radar processor will report reflectivity values interpolated onto a regular, three-dimensional array of grid points. Any such grid point within the specified volume is included in the average if and only if it has a radar reflectivity equal to or greater than 0 dBZ.

-and where-

The average cloud thickness is the altitude difference (in kilometers) between the average top and the average base of all clouds within the specified volume. The cloud base to be averaged is the higher of (1) the 0 degree C level and (2) the lowest extent (in altitude) of all cloud radar reflectivities 0 dBZ or greater. Similarly, the cloud top to be averaged is the highest extent (in altitude) of all cloud radar reflectivities 0 dBZ or greater. Given the grid-point representation of a typical radar processor, allowance must be made for the vertical separation of grid points in computing cloud thickness: The cloud base at any horizontal position shall be taken as the altitude of the corresponding base grid point minus half of the grid-point vertical separation.

Similarly, the cloud top at that horizontal position shall be taken as the altitude of the corresponding top grid point plus half of this vertical separation. Thus, a cloud represented by only a single grid point having a radar reflectivity equal to or greater than 0 dBZ within the specified volume would have an average cloud thickness equal to the vertical grid-point separation in its vicinity.

Note: The VAHIRR measurement must be made in the absence of significant attenuation by intervening storms or by water or ice on the radome itself. The VAHIRR measurement is invalid at any point on the flight track that is within 20 km of any radar reflectivity of 35 dBZ or greater at altitudes of 4 kilometers above mean sea level or greater, and at any point that is within 20 km of any type of lightning that has occurred in the previous 5 minutes. The specified volume must not contain any portion of the cone of silence above the radar, nor any portion of any sectors that may have been blocked out for payload-safety reasons. The individual grid-point reflectivities used to determine either the VARR or the average cloud thickness must be meteorological reflectivities.

Subsidiary Definition: Specified Volume
Note: See below "Interim Instructions for Implementation of VAHIRR"

20. Weather Disturbance: A weather system where dynamical processes destabilize the air on a scale larger than the individual clouds or cells. Examples of disturbances are fronts, troughs and squall lines.

21. Within: Used as a function word to specify a margin in all directions (horizontal, vertical, and slant separation) between the cloud edge, base, or top and the flight path. For example, "Within 10 NM of a thunderstorm cloud" means that there must be a 10 NM margin between every part of a thunderstorm cloud and the flight path.
Subsidiary Definitions: Cloud Base, Cloud Edge, Cloud Top, Flight Path

J. Interim Instructions for Implementation of VAHIRR
The VAHIRR quantity referred to in section 1.4.C. and the definitions require computation of both a volume average reflectivity and an average cloud thickness. These quantities are then multiplied to produce the VAHIRR. Neither of these quantities is available yet as a product on the WSR-88D and WSR-74C radar systems used to support launch operations. This instruction provides a methodology for evaluating VAHIRR criteria with currently available radar products. The methodology provides a result that is more conservative than a direct VAHIRR computation, but it should still permit the launch users to achieve much of the benefit of the VAHIRR feature.

1. Part I. Determination of average cloud thickness. The definition of VAHIRR requires determination of the average cloud top and the average cloud base above the height of the 0 °C isotherm within a square having sides 5.5 km (3 NM) north, east, south, and west of the ground projection of each point in the flight track. Average cloud thickness is defined as the difference of these two numbers. If the average cloud thickness cannot be determined at each point on the flight track, the maximum thickness within 5.5 km (3 NM) of the flight track may be used.

To determine the average cloud top height, the launch weather team may use any existing radar product that gives the height of the 0 dBZ cloud top, including "maximum height of reflectivity" and cross section products. If the average height cannot be determined, the maximum height of the 0 dBZ reflectivity may be used. If the maximum height cannot be determined, use 18 km (60 kft) for the average cloud top.

The average height of the cloud base should be derived from radar data. It is the average of the higher of (a) the bottom of the portion of the cloud producing a radar reflectivity of 0 dBZ or greater or (b) the height of the 0 °C isotherm where the 0 dBZ reflectivity extends below that level. If the average height of the cloud base cannot be determined, use the higher of (a) the height of the 0 degree C isotherm or (b) the lowest portion of the cloud producing a radar reflectivity of 0 dBZ or greater anywhere within each 5.5 km (3 n. mi.) square defined above.

2. Part II. Volume Averaged Radar Reflectivity (VARR). There is no operationally feasible way to use existing radar products to compute a volume average reflectivity. A conservative substitute is the maximum reflectivity since the volume average will always be smaller than the maximum. The WSR-88D has a "User Selectable Layer Composite Reflectivity (URL)" product and the WSR-74C has a "Max" product with user selectable base and top. The Launch Weather Team should configure these products with the bottom of the layer at the height of the 0 degree C isotherm and the top above the height of the highest radar beam within 7.8 km (4.2 n. mi.) of the ground-projected flight track in the scan strategy being used. The WSR-88D product will have to be configured at the Radar Product Generator (RPG) and included in the product scheduler for the Principal User Processor (PUP) used by the launch weather team.

3. Part III. Evaluating the constraint. The VAHIRR constraint is satisfied for a point on the flight track if the "URL" or "Max" radar product (see Part II above) everywhere within the corresponding square (defined above) is less than 10 dBZ-km divided by the average cloud thickness in km within the same square (see Part I above). (In English units, this threshold would become 33 dBZ-kft divided by the average cloud thickness in kft.) This constraint must be satisfied for every point on the flight track.

1.4     Natural and Triggered Lightning Constraints.

Notice:  The lightning definitions, explanations and examples are a critical part of the lightning constraints and must be used when evaluating all lightning constraints in this section.

Notice:  Any changes to this section will require coordination with the 45th Space Wing Range Safety Office (RSO).

The Launch Weather Team (LWT) must have clear and convincing evidence that the following hazard avoidance criteria are not violated. Even when these criteria are not violated, if any other hazardous condition exists, the LWT will report the threat to the Launch Director (LD). The LD may Hold at any time based on the instability of the weather.

A.     Surface Electric Fields

    1.     Do not launch for 15 minutes after the absolute value of any "electric field measurement" less than or equal to 5 Nautical Miles (NM) from the "flight path" has been greater than or equal to 1500 Volts/meter (V/m).

    2.     Do not launch for 15 minutes after the absolute value of any "electric field measurement" less than or equal to 5 NM of the "flight path" has been greater than or equal to 1000 V/m unless subparagraph 1.4.A.2.a. or subparagraph 1.4.2.b. is satisfied:

        a.     All "clouds" less than or equal to 10 NM from the "flight path" are "transparent";

                -or-

        b.     All "clouds" less than or equal to 10 NM from the "flight path" have "cloud tops" at altitudes where the temperature is warmer than +5 degrees Celsius and have not been part of convective "clouds" with "cloud tops" at altitudes where the temperature is colder than or equal to −10 degrees Celsius during the last 3 hours.

B.     Lightning

    1.     Do not launch for 30 minutes after any type of lightning occurs in a "thunderstorm" if the "flight path" will carry the SSV less than or equal to 10 NM from the "thunderstorm". An attached "anvil cloud" is not considered part of its parent "thunderstorm" but is covered instead by the Attached Anvil Cloud Rule.

    2.     Do not launch for 30 minutes after any type of lightning occurs less than or equal to 10 NM from the "flight path" unless all three of the following conditions are satisfied:

        a.     The "cloud" that produced the lightning is greater than 10 NM from the "flight path";

                -and-

        b.     There is at least one working "field mill" less than 5 NM from each such lightning discharge;

                -and-

c.    The absolute values of all "electric field measurements" less than or equal to 5 NM from the "flight path", and at each "field mill" specified in subparagraph 1.4.B.2.b., have been less than 1000 V/m for 15 minutes or longer.

C.    Cumulus Clouds

For the purposes of this section, "cumulus 'clouds'" do not include cirrocumulus, altocumulus, or stratocumulus "clouds". An attached "anvil cloud" is never considered part of its parent cumulus "cloud," but is covered instead by subparagraph 1.4.D.1., Attached "Anvil Clouds". Subparagraph 1.4.D.2., Detached "Anvil Clouds" applies to any detached "anvil cloud". Section 1.4.E, "Debris Clouds" applies to "debris clouds".

1.    Do not launch if the "flight path" will carry the SSV less than or equal to 10 NM from any cumulus "cloud" that has a "cloud top" at an altitude where the temperature is colder than or equal to -20 degrees Celsius.

2.    Do not launch if the "flight path" will carry the SSV less than or equal to 5 NM from any cumulus "cloud" that has a "cloud top" at an altitude where the temperature is colder than or equal to -10 degrees Celsius.

3.    Do not launch if the "flight path" will carry the SSV through any cumulus "cloud" with its "cloud top" at an altitude where the temperature is colder than or equal to -5 degrees Celsius.

4.    Do not launch if the "flight path" will carry the SSV through any cumulus "cloud" that has a "cloud top" at an altitude where the temperature lies in the range from warmer than -5 degrees Celsius to colder than or equal to +5 degrees Celsius unless all three of the following conditions are satisfied:

   a.    The "cloud" is not producing "precipitation";

-and-

   b.    The horizontal distance from the center of the "cloud top" to at least one working "field mill" is less than 2 NM;

-and-

   c.    All "electric field measurements" less than or equal to 5 NM from the "flight path" and at each "field mill" specified in subparagraph 1.4.B.4.b. above have been greater than -100 V/m, but less than +500 V/m for 15 minutes or longer.

D.    Anvil Clouds

1.    Attached "Anvil Clouds"

For the purpose of this section, if there has never been lightning in or from the parent "cloud" or "anvil cloud", subparagraphs 1.4.D.1.a and 1.4.D.1.b. shall be considered satisfied, but subparagraph 1.4.D.1.c. shall still apply.

   a.    Do not launch if the "flight path" will carry the SSV less than 10, but greater than 5 NM from any attached "anvil cloud" for the first 30 minutes after the last lightning discharge in or from the parent "cloud" or "anvil cloud" unless the portion of the attached "anvil cloud"

less than or equal to 10 NM from the "flight path" is located entirely at altitudes where the temperature is colder than 0 degrees Celsius.

b.      Do not launch if the "flight path" will carry the SSV less than or equal to 5, but greater than 3 NM from any attached "anvil cloud" for the first 3 hours after the last lightning discharge in or from the parent "cloud" or "anvil cloud" unless the portion of the attached "anvil cloud" less than or equal to 5 NM from the "flight path" is located entirely at altitudes where the temperature is colder than 0 degrees Celsius.

c.      Do not launch if the "flight path" will carry the SSV less than or equal to 3 NM from any attached "anvil cloud" unless all three of the following conditions are satisfied:

1)      The portion of the attached "anvil cloud" less than or equal to 5 NM from the "flight path" is located entirely at altitudes where the temperature is colder than 0 degrees Celsius;

-and-

2)      The Volume Averaged Height Integrated Radar Reflectivity ("VAHIRR") is less than +10 dBZ-km (+33 dBZ-kft) at every point less than or equal to 1 NM from the "flight path";

-and-

3)      All of the "VAHIRR application criteria" are satisfied.

2.      Detached "Anvil Clouds".

For the purposes of this section, detached "anvil clouds" are never considered "debris clouds".

For the purpose of this section, if there has never been lightning in or from the parent "cloud" or "anvil cloud", subparagraphs 1.4.D.2.a, 1.4.D.2.b, 1.4.D.2.c and 1.4.D.2.d.1.a shall be considered satisfied, but subparagraphs 1.4.D.2.d.1.b and 1.4.D.2.d.2 shall still apply.

a.      Do not launch if the "flight path" will carry the SSV less than or equal to 10, but greater than 3 NM from a detached "anvil cloud" for the first 30 minutes after the last lightning discharge in or from the parent "cloud" or "anvil cloud" before detachment or after the last lightning discharge in or from the detached "anvil cloud" after detachment unless the portion of the detached "anvil cloud" less than or equal to 10 NM from the "flight path" is located entirely at altitudes where the temperature is colder than 0 degree Celsius."

b.      Do not launch if the "flight path" will carry the SSV between 0 (zero) and 3 NM inclusive, from a detached "anvil cloud" for the first 30 minutes after the time of the last lightning discharge in or from the parent "cloud" or "anvil cloud" before detachment or after the last lighting discharge in or from the detached "anvil cloud" after detachment unless all three of the following conditions are met:

1)      The portion of the detached "anvil cloud" less than or equal to 5 NM from the "flight path" is located entirely at altitudes where the temperature is colder than 0 degrees Celsius;

-and-

2)        The "VAHIRR" is less than +33 dBZ-kft (+10 dBZ-km) at every point less than or equal to 1 NM from the "flight path";

-and-

3)        All of the "VAHIRR application criteria" are satisfied.

c.        Do not launch if the "flight path" will carry the SSV between 0 (zero) and 3 NM, inclusive, from a detached "anvil cloud" less than or equal to 3 hours, but greater than 30 minutes, after the time of the last lightning discharge in or from the parent "cloud" or "anvil cloud" before detachment or after the last lighting discharge in or from the detached "anvil cloud" after detachment unless subparagraph 1.4.D.2.c.1) or subparagraph 1.4.D.2.c.2) is satisfied:

1) This section is satisfied if all three of the following conditions are met:

a)        There is at least one working "field mill" less than 5 NM from the detached "anvil cloud";

-and-

b)        The absolute values of all "electric field measurements" less than or equal to 5 NM from the "flight path" and at each "field mill" specified in subparagraph 1.4.D.2.c.1)a) have been less than 1000 V/m for 15 minutes;

-and-

c) The maximum radar reflectivity from any part of the detached "anvil cloud" less than or equal to 5 NM from the "flight path" has been less than +10 dBZ for 15 minutes.

2)        This section is satisfied if all three of the following conditions are met:

a)        The portion of the detached "anvil cloud" less than or equal to 5 NM from the "flight path" is located entirely at altitudes where the temperature is colder than 0 degrees Celsius;

-and-

b)        The "VAHIRR" is less than +33 dBZ-kft (+10 dBZ-km) at every point less than or equal to 1 NM from the "flight path";

-and-

c)        All of the "VAHIRR application criteria" are satisfied.

d.        Do not launch if the "flight path" will carry the SSV through a detached "anvil cloud" unless subparagraph 1.4.D.2.d.1) or subparagraph 1.4.D.2.d.2) is satisfied:

1)      This section is satisfied if both of the following conditions are met:

a)      At least 4 hours have passed since the last lightning discharge in or from the detached "anvil cloud";

-and-

b)      At least 3 hours have passed since the time that the "anvil cloud" is observed to be detached from the parent "cloud".

2)      This section is satisfied if all three of the following conditions are met:

a)      The portion of the detached "anvil cloud" less than or equal to 5 NM from the "flight path" is located entirely at altitudes where the temperature is colder than 0 degrees Celsius;

-and-

b)      The "VAHIRR" is less than +33 dBZ-kft (+10 dBZ-km) everywhere along the "flight path";

-and-

c)      All of the "VAHIRR application criteria" are satisfied.

E.      Debris Clouds

The 3-hour time period defined in this section must begin again at the time of any lightning discharge in or from the "debris cloud".

1.      Do not launch if the "flight path" will carry the SSV between 0 and 3 NM, inclusive, from a debris cloud for 3 hours after the debris cloud is observed to be detached from the parent cloud or after the debris cloud is observed to have formed from the collapse of the parent cloud top to an altitude where the temperature is warmer than -10 degrees Celsius unless subparagraph 1.4.E.1.a. or subparagraph 1.4.E.1.b. is satisfied:

a.      This section is satisfied if all three of the following conditions are met:

1) There is at least one working "field mill" less than 5 NM from the debris cloud;

-and-

2) The absolute values of all "electric field measurements" less than or equal to 5 NM from the "flight path" and at each "field mill(s)" employed by subparagraph 1.4.E.1.a.1) of this section has been less than 1000 V/m for 15 minutes or longer;

-and-

3) The maximum radar reflectivity from any part of the debris cloud less than or equal to 5 NM of the "flight path" has been less than +10 dBZ for 15 minutes or longer.

b.      This section is satisfied if all three of the following conditions are met:

1) The portion of the debris cloud less than or equal to 5 NM from the "flight path" is located entirely at altitudes where the temperature is colder than 0 degrees Celsius;

-and-

2) The "VAHIRR" is less than +33 dBZ-kft (+10 dBZ-km) at every point less than or equal to 1 NM from the "flight path";

-and-

3) All of the "VAHIRR application criteria" are satisfied.

2.      Do not launch if the "flight path" will carry the SSV through any "debris cloud" during the 3-hour period defined in Section 1.4.E.1, unless all three of the following conditions are met:

a.      The portion of the "debris cloud" less than or equal to 5 NM from the "flight path" is located entirely at altitudes where the temperature is colder than 0 degrees Celsius;

-and-

b.      The "VAHIRR" is less than +33 dBZ-kft (+10 dBZ-km) everywhere along the "flight path";

-and-

c.      All of the "VAHIRR application criteria" are satisfied.

F.      Disturbed Weather

1.      Do not launch if the "flight path" will carry the SSV through a "cloud" associated with "disturbed weather" that has "clouds" with "cloud tops" at altitudes where the temperature is colder than or equal to 0 degrees Celsius and that contains, less than or equal to 5 NM from the "flight path", either:

a.      "Moderate precipitation" or greater;

-or-

b.      Evidence of melting "precipitation" such as a "radar bright" band.

G.      Thick Cloud Layers

For the purpose of this section neither attached nor detached "anvil clouds" are considered "thick cloud layers".

1.      Do not launch if the "flight path" will carry the SSV through a "cloud layer" that is either:

a.  Greater than or equal to 4,500 feet thick and any part of the "cloud layer" along "flight path" is located at an altitude where the temperature is between 0 degrees Celsius and -20 degrees Celsius, inclusive;

-or-

b.  Connected to a "cloud layer" that, less than or equal to 5 NM from the "flight path", is greater than or equal to 4,500 feet thick and has any part located at an altitude where the temperature is between 0 degrees Celsius and -20 degrees Celsius, inclusive.

2.  A launch operator need not apply the lightning commit criteria in subparagraph 1.4.G.1.a. and subparagraph 1.4.G.1.b. of this section if the "thick cloud layer" is a cirriform "cloud layer" that has never been "associated" with convective "clouds", is entirely at altitudes where the temperature is colder than or equal to -15 degrees Celsius and shows no evidence of containing liquid water.

H.  Smoke Plumes
1.  Do not launch if the "flight path" will carry the SSV through any cumulus "cloud" that has developed from a smoke plume while the "cloud" is attached to the smoke plume, or for the first 60 minutes after the cumulus "cloud" is observed to be detached from the smoke plume.

2.  Section 1.4.C. Cumulus "Clouds" applies to cumulus "clouds" that have formed above a fire but have been detached from the smoke plume for more than 60 minutes.

I.  Definitions, Explanations and Examples:

For the purpose of the Natural and Triggered Lightning LCC, distance from an electric "field mill" is measured differently than distance from any other object or measurement point:  Distance between a "radar reflectivity" or "VAHIRR" measurement point and any object or the "flight path" is the SHORTEST SEPARATION (horizontal, vertical, or slant range) between that point and the NEAREST PART of the object or "flight path". Similarly, distance between the "flight path" and any object is the SHORTEST SEPARATION between any point on the "flight path" the NEAREST PART of that object. For example, "every point less than or equal to 1 NM from the 'flight path'" [see subparagraph 1.4.D.1.c.2) Attached Anvil Clouds] means that the "VAHIRR" threshold must be satisfied at every point throughout the entire volume defined by a 1 NM radius from every point on the "flight path". In contrast, distance between a "field mill" or an "electric field measurement" and any object or the "flight path" is always measured HORIZONTALLY between that mill or measurement point and the nearest part of the VERTICAL PROJECTION of the object or "flight path" onto the surface of Earth. For example, "from the center of the 'cloud top' to at least one working 'field mill'" [see subparagraph 1.4.C.4.b. Cumulus "Clouds"] means that the HORIZONTAL distance between the "field mill" and a point on the surface directly beneath the center of the "cloud top" must be less than 2 NM.

1.  Anvil Cloud:  Stratiform or fibrous "cloud" produced by the upper level outflow or blow-off from "thunderstorms" or "convective clouds" having tops at altitudes where the temperature is colder than or equal to -10 degrees Celsius.

2.  Associated:  When two or more "clouds" are causally related to the same "disturbed weather" system or are physically connected. "Clouds" occurring at the same time are not necessarily "associated". A cumulus "cloud" formed locally and a cirrus layer that is physically separated from that cumulus "cloud" and that is generated by a distant source are not "associated," even if they occur over or near the launch point at the same time.

3.      Average cloud thickness is the altitude difference (in kilometers, km hereafter) between the average top and the average base of all "clouds" in the "specified volume." The cloud base to be averaged is the higher of (1) the 0 degree Celsius level and (2) the lowest extent (in altitude) of all "radar reflectivity" measurements of 0 dBZ or greater. Similarly, the cloud top to be averaged is the highest extent (in altitude) of all "radar reflectivity" measurements of 0 dBZ or greater. Given the grid-point representation of a typical radar processor, allowance must be made for the vertical separation of grid points in computing "average cloud thickness": The cloud base at any horizontal position shall be taken as the altitude of the corresponding base grid point minus half of the grid-point vertical separation. Similarly, the cloud top at that horizontal position shall be taken as the altitude of the corresponding top grid point plus half of this vertical separation. Thus, a cloud represented by only a single grid point having a "radar reflectivity" equal to or greater than 0 dBZ in the "specified volume" would have an "average cloud thickness" equal to the vertical grid-point separation in its vicinity.

4.      Bright Band: An enhancement of radar reflectivity caused by frozen hydrometeors falling and beginning to melt at any altitude where the temperature is 0 degrees Celsius or warmer.

5.      Cloud: A visible mass of suspended water droplets or ice crystals. The "cloud" is considered to be the entire volume enclosed by the visible, "nontransparent cloud" boundary as seen by an observer, or, in the absence of a visual observation, by the 0 dBZ "radar reflectivity" boundary. A visual evaluation of transparency is preferred whenever possible.

Distance from the "cloud" to a point in question refers to the separation between the point and the nearest part of that "cloud." Specifically, the wording, "less than or equal to 10 NM from any cumulus 'cloud'" means that the "flight path" must not penetrate either the INTERIOR of the "cloud" itself or the volume between 0 and 10 NM, inclusive, OUTSIDE the "cloud" boundary [for example, see Section 1.4.C., Cumulus "Clouds"]. On the other hand, "between 0 and 3 NM, inclusive, from" refers ONLY to the volume at a distance that is greater than or equal to 0, but less than or equal to 3, NM OUTSIDE the "cloud" boundary, specifically omitting the interior of the "cloud" itself [for example, see Section 1.4.E, "Debris Clouds"].

6.      Cloud Layer: A vertically continuous array of clouds, not necessarily of the same type, whose bases are approximately at the same level.

7.      Cloud Top: The visible top of the cloud, or, in the absence of a visual observation, the 0 dBZ radar top. A visual evaluation of "cloud top" is preferred whenever possible.

8.      Cone of Silence: The volume in an inverted circular cone centered on the radar that is generated by all elevation angles greater than the maximum elevation angle used in the radar scan strategy. For the purpose of "VAHIRR" calculation this volume is capped by the observed maximum "cloud top" height, the observed tropopause height, or an altitude of 20 km (66 kft), whichever is lowest.

9.      Cumulonimbus Cloud: Any convective "cloud" with any part at an altitude where the temperature is colder than –20 degrees Celsius.

10.     Debris Cloud: Any "cloud", except an "anvil cloud", that has become detached from a parent "cumulonimbus cloud" or "thunderstorm", or that results from the decay of a parent "cumulonimbus cloud" or "thunderstorm".

11.     Disturbed Weather:  A weather system where dynamical processes destabilize the air on a scale larger than the individual "clouds" or cells. Examples of disturbances are fronts, troughs and squall lines.

12.     Electric Field Measurement at the Surface:  The one-minute arithmetic average of the vertical electric field ($E_z$) at the surface of Earth, measured by a ground-based "field mill". The polarity of the electric field is the same as that of the potential gradient; that is, the polarity of the field at Earth's surface is the same as the dominant charge overhead. Do not use interpolated electric field contours for purposes of this appendix. An "'electric field measurement' less than or equal to 5 NM from the 'flight path'" [e.g., Section 1.4.A. Surface Electric Fields] is not applicable if the altitude of the "flight path" everywhere above the 5 NM circle around the "field mill" in question is greater than 20 km (66 kft).

13.     Field Mill: A specific class of electric-field sensor that uses a moving, grounded conductor to induce a time-varying electric charge on one or more sensing elements in proportion to the ambient electrostatic field.

14.     Flight Path: The planned nominal flight trajectory, including its vertical and horizontal uncertainties specified by the three-sigma guidance and performance deviations.

15.     Moderate Precipitation:  A precipitation rate of 0.1 inches/hour or a radar reflectivity factor of 30 dBZ.

16.     Nontransparent:  "Cloud" cover is "nontransparent" if one or more of the following conditions is present:
        (a) Objects above, including higher "clouds", blue sky, and stars, are blurred, indistinct, or obscured as viewed from below; or objects below, including terrain, buildings, and lights on the ground, are blurred, indistinct, or obscured as viewed from above; when looking through the "cloud" cover at visible wavelengths (the sun and moon may not be used to evaluate transparency);
        (b) Such objects are seen distinctly only through breaks in the "cloud" cover; or
        (c) The "cloud" cover has a "radar reflectivity" factor of 0 dBZ or greater.

17.     Precipitation:  Detectable rain, snow, hail, graupel, or sleet at the ground; virga; or a "radar reflectivity" factor greater than 18 dBZ at any altitude above the ground.

18.     Radar Reflectivity: The radar return from hydrometeors, in dBZ, measured by a meteorological radar operating at a wavelength greater than or equal to 5 cm. A "radar reflectivity" measurement is valid only in the absence of significant attenuation by intervening "precipitation" or by water or ice on the radome.

19.     Specified Volume: The volume bounded in the horizontal by vertical planes with perpendicular sides located 5.5 km (3 NM) north, east, south, and west of the point at which "VAHIRR" is being computed. The volume is bounded on the bottom at the altitude where the temperature is 0 degrees Celsius, and on the top by a fixed altitude of 20 km (66 kft).

20.     Thick Cloud Layer: One or more "cloud layers" whose combined vertical extent from the base of the bottom layer to the "cloud top" of the uppermost layer exceeds a thickness of 4,500 feet. "Cloud layers" are combined with neighboring layers for determining total thickness only when they are physically connected by vertically continuous "clouds", as, for example, when towering "clouds" in one layer contact or merge with "clouds" in a layer (or layers) above.

21. Thunderstorm: Any convective cloud that produces lightning.

22. Transparent: Any "cloud" that is not "nontransparent" is "transparent".

23. (reserved)

24. (reserved)

25. Volume-Averaged, Height–Integrated Radar Reflectivity (VAHIRR): The product of the "volume-averaged radar reflectivity" and the "average cloud thickness" in a "specified volume" surrounding any point at which "VAHIRR" is being computed (units of dBZ-km). The "specified volume" must not contain any portion of the "cone of silence" above the radar, nor any portion of any sectors that may have been blocked out for payload-safety reasons.

26. VAHIRR Application Criteria: The individual grid-point reflectivity measurements used to determine either the "volume-averaged radar reflectivity" or the "average cloud thickness" must be meteorological "radar reflectivity" measurements. For "VAHIRR"-evaluation points along the "flight path" itself (*not* those at a prescribed distance away from the "flight path"), the "volume-averaged, height-integrated radar reflectivity" is not applicable at any point that is less than or equal to 10 NM from any "radar reflectivity" of 35 dBZ or greater at altitudes of 4 km (13 kft) or greater above mean sea level, nor is it applicable at any point that is less than or equal to 10 NM from any type of lightning that has occurred in the previous 5 minutes.

27. Volume-Averaged Radar Reflectivity: the arithmetic average (in dBZ) of the "radar reflectivity" in the "specified volume." Normally, a radar processor will report reflectivity values interpolated onto a regular, three-dimensional array of grid points. Any such grid point in the "specified volume" is included in the average if and only if it has a "radar reflectivity" equal to or greater than 0 dBZ. If fewer than 10% of the grid points in the "specified volume" have "radar reflectivity" measurements equal to or greater than 0 dBZ, then the "volume-averaged radar reflectivity" is either the maximum "radar reflectivity" (in dBZ) in the "specified volume," or 0 dBZ, whichever is greater.

J.      Interim Instructions for Implementation of VAHIRR

The "VAHIRR" quantity referred to in Section 1.4.C. and the definitions require computation of both a volume average reflectivity and an average cloud thickness. These quantities are then multiplied to produce the "VAHIRR". Neither of these quantities is available yet as a product on the WSR-88D and WSR-74C radar systems used to support launch operations. This instruction provides a methodology for evaluating "VAHIRR" criteria with currently available radar products. The methodology provides a result that is more conservative than a direct "VAHIRR" computation, but it should still permit the launch users to achieve much of the benefit of the "VAHIRR" feature.

1.      Part I. Determination of average cloud thickness. The definition of "VAHIRR" requires determination of the average cloud top and the average cloud base above the height of the 0C isotherm within a square having sides 5.5 km (3 NM) north, east, south, and west of the ground projection of each point in the flight track. Average cloud thickness is defined as the difference of these two numbers. If the average cloud thickness cannot be determined at each point on the flight track, the maximum thickness within 5.5 km (3 NM) of the flight track may be used.

To determine the average cloud top height, the launch weather team may use any existing radar product that gives the height of the 0 dBZ cloud top, including "maximum height of reflectivity" and cross section products. If the average height cannot be determined, the maximum height of the 0 dBZ reflectivity may be used. If the maximum height cannot be determined, use 18 km (60 kft) for the average cloud top.

The average height of the cloud base should be derived from radar data. It is the average of the higher of (a) the bottom of the portion of the cloud producing a radar reflectivity of 0 dBZ or greater or (b) the height of the 0C isotherm where the 0 dBZ reflectivity extends below that level. If the average height of the cloud base cannot be determined, use the higher of (a) the height of the 0 degree C isotherm or (b) the lowest portion of the cloud producing a radar reflectivity of 0 dBZ or greater anywhere within each 5.5 km (3 NM) square defined above.

2.      Part II. Volume Averaged Radar Reflectivity (VARR). There is no operationally feasible way to use existing radar products to compute a volume average reflectivity. A conservative substitute is the maximum reflectivity since the volume average will always be smaller than the maximum. The WSR-88D has a "User Selectable Layer Composite Reflectivity (URL)" product and the WSR-74C has a "Max" product with user selectable base and top. The Launch Weather Team should configure these products with the bottom of the layer at the height of the 0 degree C isotherm and the top above the height of the highest radar beam within 7.8 km (4.2 NM) of the ground-projected flight track in the scan strategy being used. The WSR-88D product will have to be configured at the Radar Product Generator (RPG) and included in the product scheduler for the Principal User Processor (PUP) used by the launch weather team.

3.      Part III. Evaluating the constraint. The "VAHIRR" constraint is satisfied for a point on the flight track if the "URL" or "Max" radar product (see Part II above) everywhere within the corresponding square (defined above) is less than 10 dBZ-km divided by the average cloud thickness in km within the same square (see Part I above). (In English units, this threshold would become 33 dBZ-kft divided by the average cloud thickness in kft.) This constraint must be satisfied for every point on the flight track.

*A1.12 Current (through 25 March 2010) FAA LLCC (14 CFR 417, Appendix G)*
[Editor's note 1: The date shown here is the date which appears (at the time of this writing) when the rules are accessed on line in the Code of Federal Regulations (CFR). This version of the FAA rules is based on the 2005 LAP recommendations that were adopted by the Space Shuttle program as presented above in A1.10]

[Editor's note 2: The requirement that the launch operator must have **clear and convincing evidence** that the LFCC are satisfied, and the requirement (informally called the "Good Sense Rule") that any potentially hazardous conditions that are not covered by the LFCC be brought to the attention of the Launch Director do not appear in Appendix G and are not reproduced here. They appear instead in 14 CFR 417.113(c)(ii), and, like Appendix G, are mandatory.]

G417.1  General
For purposes of this section, the requirement for any weather monitoring and measuring equipment needed to satisfy the lightning flight commit criteria limits the equipment to only that which is needed. Accordingly, the equipment could include a ground-based, or airborne field mill, or a weather radar, but may or may not be limited to those items. Certain equipment, such as a field mill, when utilized with the lightning flight commit criteria, may increase launch opportunities because of the ability to verify the electric field in any cloud within 5 nautical miles of the flight path. However, a field mill is not required in order to satisfy the lightning flight commit criteria.

(a) This appendix provides flight commit criteria to protect against natural lightning and lightning triggered by the flight of a launch vehicle. A launch operator must apply these criteria under §417.113 (c) for any launch vehicle that utilizes a flight safety system.
(b) The launch operator must employ:
(1) Any weather monitoring and measuring equipment needed to satisfy the lightning flight commit criteria.
(2) Any procedures needed to satisfy the lightning flight commit criteria.

(c) If a launch operator proposes any alternative lightning flight commit criteria, the launch operator must clearly and convincingly demonstrate that the alternative provides an equivalent level of safety.

G417.3  Definitions, Explanations and Examples
For the purpose of appendix G417:

*Anvil cloud* means a stratiform or fibrous cloud produced by the upper level outflow or blow-off from thunderstorms or convective clouds.

*Associated* means that two or more clouds are causally related to the same weather disturbance or are physically connected. Associated does not have to mean occurring at the same time. A cumulus cloud formed locally and a cirrus layer that is physically separated from that cumulus cloud and that is generated by a distant source are not associated, even if they occur over or near the launch point at the same time.

*Bright band* means an enhancement of radar reflectivity caused by frozen hydrometeors falling and beginning to melt at any altitude where the temperature is 0 degrees Celsius or warmer.

*Cloud* means a visible mass of water droplets or ice crystals produced by condensation of water vapor in the atmosphere.

*Cloud edge* means the visible boundary, including the sides, base, and top, of a cloud as seen by an observer. In the absence of a visible boundary as seen by an observer, the 0 dBZ radar reflectivity boundary defines a cloud edge.

*Cloud layer* means a vertically continuous array of clouds, not necessarily of the same type, whose bases are approximately at the same level.

*Cumulonimbus cloud* means any convective cloud with any part at an altitude where the temperature is colder than -20 degrees Celsius.

*Debris cloud* means any cloud, except an anvil cloud, that has become detached from a parent cumulonimbus cloud or thunderstorm, or that results from the decay of a parent cumulonimbus cloud or thunderstorm.

*Disturbed Weather* means a weather system where dynamical processes destabilize the air on a scale larger than the individual clouds or cells. Examples of disturbed weather include fronts and troughs.

*Electric field measurement aloft* means the magnitude of the instantaneous vector electric field (E) at a known position in the atmosphere, such as measured by a suitably instrumented, calibrated, and located airborne-field-mill aircraft.

*Electric field measurement at the surface of Earth* means the 1-minute arithmetic average of the vertical electric field (Ez) at the ground measured by a ground-based field mill. The polarity of the electric field is the same as that of the potential gradient; that is, the polarity of the field at Earth's surface is the same as the dominant charge overhead. An interpolation based on electric field contours is not a measurement for purposes of this appendix.

*Field mill* is a specific class of electric-field sensor that uses a moving, grounded conductor to induce a time-varying electric charge on one or more sensing elements in proportion to the ambient electrostatic field.

*Flight path* means the planned normal flight trajectory, including its vertical and horizontal uncertainties to include the sum of the wind effects and the three-sigma guidance and performance deviations.

*Moderate precipitation* means a precipitation rate of 0.1 inches/hr or a radar reflectivity factor of 30 dBZ.

*Nontransparent* means cloud cover is nontransparent if (1) forms seen through it are blurred, indistinct, or obscured; or (2) forms are seen distinctly only through breaks in the cloud cover. Clouds with a radar reflectivity factor of 0 dBZ or greater are also nontransparent.

*Ohms/Square* means the surface resistance in ohms when a measurement is made from an electrode on one surface extending the length of one side of a square of any size to an electrode on the same surface extending the length of the opposite side of the square. The resistance measured in this way is independent of the area of a square.

*Precipitation* means detectable rain, snow, hail, graupel, or sleet at the ground; virga, or a radar reflectivity factor greater than 18 dBZ at altitude.

*Specified Volume* means the volume bounded in the horizontal by vertical plane, perpendicular sides located 5.5 km (3 NM) north, east, south, and west of the point on the flight track, on the bottom by the 0 degree C level, and on the top by the upper extent of all clouds.

*Thick cloud layer* means one or more cloud layers whose combined vertical extent from the base of the bottom layer to the top of the uppermost layer exceeds a thickness of 4,500 feet. Cloud layers are combined with neighboring layers for determining total thickness only when they are physically connected by vertically continuous clouds, as, for example, when towering clouds in one layer contact or merge with clouds in a layer (or layers) above.

*Thunderstorm* means any convective cloud that produces lightning.

*Transparent Cloud* cover is transparent if objects above, including higher clouds, blue sky, and stars can be distinctly seen from below; or objects, including terrain, buildings, and lights on the ground, can be distinctly seen from above. Transparency is only defined for the visible wavelengths.

*Triboelectrification* means the transfer of electrical charge from ice particles to the launch vehicle when the ice particles rub the vehicle during impact.

*Volume-Averaged, Height-Integrated Radar Reflectivity* (units of dBZ-kilometers) means the product of the volume-averaged radar reflectivity and the average cloud thickness within a specified volume relative to a point along the flight track.

*Within* is a function word used to specify a distance in all directions (horizontal, vertical, and slant separation) between a cloud edge and a flight path. For example, "within 10 nautical miles of a thunderstorm cloud" means that there must be a 10 nautical mile margin between every part of a thunderstorm cloud and the flight path.

G417.5  Lightning

(a) A launch operator must not initiate flight for 30 minutes after any type of lightning occurs in a thunderstorm if the flight path will carry the launch vehicle within 10 nautical miles of that thunderstorm.

(b) A launch operator must not initiate flight for 30 minutes after any type of lightning occurs within 10 nautical miles of the flight path unless:
(1) The cloud that produced the lightning is not within 10 nautical miles of the flight path;
(2) There is at least one working field mill within 5 nautical miles of each such lightning flash; and
(3) The absolute values of all electric field measurements made at the Earth's surface within 5 nautical miles of the flight path and at each field mill specified in paragraph (b)(2) of this section have been less than 1000 volts/meter for 15 minutes or longer.

(c) If a cumulus cloud remains 30 minutes after the last lightning occurs in a thunderstorm, section G417.7 applies. Sections G417.9 and G417.11 apply to any anvil or detached anvil clouds. Section G417.13 applies to debris clouds.

G417.7  Cumulus Clouds
For the purposes of this section, "cumulus clouds" do not include altocumulus, cirrocumulus, or stratocumulus clouds.

(a) A launch operator must not initiate flight if the flight path will carry the launch vehicle within 10 nautical miles of any cumulus cloud that has a cloud top at an altitude where the temperature is colder than −20 degrees Celsius.

(b) A launch operator must not initiate flight if the flight path will carry the launch vehicle within 5 nautical miles of any cumulus cloud that has a cloud top at an altitude where the temperature is colder than −10 degrees Celsius.

(c) A launch operator must not initiate flight if the flight path will carry the launch vehicle through any cumulus cloud with its cloud top at an altitude where the temperature is colder than −5 degrees Celsius.

(d) A launch operator must not initiate flight if the flight path will carry the launch vehicle through any cumulus cloud that has a cloud top at an altitude where the temperature is between +5 degrees Celsius and −5 degrees Celsius unless:

(1) The cloud is not producing precipitation;

(2) The horizontal distance from the center of the cloud top to at least one working field mill is less than 2 nautical miles; and

(3) All electric field measurements made at the Earth's surface within 5 nautical miles of the flight path and at each field mill used as required by paragraph (d)(2) of this section have been between −100 volts/meter and +500 volts/meter for 15 minutes or longer.

G417.9  Attached Anvil Clouds

(a) A launch operator must not initiate flight if the flight path will carry the launch vehicle through, or within 10 nautical miles of, a nontransparent part of any attached anvil cloud for the first 30 minutes after the last lightning discharge in or from the parent cloud or anvil cloud.

(b) A launch operator must not initiate flight if the flight path will carry the launch vehicle through, or within 5 nautical miles of, a nontransparent part of any attached anvil cloud between 30 minutes and three hours after the last lightning discharge in or from the parent cloud or anvil cloud unless:

(1) The portion of the attached anvil cloud within 5 nautical miles of the flight path is located entirely at altitudes where the temperature is colder than 0 degrees Celsius; and

(2) The volume-averaged, height-integrated radar reflectivity is less than +33 dBZ-kft everywhere along the portion of the flight path where any part of the attached anvil cloud is within the volume.

(c) A launch operator must not initiate flight if the flight path will carry the launch vehicle through a nontransparent part of any attached anvil cloud more than 3 hours after the last lightning discharge in or from the parent cloud or anvil cloud unless:

(1) The portion of the attached anvil cloud within 5 nautical miles of the flight path is located entirely at altitudes where the temperature is colder than 0 degrees Celsius; and

(2) The volume-averaged, height-integrated radar reflectivity is less than +33 dBZ-kft everywhere along the portion of the flight path where any part of the attached anvil cloud is within the specified volume.

G417.11  Detached Anvil Clouds

For the purposes of this section, detached anvil clouds are never considered debris clouds.

(a) A launch operator must not initiate flight if the flight path will carry the launch vehicle through or within 10 nautical miles of a nontransparent part of a detached anvil cloud for the first 30 minutes after the last lightning discharge in or from the parent cloud or anvil cloud before detachment or after the last lightning discharge in or from the detached anvil cloud after detachment.

(b) A launch operator must not initiate flight if the flight path will carry the launch vehicle within 5 nautical miles of a nontransparent part of a detached anvil cloud between 30 minutes and 3 hours after the time of the last lightning discharge in or from the parent cloud or anvil cloud before detachment or after the last lightning discharge in or from the detached anvil cloud after detachment unless section (1) or (2) is satisfied:

(1) This section is satisfied if all three of the following conditions are met:

(i) There is at least one working field mill within 5 nautical miles of the detached anvil cloud; and

(ii) The absolute values of all electric field measurements at the surface within 5 nautical miles of the flight path and at each field mill specified in (1) above have been less than 1000 V/m for 15 minutes; and

(iii) The maximum radar return from any part of the detached anvil cloud within 5 nautical miles of the flight path has been less than 10 dBZ for 15 minutes.

(2) This section is satisfied if both of the following conditions are met:

(i) The portion of the detached anvil cloud within 5 nautical miles of the flight path is located entirely at altitudes where the temperature is colder than 0 degrees Celsius; and
(ii) The volume-averaged, height-integrated radar reflectivity is less than +33 dBZ-kft everywhere along the portion of the flight path where any part of the detached anvil cloud is within the specified volume.

(c) A launch operator must not initiate flight if the flight path will carry the launch vehicle through a nontransparent part of a detached anvil cloud unless Section (1) or (2) is satisfied.
(1) This section is satisfied if both of the following conditions are met:
(i) At least 4 hours have passed since the last lightning discharge in or from the detached anvil cloud; and
(ii) At least 3 hours have passed since the time that the anvil cloud is observed to be detached from the parent cloud.
(2) This section is satisfied if both of the following conditions are met.
(i) The portion of the detached anvil cloud within 5 nautical miles of the flight path is located entirely at altitudes where the temperature is colder than 0 degrees Celsius; and
(ii) The volume-averaged, height-integrated radar reflectivity is less than +33 dBZ-kft everywhere along the portion of the flight path where any part of the detached anvil cloud is within the specified volume.

## G417.13  Debris Clouds

(a) A launch operator must not initiate flight if the flight path will carry the launch vehicle through any nontransparent part of a debris cloud for 3 hours after the debris cloud is observed to be detached from the parent cloud or after the debris cloud is observed to have formed from the decay of the parent cloud top to an altitude where the temperature is warmer than −10 degrees Celsius. The 3-hour period must begin again at the time of any lightning discharge in or from the debris cloud.

(b) A launch operator must not initiate flight if the flight path will carry the launch vehicle within 5 nautical miles of a nontransparent part of a debris cloud during the 3-hour period defined in paragraph (a) of this section, unless:
(1) There is at least one working field mill within 5 nautical miles of the debris cloud;
(2) The absolute values of all electric field measurements at the Earth's surface within 5 nautical miles of the flight path and measurements at each field mill employed required by paragraph (b)(1) of this section have been less than 1000 volts/meter for 15 minutes or longer; and
(3) The maximum radar return from any part of the debris cloud within 5 nautical miles of the flight path has been less than 10 dBZ for 15 minutes or longer.

## G417.15  Disturbed Weather

(a) A launch operator must not initiate flight if the flight path will carry the launch vehicle through a nontransparent cloud associated with disturbed weather that has clouds with cloud tops at altitudes where the temperature is colder than 0 degrees Celsius and that contains, within 5 nautical miles of the flight path:
(1) Moderate or greater precipitation; or
(2) Evidence of melting precipitation such as a radar bright band.

## G417.17  Thick Cloud Layers

(a) A launch operator must not initiate flight if the flight path will carry the launch vehicle through a nontransparent part of a cloud layer that is:
(1) Greater than 4,500 feet thick and any part of the cloud layer along the flight path is located at an altitude where the temperature is between 0 degrees Celsius and −20 degrees Celsius; or

(2) Connected to a thick cloud layer that, within 5 nautical miles of the flight path, is greater than 4,500 feet thick and has any part located at any altitude where the temperature is between 0 degrees Celsius and −20 degrees Celsius.

(b) A launch operator need not apply the lightning commit criteria in paragraphs (a)(1) and (a)(2) of this section if the thick cloud layer is a cirriform cloud layer that has never been associated with convective clouds, is located only at temperatures of −15 degrees Celsius or colder, and shows no evidence of containing liquid water.

G417.19  Smoke Plumes

(a) A launch operator must not initiate flight if the flight path will carry the launch vehicle through any cumulus cloud that has developed from a smoke plume while the cloud is attached to the smoke plume, or for the first 60 minutes after the cumulus cloud is observed to be detached from the smoke plume.

(b) Section G417.7 applies to cumulus clouds that have formed above a fire but have been detached from the smoke plume for more than 60 minutes.

G417.21  Surface Electric Fields

(a) A launch operator must not initiate flight for 15 minutes after the absolute value of any electric field measurement at the Earth's surface within 5 nautical miles of the flight path has been greater than 1500 volts/meter.

(b) A launch operator must not initiate flight for 15 minutes after the absolute value of any electric field measurement at the Earth's surface within 5 nautical miles of the flight path has been greater than 1000 volts/meter unless:
(1) All clouds within 10 nautical miles of the flight path are transparent; or
(2) All nontransparent clouds within 10 nautical miles of the flight path have cloud tops at altitudes where the temperature is warmer than +5 degrees Celsius and have not been part of convective clouds that have cloud tops at altitudes where the temperature is colder than −10 degrees Celsius within the last 3 hours.

G417.23  Triboelectrification

(a) A launch operator must not initiate flight if the flight path will go through any part of a cloud at an altitude where the temperature is colder than −10 degrees Celsius up to the altitude at which the launch vehicle's velocity exceeds 3000 feet/second; unless
(1) The launch vehicle is "treated" for surface electrification; or
(2) A launch operator demonstrates by test or analysis that electrostatic discharges on the surface of the launch vehicle caused by triboelectrification will not be hazardous to the launch vehicle or the spacecraft.

(b) A launch vehicle is treated for surface electrification if
(1) All surfaces of the launch vehicle susceptible to ice particle impact are such that the surface resistivity is less than $10^9$ ohms/square; and
(2) All conductors on surfaces (including dielectric surfaces that have been treated with conductive coatings) are bonded to the launch vehicle by a resistance that is less than $10^5$ ohms.

*A1.13 LAP Recommendation (2008) in FAA format.*
[Editor's Note: These are essentially the same rules as those adopted by the Space Shuttle program in 2009 and presented above in A1.11. As of May 2010, the FAA has not yet adopted these recommended changes to the FAA LFCC.]

[Editor's Note: The LAP expects section 417.113(c)(ii) (see Editor's Note 2 to A1.12 above) to remain in effect if these proposed LFCC are adopted. Its content is an integral part of the LAP's recommendation.]

## A. G417.1 General

Each of the Lightning Flight Commit Criteria (LFCC) requires **clear and convincing evidence** to trained weather personnel that its constraints are not violated. A launch operator must not initiate flight unless the constraints of **all** LFCC are satisfied. Whenever there is ambiguity about which of several LFCC applies to a particular situation, all potentially applicable LFCC must be applied. Under some conditions trained weather personnel can make a clear and convincing determination that the LFCC are not violated based on visual observations alone. However, if the weather personnel have access to additional information such as measurements from weather radar, lightning sensors, electric "field mills," and/or aircraft, this information can be used to increase both safety and launch availability. If the additional information is within the criteria outlined in the LFCC, it would allow a launch to take place where a visual observation alone would not.

    (a) This appendix provides flight commit criteria to protect against natural lightning and lightning triggered by the flight of a launch vehicle. A launch operator must apply these criteria under § 417.113 (c) for any launch vehicle that utilizes a flight safety system.

    (b) The launch operator must employ:

    (1) Any weather monitoring and measuring equipment needed to satisfy the lightning flight commit criteria.

    (2) Any procedures needed to satisfy the lightning flight commit criteria.

    (c) If a launch operator proposes any alternative lightning flight commit criteria, the launch operator must clearly and convincingly demonstrate that the alternative provides an equivalent level of safety.

## B. G417.3 Definitions, Explanations and Examples

For the purpose of this appendix, distance from an electric "field mill" is measured differently than distance from any other object or measurement point: Distance between a "radar reflectivity" or "VAHIRR" measurement point and any object or the "flight path" is the *shortest separation* (horizontal, vertical, or slant range) between that point and the *nearest part* of the object or "flight path." Similarly, distance between the "flight path" and any object is the *shortest separation* between any point on the "flight path" the *nearest part* of that object. For example, "every point less than or equal to 1 nautical mile from the 'flight path'" [see F. G417.11(c)(2) Attached "Anvil Clouds"] means that the "VAHIRR" threshold must be satisfied at every point throughout the entire volume defined by a 1 nautical mile radius from every point on the "flight path." (See also the additional explanation beneath the definition of "cloud.") In contrast, distance between a "field mill" or an "electric field measurement" and any object or the "flight path" is always measured *horizontally* between that mill or measurement point and the nearest part of the *vertical projection* of the object or "flight path" onto the surface of Earth. For example, "from the center of the 'cloud top' to at least one working 'field mill'" [see E. G417.9(d)(2) Cumulus "Clouds"] means that the *horizontal* distance between the "field mill" and a point on the surface directly beneath the center of the "cloud top" must be less than 2 nautical miles.

The following **bold-face** terms are defined here and appear in quotes wherever they are used in accordance with these definitions elsewhere in this appendix:

**Anvil cloud** means a stratiform or fibrous "cloud" produced by the upper outflow or blow-off from "thunderstorms" or convective "clouds" having tops at altitudes where the temperature is colder than or equal to -10 degrees Celsius.

**Associated** means that two or more "clouds" are causally related to the same "disturbed weather" system or are physically connected. "Clouds" occurring at the same time are not necessarily "associated." A cumulus "cloud" formed locally and a cirrus layer that is physically separated from that cumulus "cloud" and that is generated by a distant source are not "associated," even if they occur over or near the launch point at the same time.

**Average cloud thickness** is the altitude difference (in kilometers, km hereafter) between the average top and the average base of all clouds in the "specified volume." The cloud base to be averaged is the higher of (1) the 0 degree Celsius level and (2) the lowest extent (in altitude) of all "radar reflectivity" measurements of 0 dBZ or greater. Similarly, the cloud top to be averaged is the highest extent (in altitude) of all "radar reflectivity" measurements of 0 dBZ or greater. Given the grid-point representation of a typical radar processor, allowance must be made for the vertical separation of grid points in computing "average cloud thickness": The cloud base at any horizontal position shall be taken as the altitude of the corresponding base grid point minus half of the grid-point vertical separation. Similarly, the cloud top at that horizontal position shall be taken as the altitude of the corresponding top grid point plus half of this vertical separation. Thus, a cloud represented by only a single grid point having a "radar reflectivity" equal to or greater than 0 dBZ in the "specified volume" would have an "average cloud thickness" equal to the vertical grid-point separation in its vicinity.

**Bright band** means an enhancement of "radar reflectivity" caused by frozen hydrometeors falling and beginning to melt at any altitude where the temperature is 0 degrees Celsius or warmer.

**Cloud** means a visible mass of suspended water droplets or ice crystals. The "cloud" is considered to be the entire volume enclosed by the visible, "nontransparent cloud" boundary as seen by an observer, or, in the absence of a visual observation, by the 0 dBZ "radar reflectivity" boundary. A visual evaluation of transparency is preferred whenever possible.

Distance from the "cloud" to a point in question refers to the separation between the point and the nearest part of that "cloud." Specifically, the wording, "less than or equal to 10 nautical miles from any cumulus 'cloud'" means that the "flight path" must not penetrate either the *interior* of the "cloud" itself or the volume between 0 and 10 nautical miles, inclusive, *outside* the "cloud" boundary [for example, see E. G417.9(a), Cumulus "Clouds"]. On the other hand, "between 0 and 3 nautical miles, inclusive, from" refers *only* to the volume at a distance that is greater than or equal to 0, but less than or equal to 3, nautical miles *outside* the "cloud" boundary, specifically omitting the interior of the "cloud" itself [for example, see H. G417.15(a), "Debris Clouds"].

**Cloud layer** means a vertically continuous array of "clouds," not necessarily of the same type, whose bases are approximately at the same level.

**Cloud top** means the visible top of the cloud, or, in the absence of a visual observation, the 0 dBZ radar top. A visual evaluation of "cloud top" is preferred whenever possible.

**Cone of silence** means the volume in an inverted circular cone centered on the radar that is generated by all elevation angles greater than the maximum elevation angle used in the radar scan strategy. For the purpose of

"VAHIRR" calculation this volume is capped by the observed maximum "cloud top" height, the observed tropopause height, or an altitude of 20 km (66 kft), whichever is lowest.

**Cumulonimbus cloud** means any convective "cloud" with any part at an altitude where the temperature is colder than –20 degrees Celsius.

**Debris cloud** means any "cloud," except an "anvil cloud," that has become detached from a parent "cumulonimbus cloud" or "thunderstorm," or that results from the decay of a parent "cumulonimbus cloud" or "thunderstorm."

**Disturbed weather** means a weather system where dynamical processes destabilize the air on a scale larger than individual "clouds" or cells. Examples of "disturbed weather" include fronts, troughs, and squall lines.

**Electric field measurement** means the 1-minute arithmetic average of the vertical electric field ($E_z$) at the surface of Earth, measured by a ground-based "field mill." The polarity of the electric field is the same as that of the potential gradient; that is, the polarity of the field at Earth's surface is the same as the dominant charge overhead. Do not use interpolated electric field contours for purposes of this appendix. An "'electric field measurement' less than or equal to 5 nautical miles from the 'flight path'" [*e.g.*, C. G417.5(a) Surface Electric Fields] is not applicable if the altitude of the flight path everywhere above the 5 nautical mile circle around the "field mill" in question is greater than 20 km (66 kft).

**Field mill** is a specific class of electric-field sensor that uses a moving, grounded conductor to induce a time-varying electric charge on one or more sensing elements in proportion to the ambient electrostatic field.

**Flight path** means the planned nominal flight trajectory, including its vertical and horizontal uncertainties specified by the three-sigma guidance and performance deviations.

**Moderate precipitation** means a "precipitation" rate of 0.1 inches/hr or a "radar reflectivity" factor of 30 dBZ.

**Nontransparent.** "Cloud" cover is "nontransparent" if **one or more** of the following conditions is present:

(a) Objects above, including higher "clouds," blue sky, and stars, are blurred, indistinct, or obscured as viewed from below; or objects below, including terrain, buildings, and lights on the ground, are blurred, indistinct, or obscured as viewed from above; when looking through the "cloud" cover at visible wavelengths (the sun and moon may not be used to evaluate transparency);

(b) Such objects are seen distinctly only through breaks in the "cloud" cover; or

(c) The "cloud" cover has a "radar reflectivity" factor of 0 dBZ or greater.

**Precipitation** means detectable rain, snow, hail, graupel, or sleet at the ground; virga; or a "radar reflectivity" factor greater than 18 dBZ at any altitude above the ground.

**Radar reflectivity** means the radar return from hydrometeors, in dBZ, measured by a meteorological radar operating at a wavelength greater than or equal to 5 cm. A "radar reflectivity" measurement is valid only in the absence of significant attenuation by intervening "precipitation" or by water or ice on the radome.

**Specified volume.** The volume bounded in the horizontal by vertical planes with perpendicular sides located 5.5 km (3 nautical miles) north, east, south, and west of the point at which "VAHIRR" is being computed. The volume is bounded on the bottom at the altitude where the temperature is 0 degrees Celsius, and on the top by a fixed altitude of 20 km (66 kft).

**Thick cloud layer** means one or more "cloud layers" whose combined vertical extent from the base of the bottom layer to the "cloud top" of the uppermost layer exceeds a thickness of 4,500 feet. "Cloud layers" are combined with neighboring layers for determining total thickness only when they are physically connected by vertically continuous "clouds," as, for example, when towering "clouds" in one layer contact or merge with "clouds" in a layer (or layers) above.

**Thunderstorm** means any convective "cloud" that produces lightning.

**Transparent**. Any "cloud" that is **not** "nontransparent" is "transparent."

**Treated** means that a launch vehicle satisfies both of the following conditions:

(a) All surfaces of the launch vehicle susceptible to ice particle impact are such that the surface resistivity is less than $10^9$ "Ohms per square;" **and**

(b) All conductors on surfaces (including dielectric surfaces that have been coated with conductive materials) are bonded to the launch vehicle by a resistance that is less than $10^5$ ohms.

**Triboelectrification** means the transfer of electrical charge between ice particles and the launch vehicle when the ice particles collide with the vehicle during flight.

**Volume-Averaged, Height-Integrated Radar Reflectivity (VAHIRR)** is the product of the "volume-averaged radar reflectivity" and the "average cloud thickness" in a "specified volume" surrounding any point at which "VAHIRR" is being computed (units of dBZ-km). The "specified volume" must not contain any portion of the "cone of silence" above the radar, nor any portion of any sectors that may have been blocked out for payload-safety reasons.

**VAHIRR application criteria**: The individual grid-point reflectivity measurements used to determine either the "volume-averaged radar reflectivity" or the "average cloud thickness" must be meteorological "radar reflectivity" measurements. For "VAHIRR"-evaluation points along the "flight path" itself (*not* those at a prescribed distance away from the "flight path"), the "volume-averaged, height-integrated radar reflectivity" is not applicable at any point that is less than or equal to 10 nautical miles from any "radar reflectivity" of 35 dBZ or greater at altitudes of 4 km (13 kft) or greater above mean sea level, nor is it applicable at any point that is less than or equal to 10 nautical miles from any type of lightning that has occurred in the previous 5 minutes.

**Volume-averaged radar reflectivity** is the arithmetic average (in dBZ) of the "radar reflectivity" in the "specified volume." Normally, a radar processor will report reflectivity values interpolated onto a regular, three-dimensional array of grid points. Any such grid point in the "specified volume" is included in the average if and only if it has a "radar reflectivity" equal to or greater than 0 dBZ. If fewer than 10% of the grid points in the "specified volume" have "radar reflectivity" measurements equal to or greater than 0 dBZ, then the "volume-averaged radar reflectivity" is either the maximum "radar reflectivity" (in dBZ) in the "specified volume," or 0 dBZ, whichever is greater.

## C. G417.5 Surface Electric Fields

(a) A launch operator must not initiate flight for 15 minutes after the absolute value of any "electric field measurement" less than or equal to 5 nautical miles from the "flight path" has been greater than or equal to 1500 volts/meter.

(b) A launch operator must not initiate flight for 15 minutes after the absolute value of any "electric field measurement" less than or equal to 5 nautical miles from the "flight path" has been greater than or equal to 1000 volts/meter **unless either** Section 1 **or** Section 2 is satisfied:

(1) All clouds less than or equal to 10 nautical miles from the "flight path" are "*transparent*;" or

(2) All "clouds" less than or equal to 10 nautical miles from the "flight path" have "cloud tops" at altitudes where the temperature is warmer than +5 degrees Celsius and have not been part of convective "clouds" with "cloud tops" at altitudes where the temperature is colder than or equal to −10 degrees Celsius during the last 3 hours.

## D. G417.7 Lightning

(a) A launch operator must not initiate flight for 30 minutes after any type of lightning occurs in a "thunderstorm" if the "flight path" will carry the launch vehicle less than or equal to 10 nautical miles from that "thunderstorm." An attached "anvil cloud" is not considered part of its parent "thunderstorm," but is covered instead by Section F, Attached "Anvil Clouds."

(b) A launch operator must not initiate flight for 30 minutes after any type of lightning occurs less than or equal to 10 nautical miles from the "flight path" **unless all three** of the following conditions are satisfied:

(1) The "cloud" that produced the lightning is greater than 10 nautical miles from the "flight path;"

(2) There is at least one working "field mill" less than 5 nautical miles from each such lightning discharge; **and**

(3) The absolute values of all "electric field measurements" less than or equal to 5 nautical miles from the "flight path," and at each "field mill" specified in paragraph (b)(2) of this section, have been less than 1000 volts/meter for 15 minutes or longer.

## E. G417.9 Cumulus "Clouds"

For the purposes of this section, "cumulus 'clouds'" do not include cirrocumulus, altocumulus, or stratocumulus "clouds." An attached "anvil cloud" is never considered part of its parent cumulus "cloud," but is covered instead by Section F, Attached "Anvil Clouds." Section G, Detached "Anvil Clouds," applies to any detached "anvil cloud." Section H, "Debris Clouds," applies to "debris clouds."

(a) A launch operator must not initiate flight if the "flight path" will carry the launch vehicle less than or equal to 10 nautical miles from any cumulus "cloud" that has a "cloud top" at an altitude where the temperature is colder than or equal to -20 degrees Celsius.

(b) A launch operator must not initiate flight if the "flight path" will carry the launch vehicle less than or equal to 5 nautical miles from any cumulus "cloud" that has a "cloud top" at an altitude where the temperature is colder than or equal to -10 degrees Celsius.

(c) A launch operator must not initiate flight if the "flight path" will carry the launch vehicle through any cumulus "cloud" with its "cloud top" at an altitude where the temperature is colder than or equal to -5 degrees Celsius.

(d) A launch operator must not initiate flight if the "flight path" will carry the launch vehicle through any cumulus "cloud" that has a "cloud top" at an altitude where the temperature lies in the range from warmer than

-5 degrees Celsius to colder than or equal to +5 degrees Celsius **unless all three** of the following conditions are satisfied:

(1) The "cloud" is not producing "precipitation;"

(2) The distance from the center of the "cloud top" to at least one working "field mill" is less than 2 nautical miles; **and**

(3) All "electric field measurements" less than or equal to 5 nautical miles from the "flight path," and at each "field mill" specified in paragraph (d)(2) of this section, have been greater than $-100$ volts/meter, but less than +500 volts/meter, for 15 minutes or longer.

## F.  G417.11  Attached "Anvil Clouds"

For the purposes of this section, if there has never been lightning in or from the parent "cloud" or "anvil cloud," sub-sections (a) and (b) shall be considered satisfied, but sub-section (c) shall still apply.

(a) A launch operator must not initiate flight if the "flight path" will carry the launch vehicle less than or equal to 10, but greater than 5, nautical miles from any attached "anvil cloud" for the first 30 minutes after the last lightning discharge in or from the parent "cloud" or "anvil cloud" **unless** the portion of the attached "anvil cloud" less than or equal to 10 nautical miles from the "flight path" is located entirely at altitudes where the temperature is colder than 0 degrees Celsius.

(b) A launch operator must not initiate flight if the "flight path" will carry the launch vehicle less than or equal to 5, but greater than 3, nautical miles from any attached "anvil cloud" for the first three hours after the last lightning discharge in or from the parent "cloud" or "anvil cloud" **unless** the portion of the attached "anvil cloud" less than or equal to 5 nautical miles from the "flight path" is located entirely at altitudes where the temperature is colder than 0 degrees Celsius.

(c) A launch operator must not initiate flight if the "flight path" will carry the launch vehicle less than or equal to 3 nautical miles from any attached "anvil cloud" **unless all three** of the following conditions are satisfied:

(1) The portion of the attached "anvil cloud" less than or equal to 5 nautical miles from the "flight path" is located entirely at altitudes where the temperature is colder than 0 degrees Celsius;

(2) The "volume-averaged, height-integrated radar reflectivity" is less than +10 dBZ-km (+33 dBZ-kft) at every point less than or equal to 1 nautical mile from the "flight path;" **and**

(3) All of the "VAHIRR application criteria" are satisfied.

## G.  G417.13  Detached "Anvil Clouds"

For the purposes of this section, detached "anvil clouds" are never considered "debris clouds."

For the purposes of this section, if there has never been lightning in or from the parent "cloud" or "anvil cloud," sub-sections (a), (b), (c), and (d)(1)(i) shall be considered satisfied, but sub-sections (d)(1)(ii), and (d)(2), shall still apply.

(a) A launch operator must not initiate flight if the "flight path" will carry the launch vehicle less than or equal to 10, but greater than 3, nautical miles from a detached "anvil cloud" for the first 30 minutes after the last lightning discharge in or from the parent "cloud" or "anvil cloud" before detachment or after the last lightning discharge in or from the detached "anvil cloud" after detachment **unless** the portion of the detached "anvil cloud" less than or equal to 10 nautical miles from the "flight path" is located entirely at altitudes where the temperature is colder than 0 degrees Celsius.

(b) A launch operator must not initiate flight if the "flight path" will carry the launch vehicle between 0 (zero) and 3 nautical miles, inclusive, from a detached "anvil cloud" for the first 30 minutes after the time of the last lightning discharge in or from the parent "cloud" or "anvil cloud" before detachment or after the last lightning discharge in or from the detached "anvil cloud" after detachment **unless all three** of the following conditions are met:

> (1) The portion of the detached "anvil cloud" less than or equal to 5 nautical miles from the "flight path" is located entirely at altitudes where the temperature is colder than 0 degrees Celsius;

> (2) The "volume-averaged, height-integrated radar reflectivity" is less than +10 dBZ-km (+33 dBZ-kft) at every point less than or equal to 1 nautical mile from the "flight path;" **and**

> (3) All of the "VAHIRR application criteria" are satisfied.

(c) A launch operator must not initiate flight if the "flight path" will carry the launch vehicle between 0 (zero) and 3 nautical miles, inclusive, from a detached "anvil cloud" less than or equal to 3 hours, but greater than 30 minutes, after the time of the last lightning discharge in or from the parent "cloud" or "anvil cloud" before detachment or after the last lightning discharge in or from the detached "anvil cloud" after detachment unless Section (1) or Section (2) is satisfied:

> (1) This section is satisfied **if all three** of the following conditions are met:

>> (i) There is at least one working "field mill" less than 5 nautical miles from the detached "anvil cloud;"

>> (ii) The absolute values of all "electric field measurements" less than or equal to 5 nautical miles from the "flight path," and at each "field mill" specified in paragraph (c)(1)(i) of this section, have been less than 1000 V/m for 15 minutes; **and**

>> (iii) The maximum radar reflectivity from any part of the detached "anvil cloud" less than or equal to 5 nautical miles from the "flight path" has been less than +10 dBZ for 15 minutes.

> (2) This section is satisfied **if all three** of the following conditions are met:

>> (i) The portion of the detached "anvil cloud" less than or equal to 5 nautical miles from the "flight path" is located entirely at altitudes where the temperature is colder than 0 degrees Celsius;

>> (ii) The "volume-averaged, height-integrated radar reflectivity" is less than +10 dBZ-km (+33 dBZ-kft) at every point less than or equal to 1 nautical mile from the "flight path;" **and**

>> (iii) All of the "VAHIRR application criteria" are satisfied.

(d) A launch operator must not initiate flight if the "flight path" will carry the launch vehicle through a detached "anvil cloud" unless Section (1) or Section (2) is satisfied

> (1) This section is satisfied if **both** of the following conditions are met.

(i) At least 4 hours have passed since the last lightning discharge in or from the detached "anvil cloud;" **and**

(ii) At least 3 hours have passed since the time that the "anvil cloud" is observed to be detached from the parent "cloud."

(2) This section is satisfied **if all three** of the following conditions are met.

(i) The portion of the detached "anvil cloud" less than or equal to 5 nautical miles from the "flight path" is located entirely at altitudes where the temperature is colder than 0 degrees Celsius;

(ii) The "volume-averaged, height-integrated radar reflectivity" is less than +10 dBZ-km (+33 dBZ-kft) everywhere along the "flight path;" **and**

(iii) All of the "VAHIRR application criteria" are satisfied.

## H. G417.15 "Debris Clouds"

The 3-hour time period defined in this Section must begin again at the time of any lightning discharge in or from the "debris cloud."

(a) A launch operator must not initiate flight if the "flight path" will carry the launch vehicle between 0 and 3 nautical miles, inclusive, from a "debris cloud" for 3 hours after the "debris cloud" is observed to be detached from the parent "cloud," or after the "debris cloud" is observed to have formed by the collapse of the parent "cloud top" to an altitude where the temperature is warmer than −10 degrees Celsius **unless** Section (1) **or** Section (2) is satisfied:

(1) This section is satisfied **if all three** of the following conditions are met:

(i) There is at least one working "field mill" less than 5 nautical miles from the "debris cloud;"

(ii) The absolute values of all "electric field measurements" less than or equal to 5 nautical miles from the "flight path" and at each "field mill" employed by paragraph (a)(1)(i) of this section has been less than 1000 volts/meter for 15 minutes or longer; **and**

(iii) The maximum radar reflectivity from any part of the "debris cloud" less than or equal to 5 nautical miles from the "flight path" has been less than +10 dBZ for 15 minutes or longer.

(2) This section is satisfied **if all three** of the following conditions are met:

(i) The portion of the "debris cloud" less than or equal to 5 nautical miles from the "flight path" is located entirely at altitudes where the temperature is colder than 0 degrees Celsius;

(ii) The "volume-averaged, height-integrated radar reflectivity" is less than +10 dBZ-km (+33 dBZ-kft) at every point less than or equal to 1 nautical mile from the "flight path;" **and**

(iii) All of the "VAHIRR application criteria" are satisfied.

(b) A launch operator must not initiate flight if the "flight path" will carry the launch vehicle through any "debris cloud" during the 3-hour period defined in paragraph (a) of this section, **unless all three** of the following conditions are met:

> (1) The portion of the "debris cloud" less than or equal to 5 nautical miles from the "flight path" is located entirely at altitudes where the temperature is colder than 0 degrees Celsius;

> (2) The "volume-averaged, height-integrated radar reflectivity" is less than +10 dBZ-km (+33 dBZ-kft) everywhere along the "flight path;" **and**

> (3) All of the "VAHIRR application criteria" are satisfied.

## I. G417.17 "Disturbed Weather"

(a) A launch operator must not initiate flight if the "flight path" will carry the launch vehicle through a "cloud" "associated" with "disturbed weather" that has "clouds" with "cloud tops" at altitudes where the temperature is colder than or equal to 0 degrees Celsius and that contains, less than or equal to 5-nautical miles from the "flight path," **either**:

> (1) "Moderate precipitation" or greater; **or**

> (2) Evidence of melting "precipitation" such as a radar "bright band."

## J. G417.19 "Thick Cloud Layers"

For the purposes of this section neither attached nor detached "anvil clouds" are considered "thick cloud layers."

(a) A launch operator must not initiate flight if the "flight path" will carry the launch vehicle through a "cloud layer" that is **either**:

> (1) Greater than or equal to 4,500 feet thick and any part of the "cloud layer" along the "flight path" is located at an altitude where the temperature is between 0 degrees Celsius and −20 degrees Celsius, inclusive; **or**

> (2) Connected to a "thick cloud layer" that, less than or equal to 5 nautical miles from the "flight path," is greater than or equal to 4,500 feet thick and has any part located at an altitude where the temperature is between 0 degrees Celsius and −20 degrees Celsius, inclusive.

(b) A launch operator need not apply the lightning commit criteria in paragraphs (a)(1) and (a)(2) of this section if the "thick cloud layer" is a cirriform "cloud layer" that has never been "associated" with convective "clouds," is located entirely at altitudes where the temperature is colder than or equal to −15 degrees Celsius, and shows no evidence of containing liquid water.

## K. G417.21 Smoke Plumes

(a) A launch operator must not initiate flight if the "flight path" will carry the launch vehicle through any cumulus "cloud" that has developed from a smoke plume while the "cloud" is attached to the smoke plume, or for the first 60 minutes after the cumulus "cloud" is observed to be detached from the smoke plume.

(b) Section E, Cumulus "Clouds," applies to cumulus "clouds" that have formed above a fire but have been detached from the smoke plume for more than 60 minutes.

## L. G417.23 "Triboelectrification"

A launch operator must not initiate flight if the "flight path" will carry the launch vehicle through any part of a cloud, *specifically including all "transparent" parts*, at any altitude where **both** Section (a) and Section (b) are satisfied:
(a) The temperature is colder than or equal to –10 degrees Celsius; **and**

(b) The launch vehicle's velocity is less than or equal to 3000 feet/second;

**unless** Section (1) or Section (2) is satisfied:

> (1) The launch vehicle is "treated" for surface electrification; or

> (2) A launch operator has previously demonstrated by test or analysis that electrostatic discharges on the surface of the launch vehicle caused by "triboelectrification" will not be hazardous to the launch vehicle or the spacecraft.

Interim Instructions for Implementation of Volume Averaged Height Integrated Radar Reflectivity (VAHIRR) dated 28 October 2004

The VAHIRR quantity referred to in Rules G417.9 and G717.11 requires computation of both a volume average reflectivity and an average cloud thickness. These quantities are then multiplied to produce the VAHIRR. Neither of these quantities is currently available as a product on the WSR-88D and WSR-74C radar systems used to support launch operations. This instruction provides a methodology for evaluating VAHIRR criteria with currently available radar products. The methodology provides a result that is more conservative than a direct VAHIRR computation, but it should still permit the launch customer to reap much of the benefit of the new rule.

Part I. <u>Determination of average cloud thickness</u>. The definition of the VAHIRR requires determination of the average cloud top and the average cloud base above the height of the 0C isotherm within a square having sides 5.5 km (3 n. mi.) north, east, south, and west of the ground projection of each point in the flight track. Average cloud thickness is defined as the difference of these two numbers. If the average cloud thickness cannot be determined at each point on the flight track, the maximum thickness within 5.5 km (3 n. mi.) of the flight track may be used.

To determine the average cloud top height, the launch weather team may use any existing radar product that gives the height of the 0 dBZ cloud top, including "maximum height of reflectivity" and cross section products. If the average height cannot be determined, the maximum height of the 0 dBZ reflectivity may be used. If the maximum height cannot be determined, use 18 km (60 kft) for the average cloud top.

The average height of the cloud base should be derived from radar data. It is the average of the higher of (a) the bottom of the portion of the cloud producing a radar reflectivity of 0 dBZ or greater or (b) the height of the 0C isotherm where the 0 dBZ reflectivity extends below that level. If the average height of the cloud base cannot be determined, use the higher of (a) the height of the 0 degree C isotherm or (b) the lowest portion of the cloud producing a radar reflectivity of 0 dBZ or greater anywhere within each 5.5 km (3 n. mi.) square defined above.

Part II. <u>Volume Averaged Radar Reflectivity (VARR)</u>. There is no operationally feasible way to use existing radar products to compute a volume average reflectivity. A conservative substitute is the maximum reflectivity since the volume average will always be smaller than the maximum. The WSR-88D has a "User Selectable Layer Composite Reflectivity (URL)" product and the WSR-74C has a "Max" product with user selectable base and top. The launch weather team should configure these products with the bottom of the layer at the height of the 0 degree C isotherm and the top above the height of the highest radar beam within 7.8 km (4.2 n. mi.) of the ground-projected flight track in the scan strategy being used. The WSR-88D product will have to be configured at the RPG and included in the product scheduler for the PUP used by the launch weather team.

Part III. <u>Evaluating the constraint</u>. The VAHIRR constraint is satisfied for a point on the flight track if the "URL" or "Max" radar product (see Part II above) everywhere within the corresponding square (defined above) is less than 10 dBZ-km divided by the average cloud thickness in km within the same square (see Part I above). (In English units, this threshold would become 33 dBZ-kft divided by the average cloud thickness in kft.) This constraint must be satisfied for every point on the flight track.

# Appendix II    A Detailed LAP/LLCC Chronology 1991 – 2009

(Provided by John Willett)

The following sources were used to create this chronology of LAP deliberations and resulting LLCC changes:
1) A collection of *Shuttle* LLCC versions compiled for this purpose by John Madura.
2) All available LAP-published and/or formally recommended LLCC versions.
3) Meeting and teleconference notes from John Willett's files, supplemented by some input from other authors of this document.

All significantly changed LLCC sets are indicated by their release date, name as used throughout this document, LCN number, and Appendix I reference, beginning with the first release in 1991, LCN 00048 released on 01/28/91, immediately below. (Changes applying only to non-lightning LCC, as well as trivial changes of the LLCC themselves, are omitted.)  For each such new LLCC set, the significant changes from the previous version are briefly described. (The full text of the LLCC versions are not given here -- see the references given to Appendix I.)  Whenever possible, significant meetings, telecons, and publications by the LAP are also indicated by approximate date and described as to subject matter and important conclusions reached or recommendations made.

Note that the operational Shuttle rules that are primarily analyzed herein have never contained a Triboelectrification Rule, although it has been included in the PRC/LAP recommendations since at least 12 May 1995, reportedly because the Shuttle was already hardened against triboelectric charging. Reportedly a Triboelectrification Rule was included in all versions of the ELV LLCC since AC 67, but we have not attempted to distinguish between ELV and Shuttle rules in this appendix.

An Electric Fields Aloft Rule, originally recommended by Heritage (1988, Section 7.2) and continually endorsed by the PRC from 11 March 1994 through at least January 1999, has never been included in any operational rules since ASTP and Viking. See Section 5.5.6 for more on this topic.

28 January 1991 -- Space Shuttle LLCC (1991) [LCN 00048, see Appendix I, Section A1.9.1]. This is our starting point for future rule comparisons in this appendix. Many of these LLCC are similar or identical to the recommendations of Heritage (1988, Chapter 7) (see Appendix I, Section A1.8). Here we briefly indicate the important differences:

> The Field-Mill Rule (C):
>
>> This rule appears to use a different definition of Electric Field Intensity. The Heritage Committee specified "the absolute value of the 1-minute time average of the instantaneous, ground-level electric field intensity."  The present rule, on the other hand, refers to "the one minute average of absolute electric field intensity at the ground."  (See also the entry for 25 April 1995, Space Shuttle LLCC (1995) [LCN 00520R03] below.)
>>
>> The present rule replaces the second part of the original Heritage-Committee exception (involving electric-field measurements aloft) with "Smoke or ground fog is clearly causing the abnormal readings."
>
> The present Debris-Cloud Rule adds both an overall three-hour-after-detachment exception and a field-mill/radar-reflectivity exception to the 5 NM standoff.

The Heritage Committee's electric-fields-aloft exceptions for penetration of Cumulus Clouds warmer than -20 C (B1 and B2), for the 10 NM standoffs from Thunderstorms or Cumulus Clouds colder than -20 C (B3 and B4), and for the Debris-Cloud, Disturbed Weather, and Thick Cloud Rules (F, E, and D, respectively) have been eliminated. [See the PRC's attempt to re-institute parts of this exception in the entry for 15-17 February 1994 and in Koons and Walterscheid (1996).]

The Heritage Committee's Triboelectrification Rule has been dropped. [One of the LAP's attempts to document this rule can be found in Krider *et al.* (1999).]

23 Oct 1991 -- A telephone conference involving most current members of the LAP (then, PRC, including at that time Dye, Koons, Krider, Rust, Walterscheid, and Willett) was called by Jack Ernst and Bill Bihner (NASA/WSO) to consider relaxing the Surface-Field Rule (C). The telecon was held in the form of a vote on a specific change proposal that, not surprisingly, was defeated.

Discussion, apparently all by telephone and fax, continued through the winter, however. The help of the 45th WS was solicited and obtained in gathering supporting data, and the change below eventually evolved. This precedent began a push to enumerate specific exceptions in which a known, near-surface, electrification mechanism could be identified as responsible for elevated field readings. Eventually (see entry for Fall of 1996), Dave Rust and Tom Marshall (University of Mississippi) were involved in obtaining near-dawn field profiles with a tethered balloon to document a shallow layer of enhanced field that was apparently caused by a "sunrise effect." This eventually worked its way into a version released on 12 May 1997 [LCN 00765R03] (further below).

20 March 1992 -- Space Shuttle LLCC (1992) [LCN 00230R04, see Appendix I, Section A1.9.2]. Minor Change:

First introduction of the term, "transparent," without a definition, into the body of the rules. Here it is included in a "note" relative to rule C(1). ("Transparent" had previously been included only in the definition of "anvil.")

Added two new "notes" to the Surface-Field Rule (C):

C(1) "Unless there are no clouds within 10 nautical miles of the launch site" has been relaxed to allow "thin fibrous (optically transparent) clouds" or "less than or equal to 2/8... (CU/SC/ST) clouds with tops below or equal to the +5 degree Celsius temperature level," as long as they have not been associated with a thunderstorm or CU with tops higher than -10 C in the last 3 hr.

C(2) "Unless smoke or ground fog is clearly causing abnormal readings" has been relaxed provisionally, pending more data, to also allow "a maritime inversion" that causes affected mills near the ocean to read between +1 and +1.5 kV/m.

3-5 August 1992 -- MSFC's analysis of their summer ABFM I dataset resulted in proposed changes to the existing LLCC, and Jack Ernst and Bill Bihner (NASA/WSO) organized a face-to-face meeting of the PRC at MSFC to consider them. The PRC signed off on a revised set of rules (outlined below), which were apparently never implemented, and took the opportunity to go on record that "we believe the best way to insure safety from atmospheric electricity hazards, and also to improve launch availability, is to utilize an instrumented aircraft in conjunction with the ground-based field mill network to measure the electric field environment and its time development along and near the flight path." Significant changes that were adopted by the PRC were the following:

A "relative to" exception to cumulus clouds with tops colder than +5 C (B1) that allows penetration if both their radar reflectivity is low and they are verified safe by at least one nearby field mill.

A "relative to" exception to cumulus clouds with tops colder than -10 C (B2), that would allow flight between 3 and 5 nm from the *visible* edge if they were verified safe by the ground-based field-mill network, was proposed but not adopted, pending additional data. (This exception was eventually rejected.)

One of the "relative to" exceptions to no clouds within 10 nm in the presence of elevated fields (C1) was changed from "thin fibrous (optically transparent) clouds" to simply "optically transparent clouds."

The "relative to" exception to smoke's or ground fog's clearly causing elevated fields (C2), which had allowed surf and/or maritime-electrode-effect charge, was deleted. The PRC reminded NASA that this exception had been originally approved as an interim measure while further research was done on the phenomenon, that no new information had been forthcoming, and that recalibration of the field mills had recently eliminated most of the problems with this phenomenon.

Several new definitions were added, and "Debris Clouds" was significantly revised.

There was discussion of substituting the "signed one-minute average of the instantaneous electric field," plus the RMS deviation around that average, for the existing "one minute average of the absolute electric field" in both the rules and the measurement software, but this was deferred and never done.

Concern was expressed that the $10^{-4}$ acceptable level of risk that was desired for Shuttle by NASA/WSO might never be reached using ABFM data.

During 1993 several telecons were conducted (15 April 1993, 18 June 1993, 22 November 1993) regarding a MSFC proposal that was based on the full ABFM I dataset, now including the winter campaign. MSFC wanted to replace the thick-cloud and disturbed-weather rules (D & E) with a single, radar-based rule involving a threshold in vertically summed reflectivity ("VIR4"). Aerospace statisticians, working with the USAF, found that the available data did not justify the proposed new rule, and it was ultimately rejected by the PRC. Nevertheless, MSFC did convince the PRC that some form of radar rule was warranted, given a better statistical justification, and that such a rule could be very effective in reducing false alarms.

15-17 February 1994 -- A face-to-face PRC meeting was called by NASA at KSC. On the table was a NASA/MSFC-proposed set of rules containing several changes from the version approved by the PRC on 3-5 August 1992 (but never implemented). Notable were the reduction of standoff distances for B3, B4 and C1 clouds from 10 nm to 8 nm (which the PRC had previously questioned), retention of the exception relative to C(2) for a maritime inversion (which the PRC had specifically removed in 1992), replacement of the thick-cloud rule (D) with a radar rule involving VSR0C (a revised, vertically summed radar parameter), and specification of a radar threshold of 29 dBZ for "moderate or greater precipitation." Ultimately these four specific proposals were all rejected. There was also a wide-ranging discussion of a number of topics, some of which are mentioned below, which resulted in an extensive list of recommendations, both by and for the PRC. These recommendations were summarized in an 18 February 1994 letter from John Willett to Harry Koons.

The 45th WS (USAF) raised significant concerns about complexity, ambiguities, and inconsistencies in the LLCC and offered a number of suggested changes.

The PRC was asked to consider the flight rules for Shuttle, which are maintained and enforced at NASA/JSC/SMG.

Also at this meeting John Madura first formally presented data on the fraction of ELV scrubs due to various causes -- 32% due to weather -- and the fraction of weather scrubs for all vehicles that were due to various causes -- 38% due to LLCC.

Led primarily by Harry Koons, the PRC undertook a complete rewrite of the LLCC, primarily to address issues of clarity and ambiguity. By 11 March 1994 the PRC had signed off on a new LLCC version (Jim Dye apparently attended this meeting but was not a signatory on this document. Jim left the PRC somewhere around this time.) Noteworthy was the adoption of a new "Rule G," similar to a rule originally proposed by the Heritage (1988, Section 7.2). Rule G, as written, permitted exceptions to the thick-cloud, disturbed-weather, and debris-cloud rules (D, E, and F) if an operational ABFM showed no fields higher than a specified, altitude-dependent threshold along and near the flight path within the prior 15 minutes. This proposed version is documented by Koons and Walterscheid (1996 -- see entry for 1 February 1996, below, and Appendix I, Section A1.9.3) and was similar to the implemented version, "25 April 1995, Space Shuttle LLCC (1995) [LCN 00520R03]," listed below, except for the omission of "Rule G" and some major issues discussed with reference to John Madura's 12 May 1995 letter (below). Surprisingly it did not contain a triboelectrification rule.

6 October 1994 -- A PRC telecon reviewed and either approved or requested more information about several changes proposed by NASA/USAF. Available records do not indicate whether any of these proposals were adopted by the PRC.

12 May 1995 -- John Madura circulated a letter containing new rules ("25 April 1995, Space Shuttle LLCC (1995) [LCN 00520R03]," below) that had apparently already been implemented on the NASA side but may have been still pending on the USAF side.

This letter unleashed a fire-storm of protest from the PRC for several reasons:

"Rationales" had been added to the rules, apparently as part of the rules proper, without PRC knowledge or review. These rationales were felt to introduce the opportunity for ambiguity with, if not outright contradiction of, the carefully worded language of the rules. Further, some of these rationales were seen as misleading or incorrect.

The "note" at the end of the field-mill rule (C) was new to the PRC and appeared to contradict their intent with respect to failures of individual field mills.

The ABFM rule (G) had been dropped, apparently because no suitably instrumented aircraft was available, its cost effectiveness had not been established, and/or no detailed operations plan for its use had been worked out.

The triboelectrification rule had been dropped on the NASA side as irrelevant to Shuttle, although it was apparently still maintained on the USAF side.

A new definition of "documented" had been added without PRC knowledge or review (although it seemed basically OK).

All of this led to an official "letter of concern" from the PRC at the end of June of 1995. It also led to the PRC's raising the question of what its ongoing role might be in the rule-making process. Could an official charter be created and a definite membership established? Finally it was probably the motivation behind the Aerospace LLCC report -- see 1 February 1996, below.

25 April 1995, Space Shuttle LLCC (1995) [LCN 00520R03, see Appendix I, Section A1.9.4]. -- MAJOR Change (more in the way of a reorganization and clarification than a substantive relaxation):

Wholesale introduction of the term, "nontransparent," into several rules (B(3), B(5), and F(1), as well as retaining it in C(1)). Now it is defined in the Definitions.

Numerous minor wording changes throughout (*e.g.*, "launch site or planned flight path" was replaced with "flight path"); only substantive changes are noted below.

"Rationale" (*not* written nor reviewed by the LAP) were added to each of the rules IN ALL CAPS.

Cumulus Cloud Rule (B):

B(1) "Through cumulus clouds with tops higher than the 5 degrees Celsius level" has been separated into two temperature ranges. Between +5 and -5 C penetration is now allowed if three conditions are satisfied (the cloud is not precipitating, a working field mill is close enough to the cloud, and the relevant mill readings are between -100 and +1000 V/m). (Between -5 and -10 C penetration is still prohibited.)

B(3) "Cumulus clouds with tops higher than the -20 degrees Celsius level" has been replaced with "cumulonimbus or thunderstorm cloud, including nontransparent parts of its anvil." "Thunderstorm" is now defined as "any cloud that produces lightning." "Transparent" (but not "nontransparent") and "anvil" are also now defined. ("Cumulonimbus cloud" was already defined in terms of the -20 deg. C level.)

B(5) has been added to prohibit flight within 10 NM of a nontransparent *detached* anvil for the first 3 hr after detachment.

A "note" has been added excluding Altocumulus or Stratocumulus from this rule.

Field Mill Rule (C):

The reference to "launch site" in the body of this rule has been replaced with "flight path."

The traditional phrase, "the one minute average of absolute electric field intensity," in the body of the rule has been replaced with "the absolute value of any electric field intensity measurement," and a new definition of "Electric Field" now specifies "the one-minute arithmetic average of the vertical electric field." This would *appear* to be a substantive change from the average of the absolute value to the absolute value of the signed average.

In place of the cumbersome "Relative to..." "notes" at the end of the rule, formal exceptions have been added as follows:

C(1) "Thin fibrous (optically transparent) clouds" has been replaced with "transparent clouds," and "less than or equal to 2/8 cumulus/stratocumulus/stratus (CU/SC/ST) clouds" has been replaced with "clouds."

C(2) "Smoke or ground fog is clearly causing abnormal readings," including a maritime inversion via the "note," has been generalized to "a known source of electric field (such as ground fog)... previously determined and documented to be benign, is clearly causing abnormal readings." A lengthy and detailed definition of "documented" has also been added.

Disturbed Weather Rule (E): "Are associated with disturbed weather within..." has been expanded to "are associated with disturbed weather that is producing moderate (29 dBZ) or greater precipitation within..." The previous definition of "disturbed weather" has been deleted.

Debris Cloud Rule (F) has been clarified and slightly relaxed via introduction of several more sections:

F(1): "Do not launch through thunderstorm debris clouds" has been expanded to "Do not launch... through any nontransparent thunderstorm or cumulonimbus debris cloud..." The definition of "Debris Cloud" has also been clarified to explicitly include the decay of a parent CB or TRW.

F(2): "...or within 5 nautical miles of a thunderstorm debris cloud not monitored by a field mill network or producing radar returns greater than or equal to 10 dBZ" has been separated into three sections (a, b, and c) that specify explicitly what is meant by "monitored" and clarify what is meant by "producing a radar return."

Two definitions have been deleted, four retained with changes, and eight new ones added.
The two all-caps paragraphs at the end, which could be considered a rudimentary rationale, have been removed, presumably because they were subsumed into the newly written rationales for each rule.

1 February 1996 -- Publication of Koons and Walterscheid (1996) documents the rule version that was formally adopted by the PRC in March 1994, without any of the subsequent tweaks. The plan was to issue a new version in this report series each time the PRC approved a new rule revision, although only one such new version has been forthcoming to date -- see entry for 15 January 1999. This report lists the full membership of the PRC as Dye, Koons, Krider, Rust, Walterscheid, and Willett. It does not include a triboelectrification rule.

Summer of 1995, Summer and Fall, 1996 -- As recommended by the PRC during its 15-17 February 1994 meeting, experiments were conducted at KSC to study the conditions that had been causing "benign" violations of Rule C (elevated fair-weather electric fields at the surface) in apparently good weather.

Phil Krider [Studies of atmospheric electricity at the NASA Kennedy Space Center, Progress Report and a Supplemental Proposal for Grant NAG10-0092, 9 January 1996; Studies of atmospheric electricity at the NASA Kennedy Space Center (KSC) and the USAF Eastern Range (ER), Progress Report and Supplemental Proposal for Grant NAG10-0092, 24 February 1997] surveyed the electric-field distribution downwind from the surf zone and from high-voltage power lines, recorded electric-field enhancements during fog, and studied the electric-field and space-charge behavior during the "Indian River anomaly," or "sunrise effect." The only conditions suggested to be of possible concern were the elevated fields that sometimes occur near high-elevation, high-voltage power lines during persistent ground fog.

Marshall et al. (1999) made electric-field soundings with a tethered balloon to study the above "sunrise effect." They found that in the fields aloft are not as big as those at the ground and that enhanced fields extended less than 150 m above ground.

Based on the above reports, several of the potential sources of elevated fields in fair weather were characterized, if not fully understood, and were determined to be benign. This determination was based on the limited vertical extent and/or local source of the phenomena.

6 September 1996 telecon -- preliminary look at the Marshall and Rust balloon soundings of "sunrise effect."

Dave Rust reported that the fair-weather atmosphere was not repeatable, and that there have been no events above 1 kV/m although several above 500 V/m have occurred. The "effect" takes about an hour. Pre-sunrise there is a layer about 10m deep with fields near 100 V/m. At sunrise the charge density lowers and the sunrise layer deepens to about 300m. Typically less than 200 V/m. We don't understand the physics. Dr. Krider noted that these observations were consistent with his aspirator measurements of the charge density. Thought totally harmless to launches. Paul Krehbiel concurred. Speculations on possible mechanisms were offered by Rust and Merceret. Tom Marshall and Krider agreed that more data analysis was required. Additional discussions were held regarding the definitions of "fair weather" and "documented."

19 November 1996 telecon -- reviewed Marshall and Rust data/interpretation from tethered-balloon soundings in the "sunrise effect."

4 February 1997 -- LAP meeting at Aerospace Corp. in Los Angeles:

The current PRC members (Krider, Koons, Rust, Walterscheid, Willett) agreed to a charter for the re-named Lightning Advisory Panel (LAP), drafted by an "Interagency Review and Coordinating Committee" (IRCC, made up of 45WS, 30WS, KSC Weather Office, SMG, and AFMC/SMC), which appoints the LAP's members, renews its charter every three years, and to whom it reports.

Proposed exceptions to Rule C (surface fields) were discussed. It was agreed to allow specific exceptions for benign smoke, ground fog, and sunrise effect, with the provision that the elevated fields were positive and did not exceed 1500 V/m. Documentation was drafted and approved, defining each of these conditions and specifying how they could be recognize operationally.

Further discussion of a radar-based replacement of the thick cloud rule (D) was inconclusive because of lack of reliable radar data.

Harry Koons expressed concern about the potential vulnerability of space-launched payloads to sprite electric fields in the middle atmosphere.

12 May 1997 [LCN 00765R03] (not included in Appendix I) -- Minor Change:

In the Field Mill Rule (C), a parenthetical sentence has been added to C(2) pointing to the location of documentation relevant to benign ground fog, smoke, and sunrise effect.

8-9 January 1998 LAP Meeting in Tucson:

LAP charter was still under revision and had not yet been approved by IRCC.

First major participation of 45WS in the LAP process: 45WS proposed LLCC that were significantly reorganized (and to some extent changed in content) from the LAP original, and they requested clarification of (or changes to) the existing language/definitions. (The triboelectrification rule appears in their revision.) This degree of involvement was welcomed by the LAP, as it provided valuable feedback on the operational interpretation and usefulness of the rules. The basic idea of the 45WS reorganization appeared sound, but some of the original nuances were lost, leading to changes in content and indicating lack of understanding of the original intent. As a result the LAP undertook another major re-write of the rules.

John Willett and Mike Maier both presented estimates of the cost savings that might accrue from an operational ABFM. Although historical data on the cause of launch scrubs/delays is reliable, information on the cost of such scrubs/delays probably is not, and requests to log the frequency of rule violations (in the absence of actual launch countdowns) have not been successful.

It appears to have been in this meeting that the LAP decided to eliminate the "known cause" exception from Rule C (along with the "documentation" for benign smoke, ground fog, and sunrise effect), in favor of raising the field threshold to |1500 V/m|.

27-29 April 1998 LAP Meeting at KSC:

LAP charter was still under revision and had not yet been approved by IRCC.

This meeting was devoted primarily to working out the open issues in an LLCC version that had been drafted after the January meeting.

Issues involving the ground-based field-mill network were also addressed -- especially a gain shift and associated failure of the self-checking feature, use of measured (as opposed to interpolated) values, and maintenance of the sites.

6 May 1998 LAP Telecon: John Madura had applied considerable pressure to the LAP on 1 May by reporting that Don McMonagle (Shuttle Launch Integration Manager) wanted "the final fully staffed version [of the LLCC] briefed at the STS 91 Flight Readiness Review on 20 May." This began a frantic effort, documented in a flurry of e-mails and at least seven more revisions, both before and after this telecon, extending through the 20 May deadline. The final LAP-approved version ("LCC-K2.doc," distributed Wed 5/20/1998 10:41 AM EDT) is documented in the second Aerospace LCC Report -- see entry for 15 January 1999, below. Astonishingly (considering the extreme time pressure) this LAP-approved version differs from operational revision 824R03 (below) in only two significant ways:

Proposed Rule 9 (Electric Fields Aloft -- the old "Rule G") and its corresponding definition still are not included in the operational version.

Proposed Rule 10 (Triboelectrification) also is not included in the operational version.

4 June 1998, Space Shuttle LLCC (1998) [LCN 00824R03, see Appendix I, Section A1.9.5] -- MAJOR Change: Complete re-write of the rules to clarify and improve logic. Here we identify only the substantive (as opposed to the formal wording) changes:

All rationale have been eliminated from the rules.

Use of the term, "nontransparent" has been further extended to Rules 2(b), 8(b), 6, and 5.

Lightning Rule (A) has been split into two parts and made more explicit as follows: "...any type of lightning is detected... unless the meteorological condition that produced that lightning has moved more than 10 NM away from the flight path" has been divided into two parts:

1(a) "...any type of lightning occurs in a thunderstorm... within 10 NM of that thunderstorm" (*i.e.*, it's the thunderstorm that is to be avoided), and

1(b) "...any type of lightning occurs within 10 NM..." (*i.e.*, it's the lightning itself that is to be avoided), unless two conditions are satisfied to assure that that lightning did not deposit significant charge near the flight path.

Two "notes" are added to assure that any cumulus that remain after 1(a) are covered under the Cumulus Rule (2), and that anvils are covered under the new Anvil Rule (3) -- see below.

Cumulus Rule (B): This rule has been completely re-ordered, but the only substantive changes are as follows:

B(1) becomes 2(d), in which the three "unless" requirements are changed in two ways:

The horizontal distance between the cloud and the nearest working field mill has been changed from "from the farthest edge of the cloud top... is less than the altitude of the -5.0 deg. C level or 3 NM whichever is smaller" to "from the center of the cloud top... is less than 2 NM."

The allowable range of field measured at the relevant mills has been narrowed from (-100 V/m and +1000 V/m) to (-100 V/m and +500 V/m).

B(4) becomes 2(b), in which "cumulonimbus or thunderstorm cloud, including nontransparent parts of its anvil" has been replaced with "cumulus cloud with its cloud top higher than the -20 deg C level."

B(5) has been dropped entirely. The obvious intent of the LAP was that anvils (either attached or detached) are not to be covered by this rule.

Field Mill Rule (C) has been re-ordered and relaxed to become (8):

C(1) (no launch for 15 min after 1000 V/m readings at relevant mills) has become 8(b), the only change being that the second "unless" condition (clouds lower than +5 °C not associated with towering CU) now also refers only to nontransparent clouds.

C(2) (known sources such as ground fog) has been replaced with 8(a), which now prohibits flight only if the relevant field readings are above |1500 V/m|.

Thick Cloud Rule (D) has been re-ordered to become (6) and has been clarified and relaxed as follows:

Launch through transparent parts of the cloud layer is no longer prohibited.

6(1) This section appears to prohibit launch through parts of the cloud layer that are *both* >4500 ft thick and contain temperatures between 0 and -20 °C.

6(2) This section prohibits launch through a cloud layer that is *both* >4500 ft thick and contains temperatures between 0 and -20 °C within 5 NM of the flight path.

An "unless" clause has been added to the entire rule that exempts cirriform clouds that have never been associated with convective clouds, are entirely colder that -15 °C, and show no evidence of liquid water.

Disturbed Weather Rule (E), now (5):

A definition has been added for a "weather disturbance."

Now applies only to nontransparent clouds.

The radar threshold for "moderate precipitation" has been moved to a new definition of that term and increased slightly to 30 dBZ.

In addition to "moderate precipitation" the penetration prohibition has been extended to "a radar bright band or other evidence of melting precipitation." A new definition is added for "bright band."

Debris Cloud Rule (F) has been re-ordered to become (4): This rule has been substantially re-worded, but the only substantive change is to relax an "unless" condition from F(2)(c) "the maximum radar return from the entire debris cloud is less than 10 dBZ..." to 4(b)(3) "the maximum radar return from any part of the debris cloud within 5 NM of the flight path is less than 10 dBZ..."

Anvil Rule (3): This completely new rule has been added to cover and extend parts of previous rule sections, B(4) and B(5).

3(a) Covers attached anvils in place of the previous practice of considering them as part of the CB in B(4). This allows the rule for attached anvils to be made less restrictive than that for CBs, as follows:

3(a)(1) The former prohibition on penetrating non-transparent, attached anvils is retained, but

3(a)(2) flight between 0 and 5 NM is allowed 3 hr after the last lightning discharge in the storm, and

3(a)(3) flight between 5 and 10 NM is allowed 30 min after the last lighting.

3(b) covers detached anvils, in place of the previous B(5). This allows the rule for attached anvils to be fine-tuned. relative to that for CBs, as follows:

3(b)(1) The prohibition against penetration has been made more restrictive by increasing the waiting time from 1 hr to 3 hr (although after 1 hr a detached anvil might formerly have been considered a debris cloud and therefore still had the 3 hr prohibition.

3(b)(2) An entirely new requirement has been added not to penetrate a detached anvil for 4 hr after its last lightning discharge.

3(b)(3) The prohibition against flight between 0 and 5 NM has also been made more restrictive by increasing the waiting time from 1 hr to 3 hr *unless* three conditions are met, requiring adequate monitoring by at least one field mill and a radar return from the relevant portions less than 10 dBZ for 15 min.

3(b)(4)  The prohibition against flight between 5 and 10 NM has been relaxed by decreasing the waiting time from 1 hr to 30 min, but a lightning discharge in the parent cloud resets the clock.

It is explicitly stated that detached anvils are never to be considered debris clouds.

A totally new Smoke Plume Rule (7) prohibits launch through any CU that has developed from a smoke plume until 60 min after detachment. After 60 min the CU is covered by Rule 2.

Two of the previous definitions have been deleted or replaced, some of the others have been re-worded, and six totally new definitions have been added. The substantive changes are the following:

"Electric Field Measurement at the Surface" now explicitly requires the use of a ground-based field mill.

The radar threshold for "precipitation" (which replaced "precipitating cloud") has been raised from 13 dBZ to 18 dBz.

"Transparent" has been significantly changed to require visible assessment of transparency. A separate definition of "nontransparent" has also been introduced (which causes confusion later on).

15 January 1999 -- Publication of Krider *et al*. (1999) documents the rule version formally adopted by the LAP in May 1998. This report is the second (and currently the last) in a series intended to document each new rule revision that was approved by the LAP -- see previous entry for 1 February 1996. This report lists the full membership of the LAP as Koons, Krider, Rust, Walterscheid, and Willett. It includes both "Electric Fields Aloft" and "Triboelectrification" rules.

22-25 February 1999 LAP meeting at NASA/JSC in Houston -- Topics Covered in Depth:

Consider whether the Flight Rules protect the orbiter during landing. The LAP was briefed in detail on the flight rules and some of SMG's concerns. (It is not clear that the LAP ever submitted any coherent response to SMG.)

30WS and 40WS asked for and (mostly) received clarification of a number of details in the latest LLCC re-write (4 June 1998, Space Shuttle LLCC (1998) [LCN 00824R03], above). Open issues were held over to the next rule re-write.

Reports of sporadic reduced sensitivity on field mills in the LPLWS led to considerable LAP concern and to several recommendations for diagnosis and correction of the problem.

Preliminary planning of another ABFM II campaign specifically to develop radar/cloud-physics/electric-field relationships for development of new radar-based LLCC. Conclusion: conduct a workshop to plan that campaign in detail.

Hugh Christian became a new LAP member.

It was pointed out that the Federal Advisory Committee Act (FACA) would seriously cramp the LAP's style if it were constituted as an official advisory committee. LAP should consider becoming chartered under FACA as a subcommittee of one of the NASA Advisory Council committees.

1 December 1999 -- The LAP issued a formal recommendation to the KSC Weather Office for an ABFM II campaign to "validate physics-based relationships between the decay of cloud electric fields and cloud properties and to verify that these cloud properties, and also the presence of electric generators aloft, can be inferred from ground-based measurements."

25-28 January 2000 LAP Meeting at KSC:

The first planning meeting for the ABFM II Campaigns. (ABFM II is treated in detail elsewhere, but certain key ABFM II meetings are also listed herein because oversight of the experiment and the resulting data analysis was a LAP function.)

Discussion of Kodiak Launch Complex launch rules and measurements:  Incredibly high surface electric fields had been observed during brief graupel showers from very shallow clouds.

Miscellaneous follow-up of action items from the 22-25 February 1999 LAP meeting:  More information about LPLWS sensitivity decreases -- apparently mainly along the coast, associated with high humidity, and not reliably detected by the self-calibration function unless greater than about 25%. (It is not clear that this issue was ever resolved satisfactorily.)

12 February 2001 -- A letter of concern was sent by Phil Krider, in his capacity as LAP chairman, to the DOT Document Management System regarding new FAA-proposed "Lightning Flight Commit Criteria (LFCC)" -- Part 417, Appendix G, of Docket Number FAA-2000-7953. Harry Koons had been working with FAA/AST on their "translation" of the current LCC into FAA jargon, and he reported the inevitable changes in intent that had resulted. This was to become a major concern for the LAP over the next few years, with the eventual decision that the LAP should attempt to write its new rule versions in "FAA language," specifically to minimize editing by the FAA lawyers (see, for example, the entry for 6 October 2008, LAP 10/06/08 Recommendation in FAA Format).

31 August 2001 LAP telecon on launch rules at Kodiak. The LAP decided to draft some provisional "LLCC" specifically for Kodiak and discuss them in a second telecon.

5 September 2001 LAP telecon to mark up provisional "orographic" and disturbed-weather launch rules for Kodiak, which were submitted the same day as recommendations to the KSC Weather Office.

September 2001 through April 2008 -- Numerous ABFM II and joint LAP/ABFM II Telecons were conducted throughout this period. Only the actual meetings and telecons having significant results will be reported below.

27-28 November 2001 -- ABFM II Workshop at the Melbourne, FL, offices of ENSCO, Inc.:

Doug Mach and Bill Koshak reviewed A/C calibrations techniques and results at length.

Monte Bateman reviewed radar (WSR74C vs. NEXRAD) inter-comparisons.

Jim Dye, Eric Defer, and others reviewed status of the NCAR/ABFM Web site and some particularly interesting special cases.

Jim Dye began synthesis of particle concentration vs. field intensity in anvils.

John Willett reported on a theoretical approach to, and initial calculations of, field-decay times in anvils.

The LAP should be invited to the next ABFM II meeting as observers.

10 January 2002 ABFM II Telecon --

Hugh Christian officially transferred the title of Principal Investigator to Jim Dye.

Significant radar attenuation due to intervening precipitation was noted late on 10 June 2001: An attenuation warning should be included explicitly in next LLCC revision.

21 February 2002 ABFM II Telecon --

Monte Bateman argued for "vertically integrated reflectivity above 5 km (~0 C)" as a likely radar-threshold parameter for significant electrification.

Frank Merceret emphasized that an LLCC result is needed from ABFM II before the end of FY03 to meet funding commitments.

7 March 2002 ABFM II Telecon --

Beware radar geometrical cut off at low altitudes/long ranges.

26 March 2002 LAP Telecon -- The NASA/MSFC Altus Cumulus Electrification Study (ACES) experiment plan was outlined. MSFC requested LAP review of its lightning-strike avoidance plan.

4 April 2002 ABFM II Telecon --

Frank Merceret reported on attenuation due to wet radome.

Jim Dye advanced his conceptual model that strong fields are present when radar reflectivity is 10-15 dBZ or greater but that the converse is not true.

12-14 November 2002 -- Joint LAP/ABFM II Workshop at the Melbourne, FL, offices of ENSCO, Inc., to review the progress to date in data reduction and analysis and to discuss its applicability to modification of the LLCC. (Here we summarize only the LAP recommendations made in response to the progress review. For more information, refer to "Nov '02 Workshop/Workshop Summary" on the "Reports" page at "http://abfm.ksc.nasa.gov/".) John Willett's Summary of LAP Recommendations (edited and reported by Frank Merceret) follows:

1) The ABFM Analysis Team (AAT) urgently needs a final conclusion from the MSFC group about which analysis algorithm, the Mach (M) algorithm, the Koshak (K) algorithm, or a combination of these algorithms (under what conditions), should be used to provide the best estimate of the ambient electric (E) field in clouds that were sampled during the ABFM campaign.

2) The AAT should "sanitize" the cloud E-field/radar-parameter dataset so as to minimize the effects of the following: [These later evolved into a set of "filters" that could be applied, individually or in combination, to the full dataset in order to remove undesirable cases.]

A) Nearby lightning as a non-local source of the field -- use a 10 nm (18 km) standoff distance (the same as used in LCC Rule 1) and a standoff time TBD by the AAT;

B) Nearby cores of active thunderstorms that represent non-local sources of electric field – using altitude/radar thresholds and standoff distances TBD by Analysis Team;

C) Radar scan gaps, to the extent that they significantly influence the various radar parameters; and

D) Wet-radome and/or precipitation-attenuation effects, to the extent that they significantly influence the various radar parameters.

3) Further explore the cloud reflectivity parameters (*e.g.*, peak dBZ, average Z, average dBZ, box size) so as to optimize the potential relationships to ambient E-field in the "sanitized" dataset. Focus the analysis on thunderstorm anvils, thick clouds, and disturbed weather.

4) Further "sanitize" the cloud E-field/decay-time dataset so as to minimize, in addition to the effects listed in (2) above, any indications of a local electrical generator (*i.e.*, radar indications of local development of the cloud, TBD by the AAT).

5) Pursue, to the greatest practical extent, testing of the E-field-decay model using the measured cloud microphysical parameters and observed field decays. For this, we also need better estimates of the ambient winds and the parcel trajectories in anvils.

6) Look for relationships between the predicted decay times and the associated radar reflectivities in the further "sanitized" dataset detailed in (4) above. Determine whether there are other types of radars (or ground-based sensors) that could serve as a proxy for the decay time of the cloud electric field.

7) Check ALL anvils for the presence of liquid water.

8) Archive all data obtained in the current ABFM campaign (ABFM II).

9) Retrieve and, to the extent possible, archive data obtained during the previous ABFM campaign (ABFM I).

10) Integrate the ABFM I dataset with the ABFM II dataset in a coherent way. This may require re-calculating some of the radar parameters in ABFM II so that they match those previously calculated during ABFM I.

11) Re-examine in detail any specific cases that appear to be of concern in the ABFM I scatterplots.


8 January 2003 ABFM II Telecon -- An action item to use M-derived E fields only and to modify the quality control to set a flag if aircraft charge (presented in the same units as the electric field) is a factor of 10 or more larger than E.


6 March 2003 ABFM II Telecon -- Result of Frank Merceret's study of reflectivity averaging: Averaging of reflectivity measured in dBZ is strongly preferred to averaging in units of Z. Averaging the reflectivity before taking the logarithm results in a product excessively dominated by non-representative peak values.

**12-15 May 2003 -- Joint LAP/ABFM II Workshop at the Melbourne, FL, offices of ENSCO, Inc.:**

The ABFM II analysis was well along, although the team still hadn't selected the "best" radar parameter, nor the final suite of "filters" to remove cases clearly eliminated by other LLCC. Topics included an overview of the database and considerable discussion of individual cases, of the time-of-flight estimates vs. model decay times (including early indications of "redevelopment" of microphysics and electric fields within certain anvils), and of field-magnitude scattergrams for various radar parameters.

The LAP considered taking positions on these and other issues. Examples of key LAP recommendations are:

Frank Merceret and Jennifer Ward should correlate their objective determination of cloud edge with both visible transparency and radar reflectivity. (Results of this investigation were published in the January-February 2006 issue of *J. Spacecraft and Rockets*. They showed consistency among the cloud-physical, 0 dBZ radar, and visibly non-transparent estimates of "cloud edge.")

Harry Koons should apply "extreme-value analysis" to whatever scattergram(s) is (are) selected as most promising. This kind of analysis could replace the application of large safety margins (*e.g.*, 20 dB) to the observed extreme values.

IMPORTANT: The cloud boundaries should be redefined uniformly in terms of the 0 dBZ reflectivity contour, based on some ABFM data showing fields > 3 kV/m with reflectivities < 10 dBZ (the radar definition of cloud edge in the then-current LLCC).

A capability to display volume-averaged or -integrated radar parameters should be implemented for the Launch Weather Officers (LWOs) in the Range Control Center (RCC). This recommendation was embodied in a 20 May 2003 memo from the LAP to Col. Wyse, Commander, 45WS.

Frank Merceret should use the existing "merged" ABFM II data files to test the effectiveness of including more than one predictor for field magnitude.

The NCAR team should develop and apply a subjective filter to remove cases that are within 10 nm of convective cores -- towering cumulus and cumulonimbus rule violations -- in addition to the already-existing lightning and radar-attenuation filters. They should also compute an additional radar parameter, Volume Integral Reflectivity, and both of these new parameters should be added to the "Merged Files."

Move toward a radar parameter that incorporates cloud volume and/or thickness in addition to average radar reflectivity.

Somebody should research possible remote-sensing proxies for the theoretical electrical decay time, which has been found to be dominated by particles in the size range, 0.2 - 2.0 mm. (In a "white paper dated 31 July 2003 triggering field John Willett showed, for spherical particles, that electrical decay time is proportional to optical extinction coefficient. The latter is unfortunately still a local parameter, but one more susceptible to remote measurement.)

45WS presented a number of new questions of rule interpretation to the LAP for feedback. Examples of the most urgent questions and LAP responses are:

Does moisture on the windscreen of the reconnaissance aircraft constitute "precipitation?" No.

Is the mid-level blow-off from a cumulus with its top lower that the -20 °C level considered an anvil, a debris cloud, or what? Discussion of this is in process as of the 21 May 2003 LAP telecon.

Does the thick-cloud rule apply to cumulus clouds? Discussion of this is in process as of the 21 May 2003 LAP telecon.

Most participants left the meeting with several action items. Examples of the most critical ones not already included under the above LAP recommendations are:

The LAP should develop an interim version of its planned radar-based anvil rule that can be applied immediately, based on radar information already available to the LWOs, because it will take some time for the range to implement calculation/display software for whatever radar parameter is finally selected for the new rule(s). (See entry for 2 October 2003, below, for the LAP response.)

The LAP should review and comment on the latest FAA re-write of the "LFCC" for accuracy and logical consistency *re.* the existing LLCC. (This re-write was provided to Phil Krider in an e-mail from Denny Mathew (FAA/AST) dated 5 May 2003.)

21 May 2003 LAP Telecon -- Held to discuss rule revisions (starting from the LLCC as documented by Krider *et al.* [1999]) related to the 45WS questions and other issues from the 12-15 May 2003 meeting:

Re-define "transparent?" Harry Koons will suggest revisions...

Change "convective" to "cumulonimbus" in anvil definition?

Harry will draft a new FAA definition of "thick cloud layer."

Remaining 45WS and FAA issues will be deferred until July.

29 May 2003 LAP Telecon -- Continuation of the discussion of LLCC revisions begun on 21 May 2003 resulted in the following decisions:

Don't change the "anvil" definition because we are worried about any Cu's above -5 °C becoming electrified.

Do change all cloud boundaries to 0 dBZ. Also add a "cloud base" definition.

Defer any change in the definitions of "cloud layer" and of "thick cloud."

Make only minor changes to the "transparent" and "nontransparent" definitions. Defer a major re-examination of this issue.

Include a "note" under the definition of "precipitation" that specifically includes graupel.

Adopt a new definition of "field mill."

As a result of this discussion, Harry Koons put out another draft of the definitions on 29 May 2003. Although everyone seemed pretty happy with this version, there is apparently no record of any vote or formal approval. (Richard Walterscheid offered a couple of additional suggestions from Germany on 30 May 2003.)

17 July 2003 Joint LAP/ABFM II Telecon --

Another "filter" is added to the anvil dataset that removes those with radar bases (or falling precipitation) $\leq 5$ km.

Need to define an electrical length for Titan, hence an electric-field threshold for triggering hazard, to use in the extreme-value analysis of anvil scattergrams -- see next telecon.

Need to review the latest FAA rules for next telecon.

28 July 2003 Joint LAP/ABFM II Telecon -- Consideration of provisional conditions for triggering by launch vehicles from John Willett.

21 August 2003 -- Having received an indication of interest from the KSC Weather Office, John Willett suggested to Nickolay L. Aleksandrov, who had recently presented a lightning-triggering model at an international meeting that they both attended, that Prof. Aleksandrov and his Russian colleagues (Edward M. Bazelyan and Yuri P. Raizer -- both well-known experts on long-spark breakdown) develop a proposal for a small contract from NASA/KSC. The specific question of interest to NASA was whether it is possible to set reliable lower bounds on the electrical conditions as a function of altitude for lightning initiation by either a launch vehicle during boost phase or a re-entry vehicle (*e.g.*, the Space Shuttle) during glide. The altitude range of interest was specified as 0 - 10 km, with special attention to the triggering conditions at anvil altitudes. The final Russian proposal was submitted on 19 October 2003, but for various reasons it was not funded until 1 February 2005. (The resulting final report is referenced in the entry for 4 November 2005.)

29 September 2003 LAP Telecon defined the essential objections to the latest FAA LFCC.

21 October 2003 LAP Telecon -- Deadline is this week to review Harry's comments on the FAA "LFCC" re-write. There was discussion of the Russian proposal to determine a lower bound on the danger threshold aloft. Harry Koons presented an example extreme-value analysis.

31 October 2003 -- The LAP forwarded to the FAA its official response to the latest FAA version of the rules. Major LAP objections included the following:

Restore the "clear and convincing evidence" requirement to the "General" section.
Restore the requirement (in the definitions) for ground-based field mills to measure the surface electric field.
Restore the *caveat*, "Detached anvil clouds are never considered debris clouds," to the Detached Anvil rule.
Delete the "Electric Fields Aloft" rule. (This is the first available indication that the LAP had withdrawn its earlier strong support for this rule.)

16 December 2003 ABFM II Telecon -- Jim Dye is again a member of the LAP, having rejoined sometime in 2003 after leaving it sometime in 1994. Harry Koons presented extreme-value-analysis calculations for the

3X3X3 km cube radar averages. It was concluded that the 11X11 km column average (presented on or after 30 October 2003) was a better radar parameter than the 3X3X3 cube.

30 March 2004 Joint LAP/ABFM II Telecon --

It is concluded that radar parameters based on volume integrals are better than reflectivity averages for discriminating safe from unsafe anvils, and that the best way to approximate volume integrals is to multiply column averages by column thicknesses.

A question remains about what lower cut off to use for reflectivity in individual radar cells: -10 dBZ, as currently, or 0 dBZ.

More discussion about the FAA rules, especially the question of what equipment is required by the "General" section 417.1(b).

6 April 2004 LAP telecon with Titan folks about modeling exhaust-plume lengths.

13 April 2004 Joint LAP/ABFM II Telecon --

Apparently there was some discussion of the latest (12 April 2003) FAA LFCC version -- the LAP is still not happy with the required equipment and the requirement for field mills.

Agreement to use 3 kV/m as the E-field threshold.

29 April 2004 LAP Telecon -- To discuss latest revisions, primarily to the General section, of the FAA LFCC and to address the 45WS question about the recent change of the cloud-edge definition from 10 dBZ to 0 dBZ.

30 April 2004 LAP Telecon -- The LAP forwarded a formal response to Todd McNamara's question about the cloud-boundary definition.

14 May 2004 -- The LAP forwarded to the FAA its revisions to their latest LFCC draft, as discussed in great detail during and after several previous LAP telecons.

25 May LAP Telecon -- Harry Koons led the LAP through his preliminary extreme-value analysis for the 11X11 km column-integrated reflectivity.

21 July 2004 LAP telecon --

More on the extreme-value analysis from Harry Koons.

Yet another FAA LFCC version...

Initial stab at what a radar-based anvil-rule exception might look like. (Discussion and wordsmithing of this exception and the associated definitions continues via e-mail between telecons.)

11 August 2004 ABFM II Telecon --

Harry Koons presented his final version of the extreme-value analysis for the "11X11 km column-integrated reflectivity" (re-named VAHIRR on 26 August 2004) with the agreed-upon filters. Making worst-case assumptions, the probability of encountering fields $\geq$ 3 kV/m during a launch through an anvil having VAHIRR = 10 dBZ-km was estimated at $10^{-4}$. This result was eventually accepted by the

LAP as justifying the 10 dBZ-km threshold for Anvil Clouds. (See also the entry for 3 August 2007.) Chose a radar threshold of 0 dBZ for averages to avoid negative numbers going into averages, to minimize differences between radars, to reduce sensitivity to scan gaps, and to be consistent with the new definition of cloud edge.

The question remained whether to include in any new LLCC the radar fractional-volume filter that we planned to use in the data analysis.

Frank Merceret and Jennifer Ward discussed their cloud-edge determination in relation to field decay (standoff issue) and radar edge (transparency issue) in the ABFM II dataset.

26 August 2004 -- First introduction via e-mail of the final terminology for the chosen radar parameter (which I have been calling "11X11 km column-integrated reflectivity") by Jim Dye: "Volume-Averaged, Height-Integrated Radar Reflectivity (VAHIRR)."

26 August 2004 -- "White Paper" by John Willett recommended 6.2 kV/m as the best available estimate of triggering conditions for a large (Titan-class) space launcher at 10 km -- a typical anvil altitude. This estimate involved measured triggering conditions for rockets towing grounded wires [Willett, *et al.*, 1999], estimated electrical effective lengths of exhaust plumes (based on video observations of Titan and Shuttle launches, summarized by Frank Merceret on 5 January 2004), and estimates of the worst-case decrease in field threshold with atmospheric density (Paschen's Law).

23 September 2004 -- The LAP responded to John Madura's question about triggering fields for small launch vehicles with a caution that we don't currently know enough to drastically scale up the triggering thresholds with decreasing vehicle size.

22 October and 4 November 2004 LAP Telecons -- Inconclusive discussion of the VAHIRR exceptions, associated definitions, and an interim implementation with existing radar products. E-mail discussion continues.

6 November 2004 -- Harry Koons's first e-mail reference to truth-table analysis in connection with the current draft anvil rules. This analysis took us off into a major reorganization of the anvil rules -- mainly via extended e-mail discussion between John Willett and Harry Koons.

10 November 2004 LAP Telecon --

Discussed the latest truth-table-validated versions of the anvil rules.

Minor revisions to the VAHIRR definition.

12 November 2004 -- The LAP formally forwarded (via e-mail) its agreed versions of the new anvil rules, VAHIRR definitions, and interim implementation instruction to the KSC Weather Office (John Madura), but see the entry for 3 February 2005 below.

17 December 2004 -- 45WS (apparently in consultation with the 30 WS) submitted suggested revisions to the LAP's 12 November 1994 anvil rules.

1 February 2005 LAP Telecon -- Apparently to discuss the 45WS suggestions.

3 February 2005 -- The LAP formally forwarded (via e-mail, formal memo from chairman to follow) modified versions of the new anvil rules, VAHIRR definitions, and interim implementation instruction to the KSC

Weather Office (John Madura). No substantive changes were made to the 12 November 2004 version; most of the 45WS suggestions of 17 December 2004 were rejected. This seems to have been the final version (translated from FAA into USAF language), as faithfully reproduced on 3 June 2005 as Space Shuttle LLCC (2005) [LCN 01166R01, below.

11 May 2005 -- Harry Koons passed away suddenly and unexpectedly.

3 June 2005, Space Shuttle LLCC (2005) [LCN 01166R01, see Appendix I, Section A1.10] -- MAJOR Changes: This is the first application of both Harry Koons's truth-table analysis, and his statistical analysis of the new ABFM II dataset, to both clarify and relax the anvil rules. In spite of numerous wording changes throughout (which will be ignored here), there have been no substantive changes to any of the other rules (except through changes to the definitions, as detailed below).

Anvil Rules (now C) have been completely re-written. In particular, the order of the rule sections has been reversed so that the "least threat" (through) now comes last, and the "greatest threat" (within 10 NM) now comes first. They have also been logically re-organized to so that the times after lightning and/or detachment that are referenced in their different sections no longer overlap, and so that each standoff-distance range is explicitly specified (*e.g.*, 3(b)(3), "Do not launch... within 5 NM of nontransparent parts of a detached anvil cloud for the first 3 hours after...", has been replaced with C(2)(b), "Do not launch... between 0 (zero) and 5 NM from a nontransparent part of a detached anvil cloud between 30 minutes and 3 hours after..."). The substantive changes to these rules are less complex than the wording changes and can be appreciated most easily by studying the truth tables attached below. These substantive changes are summarized verbally here:

Attached Anvils (C(1)):

Satisfaction of all three VAHIRR conditions now allows flight >30 min after the last lightning at *all* ranges (including penetration); whereas penetration was previously prohibited at *all* times (including >3 hr), and flight between 0 and 5 NM was prohibited between 30 min and 3 hr after lightning.

No changes have been made to the launch permissions if any of the VAHIRR conditions is not satisfied.

Detached Anvils (C(2)):

Satisfaction of all three VAHIRR conditions now allows flight >30 min after the last lightning at *all* ranges (including penetration); whereas penetration previously was prohibited between 30 min and 4 hr under all conditions and after 4 hr unless the cloud had been detached for more than 3 hr, and flight between 0 and 5 NM was prohibited previously between 30 min and 3 hr after lightning unless a low-field low-radar condition was satisfied.

No changes have been made to the launch permissions if any of the VAHIRR conditions is not satisfied.

Definitions:

Two new definitions ("Volume-Averaged, Height-Integrated Radar Reflectivity" and "Specified Volume") have been added to define VAHIRR and how to compute it.

Additionally, an "Interim Instruction" has been added specifying how to conservatively approximate VAHIRR from currently available radar tools.

A new definition of "Field Mill" has been added to further emphasize that that is the only acceptable instrument for implementation of the Surface Field Rule (now H).

IMPORTANT: A new definition of "Cloud Base" has been added, and the definitions of "Cloud Edge," "Cloud Top," and "Nontransparent" have been substantively modified from the previous 10 dBZ radar threshold to 0 dBZ. This change affects all references to clouds throughout the rules, making them more conservative.

The definition of "Transparent" has been changed to remove references to the disk of the sun and distinct shadows on the ground, also making the rules somewhat more conservative.

Space Shuttle LLCC (2005) [LCN 01166R01] is substantially identical to the best available approximation of the rules that were approved previously by the LAP. This approximation resulted from starting with the LAP revisions of the LFCC that were forwarded to the FAA on 14 May 2004 (apparently the most recent, including the definition changes that were proposed on 29 May 2003) and then replacing the anvil rules therein with the VAHIRR versions of those rules (together with their associated definitions and interim instruction) that were forwarded to the KSC Weather Office on 3 February 2005 (also the most recent LAP-approved versions). Aside from numerous, apparently unimportant, changes in wording from FAA to KSC language (primarily in definitions and rules other than those applying to anvils) and some re-ordering of the definitions, potentially substantive differences between this composite and Space Shuttle LLCC (2005) [LCN 01166R01] are the following. (It is not clear whether the LAP actually approved these details.)

The triboelectrification rule and associated definitions present in the FAA version have been dropped in the NASA version.

The following "Notes" have been added to the NASA version of the "Surface Electric Fields" rule:

"i) Electric field measurements at the surface are used to increase safety by detecting electric fields due to unforeseen or unrecognized hazards."

"ii) For confirmed failure of one or more field mill sensors, the countdown and launch may continue."

The definition of "associated" is embellished with the following:

"An example of clouds that are not associated is air mass clouds formed by surface heating in the absence of organized lifting."

Expanded "Space Shuttle LLCC (1998)" [LCN 00824R03, 4 June 1998] Attached-Anvil Rule:

If ONE OR MORE VAHIRR conditions is NOT satisfied, or if VAHIRR is not available:

| Distance from Cloud, D: \ Time After Lightning, T | Through | $0 < D \leq 5$ nmi | $5 < D \leq 10$ nmi | 10 nmi $< D$ |
|---|---|---|---|---|
| $T \leq 30$ min | No Launch | No Launch | No Launch | -- |
| 30 min $< T \leq 3$ hr | No Launch | No Launch | -- | -- |
| 3 hr $< T$** | No Launch | -- | -- | -- |

If ALL 3 VAHIRR conditions ARE satisfied:

| Distance from Cloud, D: \ Time After Lightning, T | Through | $0 < D \leq 5$ nmi | $5 < D \leq 10$ nmi | 10 nmi $< D$ |
|---|---|---|---|---|
| $T \leq 30$ min | No Launch | No Launch | No Launch | -- |
| 30 min $< T \leq 3$ hr | No Launch | No Launch | -- | -- |
| 3 hr $< T$** | No Launch | -- | -- | -- |

Notes:

A cell entry of "--" means the rule does not explicitly mention this condition.
This rule does not invoke VAHIRR, so the two panels are identical.
** If there has never been a lightning discharge, then this time period applies.

**Figure A2-1 Expanded Space Shuttle LLCC (1998) Attached Anvil Rule**

"Space Shuttle LLCC (2005)" [LCN 01166R0, 13 June 2005] Attached-Anvil Rule:

IF ONE OR MORE VAHIRR conditions is NOT satisfied, or if VAHIRR is not available:

| Distance from Cloud, D: | Through | $0 < D \leq 5$ nmi | $5 < D \leq 10$ nmi | 10 nmi $< D$ |
|---|---|---|---|---|
| Time After Lightning, T | | | | |
| $T \leq 30$ min | No Launch | No Launch | No Launch | -- |
| 30 min $< T \leq 3$ hr | No Launch | No Launch | -- | -- |
| 3 hr $< T^{**}$ | No Launch | -- | -- | -- |

IF ALL 3 VAHIRR conditions ARE satisfied:

| Distance from Cloud, D: | Through | $0 < D \leq 5$ nmi | $5 < D \leq 10$ nmi | 10 nmi $< D$ |
|---|---|---|---|---|
| Time After Lightning, T | | | | |
| $T \leq 30$ min | No Launch | No Launch | No Launch | -- |
| 30 min $< T \leq 3$ hr | Launch Not Prohibited | Launch Not Prohibited | -- | -- |
| 3 hr $< T^{**}$ | Launch Not Prohibited | -- | -- | -- |

Notes:
A cell entry of "—" means the rule does not explicitly mention this condition.
** If there has never been a lightning discharge, then this time period applies.

**Figure A2-2  Space Shuttle LLCC (2005) Attached Anvil Rule**

Expanded "Space Shuttle LLCC (1998)" [LCN 00824R03, 4 June 1998] Detached-Anvil Rule:

If ONE OR MORE VAHIRR condition is NOT satisfied, or if VAHIRR is not available:

| Distance From Cloud, D: | Through | 0 < D ≤ 5 nmi | 5 < D ≤ 10 nmi | 10 nmi < D |
|---|---|---|---|---|
| Time After Lightning, T | | | | |
| T ≤ 30 min | No Launch | No Launch | No Launch | -- |
| 30 min < T ≤ 3 hr | No Launch | No Launch unless Low Field | -- | -- |
| 3 hr < T ≤ 4 hr | No Launch | -- | -- | -- |
| 4 hr < T** | No Launch unless Detached > 3 hr | | | |

If ALL 3 VAHIRR conditions ARE satisfied:

| Distance From Cloud, D: | Through | 0 < D ≤ 5 nmi | 5 < D ≤ 10 nmi | 10 nmi < D |
|---|---|---|---|---|
| Time After Lightning, T | | | | |
| T ≤ 30 min | No Launch | No Launch | No Launch | -- |
| 30 min < T ≤ 3 hr | No Launch | No Launch unless Low Field | -- | -- |
| 3 hr < T ≤ 4 hr | No Launch | -- | -- | -- |
| 4 hr < T** | No Launch unless Detached > 3 hr | | | |

Notes:

A cell entry of "--" means the rule does not explicitly mention this condition.

"Low field" means |s| ≥1 working mill <5 nmi from cloud, (b) <1000 V/m for 15 min at all mills ≤5 nmi from flight path AND at mill(s) in (a), and (c) cloud is <10 dBz ≤5 nmi from flight path for 15 min.

This rule does not invoke VAHIRR, so the two panels are identical.

* These cells are ambiguous between two rule sections, one saying No Launch and the other saying Launch Not Prohibited.

** If there has never been a lightning discharge, then this time period applies.

Figure A2-3 Expanded Space Shuttle LLCC (1998) Detached Anvil Rule

"Space Shuttle LLCC (2005)* [LCW 0166R0, 13 June 2005] Detached-Anvil Rule:

If ONE OR MORE VAHIRR condition is NOT satisfied, or if VAHIRR is not available:

| Distance from Cloud, D: | Through | $0 < D \le 5$ nmi | $5 < D \le 10$ nmi | 10 nmi $< D$ |
|---|---|---|---|---|
| Time After Lightning, T | | | | |
| $T \le 30$ min | No Launch | No Launch | No Launch | -- |
| 30 min $< T \le 3$ hr | No Launch | No Launch unless Low Field | -- | -- |
| 3 hr $< T \le 4$ hr | No Launch | -- | -- | -- |
| 4 hr $< T**$ | No Launch unless Detached > 3 hr | | | |

If ALL 3 VAHIRR conditions ARE satisfied:

| Distance from Cloud, D: | Through | $0 < D \le 5$ nmi | $5 < D \le 10$ nmi | 10 nmi $< D$ |
|---|---|---|---|---|
| Time After Lightning, T | | | | |
| $T \le 30$ min | No Launch | Launch Not Prohibited | Launch Not Prohibited | -- |
| 30 min $< T \le 3$ hr | No Launch | Launch Not Prohibited | -- | -- |
| 3 hr $< T \le 4$ hr | Launch Not Prohibited | -- | -- | -- |
| 4 hr $< T**$ | Launch Not Prohibited | | | |

Notes:

A cell entry of "--" means the rule does not explicitly mention this condition.

"Low Field" means (a) |E| working mill <5 nmi from cloud, (b) <1000 V/m for 15 min at all mills ≤5 nmi from flight path AND at mill(s) in (a), and (c) cloud is <10 dBZ ≤5 nmi from flight path for 15 min.

* This cell is ambiguous between sections (a), which says No Launch, and (c)(2), which says Launch Not Prohibited.

** If there has never been a lightning discharge, then this time period applies.

**Figure A2-4 Space Shuttle LLCC (2005) Detached Anvil Rule**

4 November 2005 -- A final report was received from the Russian group on a KSC contract to study triggering conditions aloft. (First mention of such a contract was reported in the entry for 21 August 2003.) This report was sent out for peer review starting about 21 December 2005. A detailed evaluation of the final report, and five reviews thereof, was provided to NASA/KSC by John Willett on 29 June 2006. Several recommendations for follow-on work were made, both by the report's authors and by John Willett, but nothing further has been done to date. Much of the contents of this report has now been published by Bazelyan *et al.* (2007).

16-17 November 2005 LAP Meeting at NASA/JSC in Houston -- Several topics were covered in some detail at this meeting:

Tim Oram (JSC/SMG) outlined the current SMG flight rules for Shuttle. Tim proposed a way to add a VAHIRR exception to the flight rules. This led to some LAP recommendations:

It is appropriate to use multiple NEXRAD radars to evaluate VAHIRR.

Conduct an experiment with near and distant NEXRAD radars looking at the same anvil to determine whether there is an upper range limit for accurate calculation of VAHIRR.

It is appropriate to forecast VAHIRR for evaluation of the flight rules.

It was tentatively agreed to change the level above which "core" radar reflectivities are reckoned in the VAHIRR-validity note from 4 km to a temperature level such as +5 C.

The problem of a really long, nontransparent anvil attached to a distant active core was raised again. John Willett offered a proposal to solve this problem by modifying the first section (through or within 10 nm) of the Attached Anvil rule. This and other VAHIRR-related issues led to a long e-mail deliberation after the meeting.

45WS raised a number of new questions about the most recently implemented (3 June 2005, Space Shuttle LLCC (2005) [LCN 01166R01]) version of the anvil rules. Frank Merceret raised the question of how to calculate VAHIRR for nontransparent anvils that cannot be seen on radar. Again, these issues led to protracted e-mail deliberation.

Richard Blakeslee (NASA/MSFC) summarized measurements over thunderstorms by the ER-2 and, more recently, by ACES. Large fields can occur in the 15 - 21 km altitude range (>17 kV/m at 15 km and >8 kV/m at 20 km). Concern was raised by several LAP members about the 2 nm standoff *above* thunderstorms in the flight rules.

Based on the ABFM II data, Jim Dye was optimistic about reducing standoff distances from anvils and debris clouds in the near future and, with considerably more effort, possibly applying VAHIRR to debris clouds.
Hugh Christian agreed to re-examine the question of whether it would be productive to include the ABFM I dataset in any such re-analysis. The main questions were whether the E-field data could still be read from the Exobyte (TM) tapes and whether the radar calibration could be salvaged.

Davis Sentman (University of Alaska, Fairbanks) reviewed what is known about sprites with respect to the hazard to Shuttle during landing. Mark Stanley (NMIMT) reviewed recent measurements of the charge and current moments in sprites. Although the total energies and peak currents are inferred to be large -- up to 10 MJ and 100 kA in extreme cases -- the energy and current densities are not well known. Elves and sprite halos are not a problem. Not enough is known about blue jets. The reentry plasma sheath probably protects Shuttle from sprites at high velocities/altitudes, but it was not immediately known to how low an altitude this might give protection. Would the ~100 V/m field transients after large positive lightning discharges be enough to initiate sprites from Shuttle itself?

John Willett gave a preliminary summary of results of a recent NASA/KSC contract to a group of Russian scientists. (First mention of such a contract was reported in the entry for 21 August 2003. The resulting final report is referenced in the entry for 4 November 2005, immediately above.) The bottom line: They found a reasonable, but not a conclusive, basis for asserting that a lower bound on the lightning-triggering conditions by a flying vehicle (taken to be the conditions for continuous propagation of a positive leader, once initiated) does not depend significantly on altitude between the surface and 10 km. (Subsequently, the LAP has continued to use Paschen's Law -- triggering threshold proportional to atmospheric density -- however, to retain an ample margin of safety.)

The need was recognized for a statistician to replace Harry Koons on the LAP. Richard Walterscheid agreed to ask Aerospace Corp. if it might provide a suitable candidate. (Once again it is not clear whether Tim Oram at SMG got formal feedback on the flight rules from the LAP after the meeting.)

After the Houston meeting Frank Merceret pointed out that the anvil rules with a VAHIRR exception to a 5 nm standoff guaranteed that VAHIRR would be invalid outside 3 nm so that much of the exception could not be used. John Willett made a diagram illustrating this problem on 2 August 2007. It was first addressed with new rule language during the 19-20 March 2008 LAP meeting.

30 January 2006 LAP/ABFM II Meeting at the AMS Meeting in Atlanta -- Doug Mach will review his cases from ABFM I to determine exactly what kind of clouds they were flying in.

19 August 2006 -- Phil Krider, as chair of the LAP, sent a memo to John Madura, KSC Weather Office, indicating the importance of the LPLWS (ground-based field-mill system) and how both safety and launch availability would be decreased if it were terminated.

6 November 2006 ABFM II Telecon –

Participants concluded that continued analysis of the ABFM II dataset would focus primarily on two topics: (1) possible reductions of standoff distances from anvil and debris clouds and (2) extension of VAHIRR to debris clouds. The work would be funded by the Launch Services Program (supporter of expendable vehicles).

ABFM II anvil data is apparently not adequate to resolve E-field decay with time.

The LAP apparently considered Frank Merceret's draft on standoff distances from anvils and debris clouds, and he was tasked to draft a straw-man standoff rule for anvils and debris cloud.

Jim Dye was tasked to identify key differences between anvils and debris clouds, as classified in the AMFM II dataset

There was discussion of a possible follow-on aircraft experiment to investigate thick clouds and disturbed weather (in winter) and to further study standoff distances and field-decay times in anvils/debris clouds (in summer). The Viking S-3 that is owned by NASA Glenn Research Center was a possible candidate. The FAA is currently paying to rebuild the A/C. MSFC would be a good place to conduct such an experiment.

The question of the flight rules, and results from the Houston meeting last year, was raised again. Tim Oram said he has some examples of CG flashes from 20 - 25 dBZ clouds.

14 February 2007 LAP/ABFM II Telecon -- Topic:  The Next Steps in ABFM II Analysis

A "phasing plan" must be developed to satisfy requirements of the new Launch Services Program sponsor.

Frank Merceret was concerned that the fall-off of fields with distance from "anvils" appeared much slower in the ABFM I dataset than in ABFM II. Doug Mach resolved this paradox by noting that the ABFM I data that Frank was using applied mostly to developing storms and was not really applicable to anvils. Further, the ABFM I data on debris clouds was concerned primarily with field decay times inside the clouds and would not help with standoffs.

Based on the above Frank Merceret was tasked to prepare a proposal to revise the anvil and debris cloud rules to reduce the standoff distances, and to do a statistical evaluation of their safety. (Frank responded on 26 February 2007 with draft rules but pointed out that he was not sure how to evaluate their safety and needed guidance from a statistician.)

Monte Bateman was tasked to extend VAHIRR to debris clouds.

Jim Dye was tasked to compare ABFM II and operational definitions of debris clouds.

It was decided not to pursue ABFM I data on thick clouds.

7 March 2007 LAP/ABFM II Telecon –

Jim Dye reviewed the radar definitions of anvils and debris clouds that were used to sort the ABFM II database. The latter definition is much less clear than the former. Examples of debris clouds were shown from the ABFM II Web site.

A definition of the radar cone of silence is needed for VAHIRR validity.

5 April 2007 LAP/ABFM II Telecon –

Frank Merceret pointed out that a capping altitude is needed for the "cone of silence" above the radar in the VAHIRR definitions. This led to general agreement but considerable debate over what altitude to choose. It was decided that the cone should be capped by the cloud-top height, the local tropopause height, or 20 km, whichever was verifiably lowest.

Frank Merceret led discussion of his proposed new standoff rules for attached and detached anvils and for debris clouds (reduced from 5 to 3 nm between 30 min and 3 hr) and his preliminary analysis of their risk based on Gaussian statistics.

Frank Merceret's risk analysis also included the following "*a fortiori*" argument:  VAHIRR is effectively zero at any location $\geq 3$ nm from the cloud edge. Therefore, flight in the region of clear air between 3 and 5 nm from cloud edge must be at least as safe as penetration of cloud for which VAHIRR $\leq 10$ dBZ-km, which is permitted (at least for anvils). Richard Walterschied did not accept the *a fortiori* argument and required a statistical justification for the reduced standoff distances.

There was general agreement that Frank's Gaussian statistics, while encouraging, were not sufficiently rigorous to justify a rule change. John Willett suggested including clouds containing <3 kV/m fields in the statistics to increase the degrees of freedom by about a factor of three.

Bill Roeder volunteered to compile a list of unusual events (*e.g.*, long-delayed lightning out of debris clouds) that the 45WS could track and document for the LAP. Everyone thought this was a great idea.

26 April 2007 LAP/ABFM II Telecon –

A space physicist with statistical expertise, Paul O'Brien, was apparently identified at Aerospace who could replace Harry Koons on the LAP and help us with the debris-cloud and standoff-distance analysis.

Frank Merceret agreed to repeat his standoff analysis including clouds with low fields, which were previously excluded, because this would increase the degrees of freedom in the analysis. Any Aerospace statistical analysis was deferred until this exercise was completed.

The language of the cone-of-silence definition was finalized.

Future use of the ABFM I dataset has two potentially serious problems:

> The radar data is nearly inaccessible on old data tapes in "Magill format." Frank Merceret is trying to convert these tapes. In any case, its calibration will remain uncertain by "several dBZ."

> As noted in November 2005, the electric-field data is archived on Exobyte (TM) tapes that might or might not be readable. Monte Bateman was tasked to try reading the Exobyte tapes. [Note: The E-field data from ABFM I has recently been salvaged and re-analyzed by Mach (2009) for use by Walterscheid et al. (2010).]

During May and June, 2007, Paul O'Brien complete a statistical analysis of Frank Merceret's standoff-distance dataset, and the LAP asked MSFC to investigate the source of persistent ~1kV/m fields outside the clouds in that dataset -- see O'Brien and Walterschied (2007), below.

15 June 2007 -- Publication of O'Brien and Walterscheid (2007), which gives an analysis of standoff distances from anvils and debris clouds based on 2-parameter Weibull statistics. Findings were (1) the electric field intensity does not depend on distance from the cloud edge and (2) the probability of exceeding 3 kV/m outside the cloud is less than $10^{-8}$.

21 June 2007 LAP/ABFM II Telecon –

The new cone-of-silence definition was formally accepted.

The status of the statistical analysis of standoffs and the *a fortiori* argument were revisited, but no conclusions were recorded.

Jim Dye led us through a discussion of the VAHIRR scatterplots for debris clouds.

Shouldn't *all* of the filters (lightning, core, attenuation, wet radome, and fraction of "specified volume" filled -- this last *not* currently included) be specifically mentioned in the rules?

Some discussion of differences in interpretation of "debris clouds" between the ABFM II analysis and the Range.

Now that Aerospace has identified an extreme-value-analysis expert, it would be interesting to compare VAHIRR results between anvils (repeat Harry Koons's analysis) and debris clouds.

3 August 2007 LAP/ABFM II Telecon –

Paul O'Brien discussed his preliminary statistics for VAHIRR in debris clouds. He pointed out a lot of temporal correlation in the dataset that needs to be mitigated before further analysis. (Frank Merceret had previously found a 10X decrease in degrees of freedom due to serial correlation.) [Harry Koons had dealt with this problem by requiring 10 samples to constitute a "launch" -- see entry for 11 August 2004.]

Revised definitions to account for the altitude cap on the cone of silence and the fraction-filled VAHIRR filter were accepted.

Paul's final statistics for anvil standoff distances were reviewed. Final statistics will be distributed and voted on next telecon.

14 September 2007 LAP/ABFM II Telecon –

Further discussion of the standoff statistics led to a request for Paul O'Brien to consider anvils and debris clouds separately to see if there's and evidence of a difference.

The LAP accepts the 10X decimation for degrees of freedom in the VAHIRR statistics.

Doug Mach assured us that the low, near-uniform fields measured outside these clouds, which were found by O'Brien and Walterschied (2007), are real.

Paul O'Brien's extreme-value analysis of VAHIRR in debris clouds looks good in spite of low effective number of samples. For VAHIRR < 15 dBZ-km, the probability of fields >3 kV/m is well under $10^{-4}$.

1 October 2007 LAP/ABFM II Telecon –

Paul O'Brien reviewed his separate standoff treatment for anvil and debris clouds. It still looks OK. Based on this result, the LAP approved the reduced standoff distance of 3 nm for both.

Jim Dye suggested lumping anvils (dropping the frozen requirement) and debris clouds together in a single VAHIRR rule, to be considered next time. (On 6 October 2007 John Willett distributed an analysis of the differences among these three rules that made it seem unlikely that they could be combined.)

Paul O'Brien will repeat the extreme-value analysis for anvils and compare with debris clouds and with Harry Koons's earlier analysis.

17 October 2007 -- Paul O'Brien was formally appointed a new member of the LAP.

26 October 2007 LAP/ABFM II Telecon –

Frank Merceret volunteered to act as "secretary" for the LAP telecons and meetings from here on out. This was seen as a big benefit, since note-taking had not been very reliable.

Todd McNamara volunteered to distribute three master's theses on the spatial distribution of lightning strikes. (This was done on 29 October 2007.)

The definition of "debris cloud" was revisited from the 7 March 2007 telecon. It was not clear that the AF and the ABFM II team were using the same definition. Todd McNamara will try to re-define the term.

When in doubt about which of two LLCC applies, apply both and use the most conservative result. Language is needed in the rules to codify this.

It's OK to use VAHIRR in clear air pending a change in the standoff distances in the anvil rules. The 45WS had submitted another list of LLCC questions on 23 July 2007. Some of these were easily answered by the LAP. Others were held over for the next telecon. A few created action items:

> John Willett will draft an exception to the anvil rule for Cu with tops between ± 5 °C.

> John Willett will clarify the language about starting the lightning clock in the anvil rule.

> John Willett will clarify the distinction between the attached anvil and its parent Cu. John Willett will consider further changing the definition of the cloud boundary to the nontransparent edge and summarize the implications for what is attached and what is detached.

19 November 2007 LAP/ABFM II Telecon --

The LAP resumed consideration of the 45WS questions:

> John Willett will look at the logic of the cumulus rules to see if they can be changed to language similar to that of the revised anvil rules.

> There was considerable discussion of where to place the VAHIRR-validity caveats. Frank Merceret pointed out that one pixel of 3 dBZ is certainly safe but results in VAHIRR invalidity. It was apparently decided not to worry about this prospect. Frank and Todd McNamara will make a list of where which caveat should be put.

> John Willett will try to draft an exception to the thick-cloud rule so that anvils don't get treated as thick clouds.

Revised anvil rules (with 3 nm standoffs) and associated VAHIRR-related definitions (version of 27 October 2007) were accepted by the LAP.

Revised debris-cloud rule (with 3 nm standoff -- version of 7 October 2007) was accepted by the LAP.

John Willett was to send out a new version of the rules that consolidates all of these changes, plus the other ones proposed in the previous telecon (lightning clock, "anvils" produced by shallow Cu, distinction between an anvil and its parent CB, and new cloud-edge definition) for a final vote by 30 November 2007. Although this consolidated version is apparently not available, all (except Hugh Christian, who apparently didn't vote) accepted these changes.

28 November 2007 -- John Madura distributed the rule-change history that has been used in preparing this chronology.

26 December 2007 -- Pre-release copy of Paul O'Brien's statistical analysis of VAHIRR for debris clouds [TOR-2008(1494)-1 (DRAFT)] was distributed to the LAP. It shows that debris clouds have systematically lower risk of >3 kV/m electric fields than do anvil clouds.

8 January 2008 -- Todd McNamara and Frank Merceret submitted revisions to the VAHIRR definitions, anvil rules, and debris-cloud rules for consideration in a telecon proposed for the end of the month. This telecon apparently never occurred.

11 or 13 March 2008 LAP telecon with FAA on Kodiak --

Only three members of the LAP attended (Krider, Walterscheid, and Willett).

There have been no commercial launches from Kodiak.

Description of the launch rules used at Kodiak -- came from Sandia.

They appear to be conflating safety rules for ground operations with those for launch.

Description of high surface e-field readings that occur at Kodiak (6-8 kV/m) and what little is known of the related meteorology ("cumulus precipitation events"). Natural lightning is very rare at Kodiak.

They have a weather radar (very limited capability) operating for all launches and a reconnaissance aircraft also available.

Based on this telecon and further discussion at the 19-20 March 2008 LAP Meeting (immediately below), several recommendations were sent via e-mail from the LAP to the FAA on 26 March 2008:

1) The occurrence of clouds producing graupel is a strong indicator of risk. This is supported by the occurrence of the high fields (6-8 kV/m) observed during situations where graupel showers occur.

2) Such high fields are themselves indicators of a danger, whether they are observed to occur in association with graupel or not. Smaller elevated fields exceeding somewhat 1.5 kV/m may occur in benign situations (*e.g.*, blowing snow), but further study is required to characterize the conditions that are safe.

3) The near absence of natural lightning should not be taken as an indicator of no risk.

4) A small-rocket campaign should be conducted to assess the triggering potential in clouds.

5) A survey of conditions that occur when high fields are recorded (clear skies, clouds, precipitation, etc.) should be undertaken.

6) A strategy for determining the existence of embedded convective clouds should be developed.

7) The weather conditions likely to present a hazard for triggering lightning during launches (the focus of the LAP up to now) are NOT the same as the conditions that present an electrostatic hazard to personnel during ground operations (the Sandia safety criteria).

8) For triggering lightning we recommended that the KLC take advantage of existing expertise at the University of Florida and/or New Mexico Tech rather than re-inventing that technology.

19-20 March 2008 LAP Meeting at KSC --

Reviewed an 11 March 2008 version of the LLCC that had been circulated by John Willett, which purported to include all changes previously agreed on in telecons, and agreed to most of them. Exceptions:

All agreed to move the short-parent exclusion for anvils from the rules to the definition of "anvil."

Language in the "General" section that requires all criteria to be evaluated was further clarified.

Agreed to exempt attached anvils from the cumulus rule. John Willett will add English units in parentheses wherever metric units were used.

Moved most of the VAHIRR-validity caveats into a separate definition of "VAHIRR Acceptance Criteria."

The two debris-cloud standoff-distance "outliers" were found to be in cloud, so not outliers after all.

Moved the surface-field rule to first place, before even the lightning rule.

Clarified the "through or within" language in several rules.

Finally addressed the invalid-VAHIRR problem that was caused by the previous VAHIRR exception to the 5 nm standoff from detached anvils (and also by the newly proposed VAHIRR exception to the 10 nm standoff for all anvils). Thanks to Frank Merceret, it was decided to draft language requiring VAHIRR to be evaluated within 1 nm of the flight path for the new 3 nm standoff (8 nmi for the 10 nm standoff) to ameliorate this problem.

It was agreed that section (a) of the attached-anvil rule was too conservative. Along the lines indicated above, it was decided to grant a VAHIRR exception to the 10 nm standoff from attached anvils during the first 30 min after lightning.

John Willett will draft language for a VAHIRR exception to the debris-cloud rule.

John Willett will draft for review a new version of the cumulus rule that is consistent with the new format of the anvil rules (as promised during the 19 November 2007 Telecon).

Todd McNamara pointed out that the current (LAP approved, 1 June 2005) rule permits flight up to the edge of *any* attached anvil (while respecting the Cu standoffs) before the first lightning (as for t > 3 hr). What if the first lightning extends out into the anvil, possibly beyond the Cu standoff distance? John Willett was tasked to write up the threat scenario that made this situation dangerous and a proposed search for first lightning in anvils. (Ultimately the LAP-approved, 6 October 2008 rules prohibited flight between 0 and 3 NM of an attached anvil at any time, in the absence of VAHIRR.)

The need for LAP funding to write scientific rationales for the rules was discussed. John Madura was urged to explore funding sources.

Frank Merceret prepared the following list of action items from the meeting:

Krider will contact Tim Oram at SMG regarding the status of the Shuttle flight rules.

Christian will examine time variations in fields above anvils when lightning occurs from the anvil.

Christian will send the ACES reports to the LAP.

Dye will discuss the debris-cloud cases with O'brien.

McNamara will advise the LAP what rules 45WS currently applies to an anvil-like cloud that is below the altitude that defines an anvil cloud in the rules.

Willett will reword the standoff rules to eliminate Frank Mercert's "donut" between 3 and 5 miles when the reduced 3 mile stand-off distance applies.

Roeder will contact Stano and Fuelberg regarding the probability that the first flash goes into the anvil.

25 March 2008 LAP/ABFM II Telecon --

Paul O'Brien reported that a re-do of his debris-cloud standoff statistics without the two outliers strengthens the conclusion that 3 nm is more than enough.

Paul O'Brien reported that his VAHIRR statistical analysis for debris clouds is comparable to his new re-analysis of Harry Koons's dataset for anvil clouds. The 10 dBZ-km threshold is very safe for debris clouds but less so for anvil clouds using the 10X decimation. Paul finds a few tenths of one percent (a few cases per thousand) chance of fields exceeding 3 kV/m in his wider (5 - 15 dBZ-km) bin. This result leads to much discussion and plans for another telecon on 1 April 2008 to follow up on this issue.

1 April 2008 LAP/ABFM II Telecon --

Paul O'Brien reports that expanding the VAHIRR bin to 0 - 20 dBZ-km (which should be more conservative) increases the number of points without increasing the peak fields, thus improving the statistics. For the combined anvil dataset (all decimated anvil samples having valid VAHIRR from either radar), the probability of exceeding 3 kV/m is reduced to 0.016% with an upper confidence limit of 0.22%.

John Willett was asked to summarize the uncertainties in the triggering conditions aloft. The resulting "white paper" is dated 14 April 2008 and concluded, "The degree of conservatism of the 3 kV/m triggering field at 10 km is quite uncertain by may be as large as a factor of six."

11 April 2008 -- 45WS submitted suggested revisions to the LAP's current draft rules.

15 April 2008 LAP/ABFM II Telecon --

Paul O'Brien presented two more analyses of VAHIRR statistics. Revising the algorithm for selecting the "combined" dataset (how to choose which radar to use in the presence of filter criteria when both VAHIRR estimates are valid) gains data a few points, reducing the risk for both anvils and debris clouds. Removing the decimation and correcting the error bounds later, however, makes matters worse (compared to the "midpoint" decimation technique) and exposes some "outliers" for debris clouds. Overall conclusion: It would be wise to fly another ABFM campaign to increase the degrees of freedom.

Paul looked up the outliers mentioned above for further study. These were examined by John Willett on 17 April 2008 and determined to be debris of a dying small thunderstorm but with a top still above the -10 C isotherm at the time and location of the high-field readings. Jim Dye concurred with this identification on 15 May 2008.

John Madura reported that Range Safety was willing to accept a *combined* risk (all causes) of 1/1000 for its flight-termination system.

The LAP re-affirmed its approval for the use of VAHIRR exceptions for anvils and approved them for debris clouds as well.

Paul O'Brien was tasked to work with John Willett to produce an overall estimate of the conservatism build into our VAHIRR risk analysis, which, though solid for debris clouds, appears only marginally safe for anvils. This exercise resulted in a PowerPoint presentation dated 22 April 2008. Its abbreviated conclusions were the following, justifying the LAP recommendation to use VAHIRR:

1) The 3 kV/m triggering threshold includes at least an arbitrary safety factor of two.

2) Increasing that threshold to only 4 kV/m decreases the triggering probability in anvils from a few per thousand to less than 1/10,000.

3) Further increasing the threshold to 5 kV/m decreases the upper 95% confidence limit on that probability below 1/10,000! (The risk in debris clouds remains considerably less.)

9 May 2008 LAP Telecon -- A new round of wordsmithing begins:

To eliminate the confusing and sometimes logically duplicative expression, "through and within," Jim Dye floated the idea of a new definition of "cloud" (everything inside the cloud boundary) and of referencing all standoff distances to the cloud itself.

The "non-transparent" definition should be re-written so that it is the true opposite of "transparent."

Frank Merceret suggested a clever way to re-write the VAHIRR definition to prevent invalid evaluations due to small fractional volumes with $\geq$ 0 dBZ radar returns.

Many of the 45WS comments of 11 April 2008 were considered and most of these were accepted for incorporation into the current draft LLCC. The remainder was held over to the next telecon.

15 May 2008 LAP Telecon --

The remaining 45WS comments were considered.

A new definition of "radar" was proposed to require wavelengths longer than 3 cm.

It was recognized that, for VAHIRR evaluation points outside the flight path, the core and lightning filters should not be applied.

12 June 2008 LAP Telecon --

Jim Dye and John Willett have been working together on the language and the logic of the rules, leading toward a major revision in both the definition of standoff distances and the structure of certain rules. John Willett reviewed his reconstruction of Harry Koons's original truth-table analysis of the anvil rules (see entries for 6 November 2004 and subsequent) and its extension to the debris-cloud and cumulus rules in the latest draft LLCC. Other than the necessity of avoiding contradictions between rule sections, the question of whether redundancies simplify the language or confuse the application was discussed.

Paul O'Brien suggested archival for future reference/use of the software that he developed for the VAHIRR statistical analysis. (At time of writing, both this software and the dataset that Paul used for his analysis have been acquired by Jim Dye and John Willett, with intent to post them on the NASA/KSC ABFM Web Site.)

Jim Dye and Frank Merceret reviewed problems that they had discovered with the automated VAHIRR-calculation software that was being developed for use at the Range.

8 July 2008 LAP Telecon --

On 22 June 2008 Jim Dye and John Willett had circulated a new draft of the LLCC that implemented the 45WS suggestions and other changes to which the LAP had agreed during prior telecons. More importantly, this version contained significant new changes to the definition of "cloud," to the description of standoff distances, and consequently to the language of the rules themselves. The logic of certain rules was also modified to remove one contradiction and several redundancies, as indicated by truth-table analysis. This new draft was discussed with the following results:

> Define "radar reflectivity" to incorporate all requirements for radar and echo characteristics.

> Put all words being explicitly defined in quotes in both the definitions and the body of the rules.

> Phil Krider will work on the syntax of the definitions.

> John Willett will turn a final draft on which everyone will vote at the next telecon on 22 July 2008.

The LAP also discussed its concern about the FAA-required statement in the "General" section, "If a launch operator proposes any alternative lightning flight commit criteria, the launch operator must clearly and convincingly demonstrate that the alternative provides an equivalent level of safety." John Madura was asked to report on what the LAP can do about this.

22 July 2008 LAP Telecon --

There was still no agreement on several aspects of the latest draft:

Hugh Christian didn't understand the rules.

Richard Walterscheid wanted to eliminate the term, "transparent." A contradiction was pointed out between the definitions of "transparent" and "nontransparent." John Willett was tasked to eliminate "transparent," incorporate "nontransparent" into the definition of "cloud," and eliminate all uses of "nontransparent" throughout the body of the rules.

Richard Walterscheid dropped his insistence on radar measurements for standoffs beyond 3 nm. This might require a major re-write of the anvil rules if the LAP adopts this position. Phil Krider and Richard Walterscheid will try to come to agreement on this issue.

Bill Roeder will draft a new definition and rule for disturbed weather.

New comments and clarifications that had been submitted by the 45WS on 18 July 2008 were discussed.

4 September 2008 LAP Telecon --

The question of dropping radar requirements on standoff (see entry for 22 July 2008) turns on what filters were used in the standoff-distance statistics. Frank Merceret reported that no core- or lightning-range filters, nor the frozen filter, were applied. LAP consensus was to drop the VAHIRR requirement, but not the frozen requirement, for >3 nm standoffs from anvils. John Willett was tasked to send out a corresponding re-draft.

John Madura requested that the LAP consider volcanic ash and the possibility of writing a rule to cover it.

18 September 2008 LAP Telecon -- On 10 September 2008 John Willett had circulated another draft that incorporated the changes recommended in the previous telecon -- especially the removal of a radar (but retention of a frozen) requirement for anvil standoffs >3 nm. This was the primary topic for discussion:

After introducing "large" before "precipitation" into the definition of "radar reflectivity, it was decided to remove it.

Hugh Christian suggests that all distances from field mills be measured in horizontal projection.

Richard Walterscheid suggested splitting off the definition of "cloud" from that of "distance." This led to ongoing discussion (about how to specify standoffs and whether to use non-overlapping distance ranges) with John Willett after the telecon.

The definition of "Ohms per Square" should be deleted.

Phil Krider will do a literature search on volcanic-plume electrification.

Bill Roeder decided to drop his request to modify the disturbed-weather rule this time around.

2 October 2008 LAP Telecon --

The LAP approved the rule version dated 19 September 2008 with very minor changes. This led immediately to the final version described below, which was forwarded to the KSC Weather Office on 17 October 2008.

6 October 2008, LAP 10/06/08 Recommendation in FAA Format [see Appendix I, Section A1.13] -- MAJOR Changes: This is the second application of Harry-Koons-inspired logical analysis and of the ABFM II dataset, as further analyzed statistically by Paul O'Brien and Frank Merceret.

The order of the rule sections has been changed in the following ways: The Definitions (G417.3) now come immediately after the preamble to emphasize that they are an integral part of the rules. Surface Electric Fields (G417.5) is now the first rule, to emphasize that elevated electric fields themselves are the hazard. Also, the Anvil rules have been separated into two rules, Attached Anvil Clouds (G417.11) and Detached Anvil Clouds (G417.13). In general, an effort has been made to specify whether or not the end points of a measurement range are included in a criterion (*e.g.*, "less than or equal to N, but greater than M, nautical miles from"). The distinction between a thunderstorm or cumulonimbus cloud and its attached anvil has been made even more explicit. Again we ignore minor wording changes and concentrate on substantive changes in the following.

General (G417.1): The requirement has been added that the constraints of *all* LFCC must be satisfied and that all potentially applicable LFCC must be applied. [Note that although the familiar "Good Sense Rule" is omitted from these proposed LFCC as presented here, essentially the same wording does appear in another section of the current FAA LFCC implementation where it is still a mandatory requirement. See Appendix I, A1.12 and A1.13. The LAP expects that this requirement will remain in the regulations as the LFCC are updated by the FAA.]

Definitions (G417.3): A substantial preamble has been added, explaining the new philosophy for measuring distances. (This philosophy has been further elaborated under the definition of "cloud.") In particular, two different kinds of distance measurements are defined. Distance between a field mill or measurement point and any object means the *horizontal* distance between that field point and the nearest part of the *vertical projection* of that object on Earth's surface. On the other hand, distance between a cloud and the flight path means the *shortest slant range* between any point on the flight path and *any part* of that cloud. The latter definition is intended to facilitate the separation of standoff distances in different sections of a given rule into non-overlapping ranges. For example, "if the flight path will carry the launch vehicle less than or equal to N nautical miles from any X cloud" means that the flight path must penetrate neither the *interior* of that cloud nor the volume between 0 and N nautical miles, inclusive, *outside* the cloud boundary. On the other hand, "between 0 and N nautical miles, inclusive, from a X cloud" refers only to the volume between 0 and N nautical miles, inclusive, *outside* the cloud boundary but specifically omits the *interior* of the cloud itself.

"Anvil Clouds" are now restricted to those produced by cumuli having tops colder than -10 °C.

A new definition of "Cloud" now subsumes the previous "Cloud Edge" and "Cloud Base" (but not "Cloud Top," which is re-worded but retained) and is considered to include everything inside the nontransparent cloud boundary.

The definition of "Electric field measurement" has been restricted to exclude cases in which the flight path is at or above 20 km altitude.

"Nontransparent" and "Transparent" have been reorganized so that the former is defined more clearly and the latter is defined as the opposite of the former.

New definitions have also been added for "Radar Reflectivity," "Thick Cloud Layer," "Treated," and "Triboelectrification."

"Within" has been eliminated and replaced by the new expressions of distance mentioned above.

The definition of "VAHIRR" has been reorganized to separate out the subsidiary definitions, "Average Cloud Thickness," Cone of Silence," "VAHIRR Application Criteria," and "Volume-Averaged Radar Reflectivity."

Anvil Clouds (G417.11 and G417.13): In both of these rules it has been explicitly stated that all waiting times after lightning are considered satisfied if there has never been a relevant lightning discharge. Also in both rules, the volume around the flight path in which the VAHIRR threshold (but NOT the VAHIRR Application Criteria) must be satisfied has been adjusted so that there is no case when VAHIRR is not "applicable" because too little of the cloud in question is within the Specified Volume to satisfy the "frac" criterion, which is now given at the end of the "Volume-Averaged Radar Reflectivity" definition. For other substantive changes to these rules, the reader is referred to the truth-table comparisons below. The following statements verbally summarize these changes:

Attached Anvil Clouds (G417.11):

Satisfaction of all three VAHIRR conditions now allows flight at *all* times after the last lightning at *all* ranges (including penetration); whereas flight was previously prohibited at ranges ≤ 10 NM for times ≤ 30 min.

If any of the VAHIRR conditions is *not* satisfied, flight is now permitted between 3 and 5 NM for times ≤ 3 hr and between 5 and 10 NM for times ≤ 30 min if the relevant part of the cloud is frozen; whereas flight was previously prohibited within these range and time intervals.

HOWEVER, flight is now prohibited between 0 and 3 NM for times greater than 3 hr if any of the VAHIRR conditions is not satisfied -- conditions under which flight was previously allowed.

Detached Anvil Clouds (G417.13):

Satisfaction of all three VAHIRR conditions now allows flight at *all* times after the last lightning at *all* ranges (including penetration); whereas flight was previously prohibited at ranges ≤ 10 NM for times ≤ 30 min.

If any of the VAHIRR conditions is not satisfied, flight is now permitted between 3 and 5 NM between 30 min and 3 hr; whereas a low-field/low-radar condition was previously required.

Also if any of the VAHIRR conditions is not satisfied, flight is now permitted between 3 and 10 NM during the first 30 min if the relevant part of the cloud is frozen; whereas flight was previously prohibited within this range and time interval.

Debris Clouds (G417.15):  A VAHIRR condition has now been introduced into this rule. As with the Anvil rules, the volume around the flight path in which the VAHIRR threshold must be satisfied has been adjusted so that there is no case when VAHIRR is not "applicable" because too little of the cloud in question is within the Specified Volume to satisfy the "frac" criterion. For other substantive changes, the reader is referred to the truth-table comparisons below. The following statements verbally summarize these changes:

> Satisfaction of all three VAHIRR conditions now allows flight at *all* times after detachment (independent of lightning) at *all* ranges (including penetration); whereas flight was previously prohibited at ranges ≤ 5 NM for times ≤ 3 hr, with or without conditions.

> If any of the VAHIRR conditions is not satisfied, flight is now permitted unconditionally between 3 and 5 NM for times ≤ 3 hr; whereas flight was previously permitted within this range and time interval only if a low-field low-radar condition was satisfied.

Thick Cloud Layers (G417.19):  It is now explicitly stated that anvil clouds are never considered to be thick cloud layers.

The Triobelectrification Rule (G417.23) now specifically includes transparent clouds. (NOTE that the Triboelectrification Rule does *not* originate in this version. It was omitted from the previous versions analyzed here only because they were taken from the Shuttle-Program rule collection, which did not include a Triboelectrification Rule.)

21 April 2009 LAP/ABFM III telecon -- Preliminary planning of another ABFM (III) campaign proposal, should funding be forthcoming. A final core-proposal was forthcoming from Frank Merceret on 9 July 2009.

24 June 2009 LAP Telecon -- Planning the present LAP/LLCC History/Rationale project.

Expanded "Space Shuttle LLCC (2005)" (LCN 01166R0, 13 June 2005) Attached-Anvil Rule:

If ONE OR MORE VAHIRR conditions is NOT satisfied, or if VAHIRR is not available:

| Distance from Cloud, D: | Through | 0 < D ≤ 3 nmi | 3 < D ≤ 5 nmi | 5 < D ≤ 10 nmi | 10 nmi < D |
|---|---|---|---|---|---|
| Time After Lightning, T | | | | | |
| T ≤ 30 min | No Launch | No Launch | No Launch | No Launch | --- |
| 30 min < T ≤ 3 hr | No Launch | No Launch | No Launch | --- | --- |
| 3 hr < T++ | No Launch | No Launch | --- | --- | --- |

If ALL 3 VAHIRR conditions ARE satisfied:

| Distance from Cloud, D: | Through | 0 < D ≤ 3 nmi | 3 < D ≤ 5 nmi | 5 < D ≤ 10 nmi | 10 nmi < D |
|---|---|---|---|---|---|
| Time After Lightning, T | | | | | |
| T ≤ 30 min | No Launch | No Launch | No Launch | No Launch | --- |
| 30 min < T ≤ 3 hr | Launch Not Prohibited | Launch Not Prohibited | Launch Not Prohibited | --- | --- |
| 3 hr < T** | Launch Not Prohibited | Launch Not Prohibited | --- | --- | --- |

Notes:
A cell entry of "--" means the rule does not explicitly mention this condition.
** If there has never been a lightning discharge, then this time period applies.

**Figure A2-5 Expanded Space Shuttle LLCC (2005) Attached Anvil Rule**

Expanded "LAP 10/06/08 Recommendation in FAA Format" Attached-Anvil Rule:

If ONE OR MORE VAHIRR conditions is NOT satisfied, or if VAHIRR is not available:

| Distance from Cloud, D: \ Time After Lightning, T | Through | 0 < D ≤ 3 nmi | 3 < D ≤ 5 nmi | 5 < D ≤ 10 nmi | 10 nmi < D |
|---|---|---|---|---|---|
| T ≤ 30 min | No Launch | No Launch | No Launch unless Frozen | No Launch unless Frozen | -- |
| 30 min < T ≤ 3 hr | No Launch | No Launch | No Launch unless Frozen | -- | -- |
| 3 hr < T** | No Launch | No Launch | -- | -- | -- |

If ALL 3 VAHIRR conditions ARE satisfied:

| Distance from Cloud, D: \ Time After Lightning, T | Through | 0 < D ≤ 3 nmi | 3 < D ≤ 5 nmi | 5 < D ≤ 10 nmi | 10 nmi < D |
|---|---|---|---|---|---|
| T ≤ 30 min | Launch Not Prohibited | Launch Not Prohibited | -- | -- | -- |
| 30 min < T ≤ 3 hr | Launch Not Prohibited | Launch Not Prohibited | -- | -- | -- |
| 3 hr < T** | Launch Not Prohibited | Launch Not Prohibited | -- | -- | -- |

Notes:

A cell entry of "--" means the rule does not explicitly mention this condition.

"Frozen" means cloud is colder than 0 C within prescribed distance from flight path.

Yellow cells implicitly become green in lower matrix, since "Frozen" is one of the VAHIRR conditions.

** If there has never been a lightning discharge, then this time period applies.

**Figure A2-6 Expanded LAP Recommended (10/06/08) Attached Anvil Rule**

Expanded "Space Shuttle LLCC (2005)" [LCN 0166R0, 13 June 2005] Detached-Anvil Rule:

If ONE OR MORE VAHIRR condition is NOT satisfied, or if VAHIRR is not available:

| Distance from Cloud, D: | Through | 0 < D ≤ 3 nmi | 3 < D ≤ 5 nmi | 5 < D ≤ 10 nmi | 10 nmi < D |
|---|---|---|---|---|---|
| Time After Lightning, T | | | | | |
| T ≤ 30 min | No Launch | No Launch unless Low Field | No Launch unless Low Field | No Launch | -- |
| 30 min < T ≤ 3 hr | No Launch | No Launch unless Low Field | No Launch unless Low Field | -- | -- |
| 3 hr < T ≤ 4 hr | No Launch | -- | -- | -- | -- |
| 4 hr < T** | No Launch unless Detached > 3 hr | | | | |

If ALL 3 VAHIRR conditions ARE satisfied:

| Distance from Cloud, D: | Through | 0 < D ≤ 3 nmi | 3 < D ≤ 5 nmi | 5 < D ≤ 10 nmi | 10 nmi < D |
|---|---|---|---|---|---|
| Time After Lightning, T | | | | | |
| T ≤ 30 min | No Launch* | No Launch | No Launch | No Launch | -- |
| 30 min < T ≤ 3 hr | Launch Not Prohibited | Launch Not Prohibited | -- | -- | -- |
| 3 hr < T ≤ 4 hr | Launch Not Prohibited | -- | -- | -- | -- |
| 4 hr < T** | Launch Not Prohibited | -- | -- | -- | -- |

Notes:

A cell entry of "--" means the rule does not explicitly mention this condition.

"Low Field" means (a) ≥1 working mill <5 nmi from cloud, (b) <1000 V/m for 15 min at all mills ≤5 nmi from flight path AND at mill(s) in (a), and (c) cloud is <10 dBz ≤5 nmi from flight path for 15 min.

* This cell is ambiguous between sections (3), which says No Launch, and (c)(2), which says Launch Not Prohibited.

** If there has never been a lightning discharge, then this time period applies.

**Figure A2-7  Expanded Space Shuttle LLCC (005) Detached Anvil Rule**

Expanded "LAP 10/06/08 Recommendation in FAA Format" Detached-Anvil Rule:

If ONE OR MORE VAHIRR condition is NOT satisfied, or if VAHIRR is not available:

| Distance from Cloud, D: | Through | 0 < D ≤ 3 nmi | 3 < D ≤ 5 nmi | 5 < D ≤ 10 nmi | 10 nmi < D |
|---|---|---|---|---|---|
| Time After Lightning, T: | | | | | |
| T ≤ 30 min | No Launch | No Launch | No Launch unless Frozen | No Launch unless frozen | -- |
| 30 min < T ≤ 3 hr | No Launch | No Launch unless Low Field | -- | -- | -- |
| 3 hr < T ≤ 4 hr | No Launch | -- | -- | -- | -- |
| 4 hr < T** | No Launch unless Detached > 3 hr | | | | |

If ALL 3 VAHIRR conditions ARE satisfied:

| Distance from Cloud, D: | Through | 0 < D ≤ 3 nmi | 3 < D ≤ 5 nmi | 5 < D ≤ 10 nmi | 10 nmi < D |
|---|---|---|---|---|---|
| Time After Lightning, T: | | | | | |
| T ≤ 30 min | Launch Not Prohibited | Launch Not Prohibited | -- | -- | -- |
| 30 min < T ≤ 3 hr | Launch Not Prohibited | Launch Not Prohibited | -- | -- | -- |
| 3 hr < T ≤ 4 hr | Launch Not Prohibited | -- | -- | -- | -- |
| 4 hr < T** | Launch Not Prohibited | -- | -- | -- | -- |

Notes:

A cell entry of "--" means the rule does not explicitly mention this condition.

"Frozen" means cloud is colder than 0 C within prescribed distance from flight path.

"Low Field" means (a) 21 working mil <5 nmi from cloud, (b) <1000 V/m for 15 min at all mills ≤5 nmi from flight path AND at mill(s) in (a), and (c) cloud is <10 dBZ ≤5 nmi from flight path for 15 min.

Some yellow cells implicitly become green in lower matrix, since "Frozen" is one of the VAHIRR conditions.

** If there has never been a lightning discharge, then this time period applies.

**Figure A2-8 Expanded LAP Recommended (10/06/08) Detached Anvil Rule**

4-6 August 2009 LAP Meeting at KSC -- LAP/LLCC History/Rationale first formal meeting:

Facing the probable resignation of Dave Rust, the LAP considered potential new members to invite. (As of 18 December 2009 it appears that Dave Rust will not resign, so this issue is moot.)

There is a need for the LAP to serve the entire space community beyond the retirement of Shuttle. John Madura and Frank Merceret indicated plans to continue in the KSC Weather Office until 2013.

Karen Shelton-Mur (FAA/AST) had requested to attend the LAP meeting on the LLCC rationales, apparently to ask general questions. It was decided to request salary and per diem from the FAA to add a day to the working meeting for that purpose.

Frank Merceret will set up regular recording of LLCC violations (or lack thereof) during all countdowns.

The LAP made a formal request to the KSC Weather Office and the 45WS for the above and also for recording of any high-field readings without explanation.

The History Document is due end of December 2009. The Rationale Document is due end of September 2010.

History Topic:

Three chronologies are needed: LLCC origin and evolution (and co-evolution of FAA rules), weather/atmospheric-electrical instrumentation at KSC, and organizational structure of weather support at KSC.

Numerous assignments of responsibility were made.

Rationale Topic:

A starting point might be the rationale outline in Krider *et al.* (2006, Section 3.2.2) -- needs updating to current (10/6/08) version of LLCC.

Cloud electrification -- starting point might be Krider *et al.* (2006, Section 3.2.1). Details go into an appendix.

The development of ideas about triggered lightning: When and how was it first realized that triggered lightning was a threat to space launches? What is now known about triggering conditions? (Starting point might be Krider *et al.* (2006, Sections 2.2 and 3.1) -- details go into an appendix.)

Key rationale problem will be the thick-cloud rule, because it appears to be based on and referenced to nimbostratus, but there is new data from the ABFM I experiment.

Numerous assignments of responsibility were made.

2 Sep 2009 -- John Madura relayed via e-mail a question to the LAP from Tim Oram (JSC/SMG), echoed by the 45WS, about the new anvil rules. This question was viewed as urgent, since John indicated that it was delaying implementation of the latest LLCC. Phil Krider was out of the country at the time, but the question was addressed by Jim Dye and John Willett with e-mail review by Phil but without formal discussion among

the entire LAP. An informal e-mail response was sent to John Madura and others by John Willett and Jim Dye on 18 September 2009. Below is a key quotation from that response:

> "The rationale for relaxation of the rules relative to time after lightning is essentially that the ABFM II dataset includes a large proportion of clouds in which (or in whose parent clouds) lightning was actively occurring during the aircraft penetrations. There was no apparent distinction between the clouds that did and that did not contain lightning as far as launch safety was concerned. The changes were made on the assumption that ALL of the LLCC must be satisfied as a package. There was never any intention to allow flight within 10 nmi of lightning."

John Madura replied that this response was adequate and closed the LAP issue.

24 October 2009, Current Space Shuttle LLCC (adopted 2009) [LCN WEA-1, see Appendix I, Section A1.11] -- A new version of the LLCC that is essentially equivalent to the version approved by the LAP on 2 October 2008 was implemented by the Range. Aside from numerous wording changes to translate from FAA to Range terminology, and reversal of the order to put the definitions at the end (counterbalanced by a statement at the beginning emphasizing that they are an integral part of the rules), there are only the following substantive changes:

> The LAP wording, "Whenever there is ambiguity about which of several LFCC applies to a particular situation, all potentially applicable LFCC must be applied," is not echoed by any similar wording in the Range version (although this is current practice at the Eastern Range).

> The Range wording, "Even when these criteria are not violated, if any other hazardous condition exists, the LWT will report the threat to the Launch Director (LD). The LD may Hold at any time based on the instability of the weather," does not correspond to any similar wording in the LAP/FAA version. [Note that although the familiar "Good Sense Rule" is omitted from these proposed LFCC as presented here, essentially the same wording does appear in another section of the current FAA LFCC implementation where it is still a mandatory requirement. See Appendix I, A1.12 and A1.13. The LAP expects that this requirement will remain in the regulations as the LFCC are updated by the FAA.]

> The LAP reference in the "preamble" to the definitions, "(See also the additional explanation beneath the definition of 'cloud.')," to additional explanation of distance measurements is missing from the Range-approved version.

> The LAP definition and rule for Triboelectrification has been removed by the Range (presumably because Shuttle has been hardened against triboelectric hazards).

Expanded "Space Shuttle LLCC (2005)" ILCN 0166R0, 13 June 2005 Debris-Cloud Rule:

If ONE OR MORE VAHIRR conditions is NOT satisfied, or if VAHIRR is not available:

| Distance from Cloud, D: Time After Detachment, T: | Through | 0 < D ≤ 3 nmi | 3 < D ≤ 5 nmi | 5 nmi < D |
|---|---|---|---|---|
| T ≤ 3 hr*** | No Launch* | No Launch unless Low Field | No Launch unless Low Field | --- |
| 3 hr < T | | --- | --- | --- |

If ALL 3 VAHIRR conditions ARE satisfied:

| Distance from Cloud, D: Time After Detachment, T: | Through | 0 < D ≤ 3 nmi | 3 < D ≤ 5 nmi | 5 nmi < D |
|---|---|---|---|---|
| T ≤ 3 hr*** | No Launch* | No Launch unless Low Field | No Launch unless Low Field | --- |
| 3 hr < T | | --- | --- | --- |

Notes:

A cell entry of "---" means the rule does not explicitly mention this condition.

This rule does not invoke VAHIRR, so the two panels are identical.

"Low Field" means (a) ≥1 working mill <5 nmi from cloud, (b) <1000 V/m for 15 min at all mills ≤5 nmi from flight path AND at mill(s) in (a), and (c) cloud is <10 dBZ ≤5 nmi from flight path for 15 min.

\* This cell is ambiguous between sections (a), which says No Launch, and (b), which allows a field-mill exception.

\*\*\* The three-hour waiting period begins again if lightning occurs in the debris cloud.

**Figure A2-9 Expanded Space Shuttle LLCC (2005) Debris-Cloud Rule**

Expanded "LAP 10/06/08 Recommendation in FAA Format" Debris-Cloud Rule:

If ONE OR MORE VAHIRR conditions is NOT satisfied, or if VAHIRR is not available:

| Distance from Cloud, D: | Through | 0 < D ≤ 3 nmi | 3 < D ≤ 5 nmi | 5 nmi < D |
|---|---|---|---|---|
| Time After Detachment, T | | | | |
| T ≤ 3 hr*** | No Launch | No Launch unless Low Field | -- | -- |
| 3 hr < T | -- | -- | -- | -- |

If ALL 3 VAHIRR conditions ARE satisfied:

| Distance from Cloud, D: | Through | 0 < D ≤ 3 nmi | 3 < D ≤ 5 nmi | 5 nmi < D |
|---|---|---|---|---|
| Time After Detachment, T | | | | |
| T ≤ 3 hr*** | Launch Not Prohibited | Launch Not Prohibited | -- | -- |
| 3 hr < T | -- | -- | -- | -- |

Notes:

A cell entry of "--" means the rule does not explicitly mention this condition.

"Low Field" means (a) ≥1 working mill <5 nmi from cloud, (b) <1000 V/m for 15 min at all mills ≤5 nmi from flight path AND at mill(s) in (a), and (c) cloud is <10 dBZ ≤5 nmi from flight path for 15 min.

*** The three-hour waiting period begins again if lightning occurs in the debris cloud.

**Figure A2-10 Expanded LAP Recommended (10/06/08) Debris-Cloud Rule**

# Appendix III    Members of the PRC/LAP

**Christian, Dr. Hugh J. (1999 to present)**
Principal Research Scientist
University of Alabama in Huntsville

Dr. Christian specializes in the observation and measurement of electrical phenomena in the atmosphere through in situ and remote sensing techniques. His expertise includes ground-based and airborne field mill measurements, cloud electrical structure, techniques for location of lightning using ground-based electromagnetic remote sensing, and location of lightning using space-based optical remote sensing. He has authored or co-authored numerous conference papers and peer-reviewed journal articles.

Member of Busse Committee after the AC 67 incident; Principal Investigator of ABFM I experiment; LAP member since 1999; planning and oversight of the ABFM II experiment; NASA perspective.

**Dye, Dr. James E. (2002 to present)**
Senior Scientist Emeritus
National Center for Atmospheric Research, Boulder, Colorado

Dr. Dye is an internationally recognized leader in cloud physics, atmospheric electricity, airborne instrumentation, polar stratospheric clouds, production of nitrogen oxides by lightning, and use of both intra-cloud and cloud-to-ground lightning for characterizing severe storms. He has worked extensively in the design, direction, participation and data analysis of coordinated multiple aircraft--radar studies of thunderstorms using both powered aircraft and an instrumented sailplane; and the development and evaluation of airborne cloud particle measuring devices and liquid water content sensors and has participated in or led more than 20 national and international field experiments. He has published over 90 refereed journal articles, reports or chapters in books and over 125 conference publications and reports. He is a fellow of the American Meteorological Society, a member of the American Geophysical Union, a former editor of the Journal of Geophysical research and former Associate and Acting Editor of the Journal of Atmospheric and Oceanic Technology.

Principal Investigator of ABFM II experiment; PRC member initially and LAP member since 2003; academic-research perspective.

**Krider, Dr. E. Phillip, Chairman (inception to present)**
Professor, Department of Atmospheric Sciences and Institute of Atmospheric Physics
The University of Arizona

Dr. Krider is an author or co-author of more than 130 reviewed publications, and he has 8 patents. He is a Fellow of the American Geophysical Union (AGU) and the American Meteorological Society (AMS), and a Member of the American Association of Physics Teachers and the History of Science Society. Dr. Krider received the AMS award for Outstanding Contributions to the Advance of Applied Meteorology in 1985, and he currently chairs the NASA/Air Force Lightning Advisory Panel for space launch operations. Dr. Krider is a former Co-Chief Editor and Editor of the *Journal of the Atmospheric Sciences*, former Associate Editor of the *Journal of Geophysical Research*, and he is a past President of the IAMAS/IUGG International Commission on Atmospheric Electricity (ICAE).

Participated in Apollo XII investigation; member of Aerospace (Heritage) and NAS/NRC Committees after AC 67; Chairman of the PRC/LAP since its inception; planning and oversight of all LAP-proposed field experiments and direct participation in the ABFM II at KSC; academic perspective.

**Koons, Dr. Harry C. (inception to 2005, deceased)**
Distinguished Scientist
Space Applications laboratory, Aerospace Corporation, El Segundo, CA.

Fields of specialization include magnetospheric physics, spacecraft charging, spacecraft environmental anomaly investigations, very-low-frequency propagation, wave-particle interactions, and active experiments in space. He pioneered the use of expert systems and neural networks in areas related to the space environment. He was an investigator for space flight experiments on the OV1-21, SCATHA, AMPTE and CRRES spacecraft. Dr. Koons was a member of the American Geophysical Union, the Union of Radio Science, the American Association for the Advancement of Science, the American Institute of Aeronautics and Astronautics and the Society of Sigma Xi.

Statistical analysis of ABFM II dataset; LAP member continuously from its inception until his death; Air Force-research perspective.

**O'Brien, Dr. T. Paul (2007 to present)**
Aerospace Corporation, El Segundo, CA.

Paul O'Brien received his B.A. in Space Physics and Astronomy from Rice University in 1997. Dr. O'Brien received his MS degree in 1999 from UCLA, and completed his Ph.D. in 2001 under Dr. Robert McPherron in the area of Geophysics and Space Physics. His work with Dr. McPherron focused on geomagnetic storms as space weather with focus on the ring current and trapped energetic electrons. During a 1-year post-doc at UCLA Dr. O'Brien worked part time with Dr. Didier Sornette covering a broad range of complex systems, from helicopter engines to earthquakes. At Aerospace Corporation he continues to work actively on the radiation belts, ring current, and other phenomena in the inner magnetosphere, with emphasis on the application of space science results to tools for spacecraft design and situational awareness. Dr. O'Brien's work frequently uses statistical, nonlinear and artificial intelligence methods for analyzing large data sets to describe the magnetosphere.

Statistical analysis of ABFM II dataset; LAP member continuously since 2007; Air Force-research perspective.

**Rust, Dr. W. David (inception to present)**
Physical Scientist
National Severe Storms Laboratory

Dr. Rust coauthored the graduate level text and professional reference book, "The Electrical Nature of Storms." He has pursued research that covers topics ranging from the electrical structure of storms to lightning. He is internationally recognized as an expert in making balloon-borne measurements of electrical parameters in storms using free-flying, instrumented balloons, and has significant experience in the development and use of mobile laboratories to conduct research requiring the ability to collect data while tracking with storms. He has been served on the Lightning Advisory Panel, and its various predecessor groups, addressing atmospheric electricity launch issues since Apollo-Soyuz.

Member of the Aerospace (Heritage) Committee after the AC 67 incident; PRC/LAP member since its inception around 1990; planning and oversight of all LAP-proposed field experiments and direct participation in one at KSC; NOAA perspective.

**Walterscheid, Dr. Richard L. (inception to present)**
Senior Scientist
Space Sciences Department, Aerospace Corporation, El Segundo, CA.

Dr Walterscheid has extensive experience in weather and the meteorology of cloud electrification. His experience in weather includes service as a Weather Officer and Aerospace Sciences Officer for the Air Force Air Weather Service and an Interagency Personnel Act assignment to the National Oceanic and Atmospheric Administration. His work related to lighting includes aviation forecasting of thunderstorm activity in Florida, a study of triggered lightning strikes to aircraft in Europe, and long-standing service on the NASA and Air Force Lighting Advisory Panel and its predecessors since 1987. His research work in the physics and dynamics of the atmosphere has led to approximately 130 peer reviewed publications. Dr. Walterscheid is a member of the American Geophysical Union, the American Meteorological Society, and Phi Beta Kappa.

Member of the Aerospace (Heritage) Committee after the AC 67 incident; PRC/LAP member since its inception about 1990; planning and oversight of all LAP-proposed field experiments; Air Force research perspective.

**Willett, Dr. John C. (inception to present)**
Consultant in meteorological sciences

Dr. Willett has extensive experience in lightning physics, atmospheric electricity, and boundary-layer meteorology. He has served as a senior physicist at the Air Force Research Laboratory, a NRC Senior Research Associated at NASA/GSFC, and a meteorologist at the Naval Research Laboratory; and he was an Adjunct Professor in the Meteorology Department at the University of Maryland. He has served as a member of the USAF/NASA Lightning Advisory Panel from 1992 to the present, as chairman of the American Geophysical Union Committee on Atmospheric and Space Electricity from 1992 to 1996, and as an Associate Editor for JGR Atmospheres for two four-year terms. He has also served NASA/KSC as a member of both the Review Panel on the Rocket-Triggered Lightning Program and the Advisory Panel on Upgrading the Field Mill System. He has numerous publications on atmospheric electrical and lightning research.

PRC/LAP member since its inception around 1990; planning and oversight of all LAP-proposed field experiments; Navy/Air Force perspective.

# Appendix IV    ABFM I Science Team and Key Results

Dr. Monte G. Bateman, NASA Global Hydrology and Climate Center, Marshall Spaceflight Center
Dr. Hugh J. Christian, NASA Global Hydrology and Climate Center, Marshall Spaceflight Center
Dr. Douglas M. Mach, NASA Global Hydrology and Climate Center, Marshall Spaceflight Center
Launa Maier, NASA, KSC

Key Results:
ABFM I Winter studies (MSFC Feb 92 and Oct 92 reports and Mach et al., 1992)

- For layer clouds a strong correlation was found between electric fields inside the clouds and vertically integrated radar reflectivity (VIR).
- Fields at the surface were not always correlated with fields aloft in layer clouds.
- The lowest altitude with fields >=5 kV/m was 13 kft and the highest altitude with fields >=5 kV/m was 20 kft (~-10 °C) suggesting that the main area of hazard was between 0 and -10 °C
- For disturbed weather clouds there was a strong correlation between the electric fields aloft and VIR.
- For disturbed weather clouds all penetrations with fields >5 kV/m were between 10 and 21 kft (approximately 0 to -10 °C).
- Fields measured at the surface below disturbed weather clouds correlated well with fields aloft, but there were 3 exceptions where fields at the surface were <1 kV/m and fields aloft were >5 kV/m.
- Penetrations with fields > 3kV/m were generally associated with regions of higher reflectivity (embedded convection).
- Cloud thickness and electric field strength were very poorly correlated.

ABFM I Summer studies

- The fields outside of storms with lightning fell over to <5 kV/m at a range of 5 nm.
- For cumulus clouds, the electric fields depended strongly on the cloud top height. Fields in clouds with 10 dBZ tops lower than the 0 °C level did not exceed 3 kV m$^{-1}$. Fields >3-5 kVm$^{-1}$ did not develop in the clouds until the echo tops had grown higher than the -10 °C level. The clouds did not produce lightning until the tops were higher than the -20 °C level. Fields at the edge of clouds with echo tops higher than the -20 °C level could be > 50 kV m$^{-1}$.
- When clouds were within 5-6 nm of the Ground Based Field Mill network there were no cases when the fields aloft were >5 kV/m and fields at the surface were <1 kV/m.
- Even with fields of tens of kV m$^{-1}$ inside electrically active convective clouds, the fields external to these clouds decayed to < 3 kV m$^{-1}$ within 15 km of cloud edge. This was true even for clouds that were producing lightning with tops above -20 °C.

# Appendix V    ABFM II Science Team and Key Results

All LAP Members listed in Appendix III were part of the ABFM II Science team.

Other Members:

Dr. Monte G. Bateman, USRA, Huntsville, AL

Dr. Dennis Boccippio, NASA, MSFC

Dr. Eric Defer, NCAR

Dr. Cedric A. Grainger, University of North Dakota

Dr. William. D. Hall, NCAR

Sharon A. Lewis, NCAR

Dr. Douglas M. Mach, University of Alabama at Huntsville

John Madura, NASA, Kennedy Space Center Weather Office

Dr. Francis J. Merceret, NASA, Kennedy Space Center Weather Office

Natalie Murray, University of Arizona

Michael Stewart, NASA, MSFC

Paul T. Willis, NOAA, Hurricane Research Division

Jennifer G. Wilson, NASA, Kennedy Space Center Weather Office

In addition, the winter 2001 cloud radar team from NOAA/ETL included Brooks Martner and Bruce Bartram.

Key Results:
- An extensive set of airborne measurements of electric field; particle concentrations, sizes and types were obtained in coordination with radar reflectivity measurements in anvils and debris clouds.
- The measurements showed that when electric fields were strong (> 10 kV/m) the radar reflectivity at or near the aircraft location was always high (>5 to 10 dBZ), but high reflectivity did not always mean that the electric field was strong.
- Similarly, when electric fields were strong, the particle concentrations in all size ranges from ~10 microns to several millimeters were also high.
- The transition from weak fields (< 1 kV/m) to strong fields (>10 kV/m) inside the anvil and debris clouds occurred quite abruptly and this transition usually occurred as the aircraft flew from weaker reflectivity into regions with reflectivity greater than approximately 10 dBZ.
- As a manifestation of the abrupt transition from weak to strong electric field, scatter-plots of electric field versus reflectivity for the entire anvil and debris cloud data set exhibited a threshold behavior such that for fields ~<3 kV/m the reflectivity was weak.

- Using the above finding as a basis, the radar based parameter, volume averaged height integrated radar reflectivity (VAHIRR) was developed. VAHIRR is the product of the average reflectivity times the anvil/debris cloud thickness above the 0 °C level.

- In some long-lived anvils a secondary stratiform-like development was observed in which both electric field and reflectivity became enhanced and persisted for periods more than an hour over large areas. Even in these conditions VAHIRR was able to identify regions with strong electric fields.

- Measurements near anvil or debris cloud edges clearly showed that particles extend out to or beyond the 0 dBZ radar contour and well beyond the 10 dBZ radar contour. As a result of these observations the LAP changed the definition of "anvil edge" in the LLCC rules from +10 dBZ to 0 dBZ.

- A simple model was developed to estimate the decay of electric field in the ABFM anvils based upon the actually observed particle size distributions. The estimates from the model show that the time for the electric field to decay from 50 kV/m to near zero can be as long as 1 ½ hours in the dense part of an anvil, but as short as only a few minutes in weak reflectivity near the edge of the anvil. These estimates are considered to be upper limits.

- Comparisons for case study days of electric field decay time scale from the model were generally consistent with observed decay times, but only one ABFM anvil case permitted a meaningful comparison.

- In anvil and debris clouds observed during ABFM II the electric field fell off rapidly from the cloud edge to values <1 kV/m by 8 km distance from the edge.

# Appendix VI  Apollo XII Timeline

Launch of Apollo 12 occurred on 14 November 1969 at 16:22 UT   (Returned to Earth on 24 November 1969 at 20:58 UT)

Two lightning strikes occurred after lift-off, a 2-stroke CG flash (with two ground terminations) recorded on video at T+36.5 sec and an IC discharge occurred at T+52 sec.

November 15, 1969 *NY Times* (page 22, col. 1) '…aircraft reported that there seemed to be no electricity in the clouds.'

November 17, 1969 Dr. Richard E. Orville drafted letter to *NY Times* pointing out that lightning can be 'initiated by towers and tall buildings … and perhaps by a fueled rocket launching astronauts toward the moon.'

November 24, 1969 Article by B. K. Thomas, Jr. appeared in *Aviation Week and Space Technology* with the title 'Power Source Loss Marks Apollo Launch' illustrates initial confusion about the cause of incident. Mentions possibilities like 'discharge of static electricity,' 'vibration during launch phase opening circuit breakers,' and 'vehicle had been struck by lightning. But this was considered remote because the effects would probably have been more severe than those actually experienced, according to Walter J. Kapryan, KSCs launch director.'

November 26, 1969 Informal Seminar and Discussion on <u>The Apollo 12 Lightning Flash</u> was hosted by Dr. Arthur A. Few, Jr. at Rice University on the occasion of a visit to MSC by Dr. Orville.

December 4, 1969 'Meeting on AS-507 Lightning' held at Marshall Space Flight Center (MSFC) hosted by Glenn E. Daniels (S&E-AERO-YT) to discuss: Magnitude of the lightning hazard to a Saturn V vehicle in flight; Mechanism of atmospheric electrical charge and discharge on the Saturn V vehicle in flight; and What launch operational constraints should be imposed to prevent a reoccurrence of the incident.

December 5, 1969 Open letter by Dr. Few to the scientific community working on lightning physics and related questions stating that arrangements had been made with the AGU and the NASA-MSC to have a Special Session at the upcoming Fall AGU Meeting to obtain more 'hard' information on the Apollo 12 event.

December 8, 1969 Letter by Dr. Andrew E. Potter (NASA-MSC) to specialists in relevant areas announced that there would be an Open Meeting at the Jack Tar Hotel (venue for the Fall AGU meeting) at 12:15 PM on December 16 where the facts available on the Apollo 12 lightning incident and its effects on the spacecraft will be presented. Ernie Amman, the ESSA weather forecaster working at KSC at the time of the incident, will review the meteorological situation at the time of the launch. Potter also mentioned an invitation to a 'small discussion group' would be held at 5:00 PM to cover specific questions such as: Would these discharges have occurred in the absence of the rocket launch, or were they triggered by the rocket? Can the magnitude of the vertical electric field and the intensity of rainfall at the launch site be used to indicate hazardous launch conditions? Is it possible to define a characteristic weather pattern which might yield rocket-triggered lightning discharges? Are additional and/or improved measurements necessary to estimate the probability of lightning during a launch? (A list of the people who were invited to the 'small group' discussions are acknowledged at the beginning of the Preface to the MSC-01540 Report.)

December 9, 1969 M. Brook, C. R. Holmes, and C. B. Moore of New Mexico Tech draft a Technical Note entitled 'Exploration of Some Hazards to Naval Equipment and Operations Beneath electrified Clouds' for the

Office of Naval Research (ONR) under Project Themis. This note was subsequently published as Appendix 2 in NASA Report MSC-01540, February 1970, and in the ONR *Naval Research Reviews*, April 1970.

December 16, 1969 The special session and subsequent discussions were held at the Fall AGU Meeting and were Chaired by Donald D. Arabian.

December 22, 1969 An article appeared in *Aviation Week* entitled 'Apollo 13 Launch Rule Changes Will Be Recommended to NASA' (also mentioned an article in the December 8 issue, p. 52) where D. Arabian is quoted as saying 'We have to take the state of the art and apply it.'

December 23, 1969 Rocco A. Petrone, the Apollo Program Director at NASA Headquarters, called on the MSC, MSFC, and JSC to resolve the Apollo 12 electrical discharge incidents before the Apollo 13 Flight Readiness Review (FRR) on February 5, 1970. Scheduled an LDX meeting with MSC/KSC/MSFC and Headquarters on January 22, 1970 (this may have subsequently been postponed to February 6). Petrone asked for the MSC (Don Arabian) to coordinate the efforts of the three NASA centers in drafting this report.

December 30, 1969 Memorandum from W. O. Campbell of Bellcomm, Inc. on 'Apollo 12 Launch Weather' compared Apollo 12 launch weather conditions with mission rules and noted that the candidate rules for lightning avoidance then under discussion 'suffer generally from ambiguity, lack of objectivity, and depend to some extent on local forecasting experience.'

January 7, 1970 Arabian memo 'Reporting of the Apollo 12 lightning incident' sent to W. Vaughan (MSFC) and H. Clark (KSC) suggesting eight specific chapters for the Apollo 12 report.

January 8, 1970 KSC Systems Engineering responded to request from H. J. Clark and provided:
   Description of the present Lightning Instrumentation
   A summary of the ESSA Research Studies then underway
   Proposed instrument changes for the Apollo 13 launch
   Past history of lightning strikes at the LC-39 (500F) launch site

January 13, 1970 Dr. Robert Manka (MSC) summarized recommendations for 'Methods of Measuring Electric Fields' that were discussed at the AGU Meeting on December 16. Ground-based and airborne measurements were considered. Stated that preliminary 'hazardous field levels' are between $\pm 1$ and $\pm 3$ kilovolts/meter. Need to obtain relationships between airborne and ground-based electric fields and values of the 'critical electric field' inside clouds.

January 14, 1970 Memo from W. J. Waugh, Manager of Spacecraft Engineering at Boeing-Houston, to D. Arabian on 'Apollo 12 Electrical Power Transient During Boost Anomaly.' Reviewed environmental factors at the time of the Apollo 12 launch and gave specific recommendations for 'operationally practical mission launch rules' with associated rationale.

January 14, 1970 D. Arabian invited (through E. P. Krider) M. Brook, D. Fitzgerald, G. Freier, C. B. Moore, and E. T. Pierce to meet at MSC on January 21-22, 1970, 'to discuss further the problem of forecasting lightning hazards to Apollo launches. Particular emphasis will be placed on electric field measurements, sferics detection, and radar observations.'

January 21-22, 1970 Above meetings were held at MSC.

January-February, 1970  MSC-01540 Report was written.

January 30, 1970 MSFC releases MPR-SAT-FE-70-1, 'Saturn V Launch Vehicle Flight Evaluation Report AS-507 Apollo 12 Mission.'

February, 1970 Report on 'Analysis of Apollo 12 Lightning Incident' (MSC-01540) released by the MSFC, KSC, and MSC. Contained LLCC that 'will satisfy the low risk requirements of the spacecraft and the lesser restrictions of the launch vehicle' (p. 50).

February 2, 1970 Arabian announced that an LDX meeting on the Apollo 12 lightning will be held between NASA Headquarters, MSFC, KSC, and MSC on February 6, 1970. This meeting covered
- Precisely what happened
- Cause of the incident
- Hardware and software changes
- Instrumentation to improve identification of potentially hazardous electric fields
- Adequacy of automatic abort system and ordnance device for lightning
- Recommended changes in the launch rules

February 4, 1970 Memo from GE Apollo Systems Department described some telephone contacts that were made with Drs. Brook, Freier, Moore, and Fitzgerald who were asked "Are you aware of any documented violation of the 1 to 1 cone used for protection against direct lightning strokes?" (These were initiated in response to questions about the efficacy of lightning protection at the launch site.)

March 19, 1970 Memo from Drs. R. H. Manka and A. E. Potter to D. Arabian detailing consensus suggestions for the type of electric field sensors and other ground-based instrumentation to have at KSC (and their calibration). Also, recommended a 'small, low-power radar with range-height indicator' that could provide rapid scans in elevation (3 to 5 seconds per scan) through the vertical, and slowly in azimuth (about 2 to 3 minutes per hemisphere).

April 22, 1970 Special Session held at Spring AGU Meeting on 'Triggered Lightning/Atmospheric Electricity' that was Chaired by B. Vonnegut and M. Brook. 16 abstracts of papers were published in EOS, Vol. 51, No. 4, April 1970.

May 21, 1970 Summary of a 'Special Discussion' that followed the above AGU session (prepared by Drs. Krider, Manka, and Freden) indicates that the discussion was divided into three parts:
    i) Comments on the launch rules that existed then (i.e. after Apollo 12);
    ii) Instrumentation that was required to implement the launch rules; and
    iii) Experiments which would be desirable during future launches.

May 22, 1970 James Hughes (ONR) visited D. Arabian (who was busy with Apollo 13 problems) in Houston asking whether the Kennedy Space Center could be used to study the atmospheric electrical environment at KSC, cumulus clouds as they make transitions between land and sea, and electrical phenomena that are associated with the launching of rockets.

June 19, 1970 KSC Memo from Manager, Apollo-Skylab Programs (Brigadier General, USAF, Thomas W. Morgan) to MSFC (L. B. James) and MSC (J. A. McDivitt) on 'KSC's Role – Research in Atmospheric Electricity.' Realized that additional efforts are required and that additional efforts will be 'somewhat useless to NASA and other agencies if not properly integrated such that a user (launch site, such as KSC) can utilize or implement results into present and future operational programs.'

# Appendix VII  Evolution of the Lightning Instrumentation at CCAFS

This appendix provided by Michael W. Maier, of the Range Technical Services Contractor, CSR. It includes his personal recollections as an active participant in the events described.

Mid-Late 1960's (Prior to Apollo 12)
NASA deploys eight radioactive probe electric field meters at KSC manufactured by the Sweeney Corporation. Date of initial installation unknown, but they were operating when Apollo 12 was launched in November 1969. NASA also deployed eight corona current detectors (4 ft whip antenna connected to a microampmeter) at the same sites as the radioactive probes and operated a two station sferics locating system. Five of the eight were located on the ground (1.5 m above grade) and three on buildings (VAB, MSOB which is I think now called the O&C Building, and one other unnamed building). There is a map of stations in the Apollo 12 event report (NASA TM-X-62894)

There was much discussion of the calibration of these instruments (see Calibration of Atmospheric Electric Field Meters and Determination of Form Factors at Kennedy Space Center, by H. W. Kasemir, ERL APCL-11, April 1971). Kasemir concluded the maximum field values at the surface during the one hour prior to the Apollo 12 launch were ±4 kV/m. He also noted the instrument at Site 8 located at the pad recorded fields of -3.2 kV/m from the steam cloud at the pad during launch of Apollo 12 launch and +3.7 kV/m during Apollo 13 launch on a fair weather day.

1969
November 14 - Apollo 12 struck by lightning at 36.5 seconds (6400 ft vehicle altitude) and 52 seconds (14400 ft vehicle altitude) after liftoff. Mission comes within seconds of abort due to loss of all telemetry when ground controller John Aaron suggests Astronaut Alan Bean select AUX setting on spacecraft signal conditioning equipment.

1970-1972
NASA deploys 25 field mills designed by George Freier of the University of Minnesota Physics Department at sites on KSC. These would later be designated the NASA Phase I instruments. The specification is found in KSC-SPEC-M-0007, Specification for Lightning Instrumentation System at KSC. The system begins operating on 1200 ET 28 July 1972. The initial transmission format from the site to the CIF was FSK. The FSK signals were not transmitted continuously, the CIF sent a 1.5 s long interrogation signal to all mills and then each mill responded following a preprogrammed delay so not two sites would be transmitting at the same time. Not sure what the actual sample time was.

The instruments were implemented in the original "two box" site design where the rotor, stator, and motor located in one box (at the center of the cleared area) with a second box located on the edge of the cleared area containing the signal conditioning circuit cards. The FSK signals were hard wired to the CIF PCM Data Station (Room 323) for conversion, recording and display. The data from up to eight instruments mills were converted to analog and could be plotted on Brush strip chart recorders. The PCM data was also transmitted to the Meteorological Predication Center (MPC) CIF for use by the NOAA SMG team resident at KSC.

1973
Calibration procedure for the NASA field mill is described in GP-1019 (Rev 1) dated 25 July 1973. Instruments should have a Vout of +8V when +957.6 V is applied to a plate 2.1 cm above the rotor.

<u>1972-1976</u>
Sometime during this period the CIF function was relocated to the Central Measurement Data Processing System (CMDPS) in Rooms 2P19 and 2P20 of the LCC and the first real-time computer processing of the data began. I believe this was also when the interface to the NASA RDS 500 computer was established (the LCC Environmental and Special Measurement System) and digital processing and display of the KSC field mill data began with a real-time contouring program developed by Elmer Magaziner of NOAA ERL.

The AFETR installed a network of eight electrostatic field mills of a type different than the 25 NASA instruments in the early 1970's. The system is described as the Launch Pad Lightning Warning System (LPLWS) in the July 1976 Range Instrumentation Handbook (and not in the 1965 handbook). The eight mills on CCAFS were located at Pad 41, Pad 40, Pad 13, Weather Station, Pad 36, Pad 17, Pad 25, and the Navy Port. There was also one field mill at PAFB which I recall seeing when I visited there in the 72-75 time frame. Data were processed by the AFETR CDC 3100 computer in real-time with a teletype output. There were also strip chart outputs in the Cape Canaveral Forecast Facility (CCFF) at the Range Control Center (RCC).

NASA documentation is confusing about the number of field mill sites during this period. It appears there were more than 25 instruments (or sites) but only 25 could be processed by the RDS computer. I have a couple different maps showing different configurations for manned launches versus unmanned launches. It appears NASA may have moved some of the instruments to sites on Air Force property just prior to STS 1 to provide better coverage since the ESMC mills were concentrated around the ELV pads.

<u>1974</u>
Phil Krider recommends NASA increase the calibration plate spacing from 2.1 to 10.0 cm based on his field comparisons from that summer. He reported the 2.1 cm spacing was too small to provide an accurate calibration since the rotor to stator spacing of the NASA mills varied from instrument to instrument by 4 mm (between 2 to 6 mm spacing). He also recommended that NASA improve the spacing between adjacent stator plates on the "second lot" of field mills. The second lot was made with larger stators and the spacing between adjacent stators was so small that rain drops would short the adjacent sectors during heavy rain leading to about half the network becoming inoperative during heavy rain.

<u>1975</u>
GP-1019 is revised in May 1976 (Rev 2) and still calls for a nominal Vout of 8.0V when 984.4 V is applied to the calibration plate with 2.1 cm spacing so evidently the increase in plate spacing recommended by Phil Krider was not implemented at this time.

In April NASA did field testing of mills with 10 cm calibration plate spacing. This larger spacing must have been implemented at some time between 1975 and 1986 when I worked for Pan Am since it was close to 10 cm in 1986. I wasn't aware of GP-1019 or the change to 10 cm and I didn't measure it until 1990 when it was 9.4 cm. I don't have any drawings or records of when it changed.

<u>1976</u>
Charlie Moore and Steve Friberg of New Mexico Institute of Mines and Technology conduct an independent evaluation of the absolute calibration of the 25 KSC field mills. They place a portable reference instrument next to each KSC mill and compare data under fair weather conditions. They find the mean ratio of all collocated mill readings was 0.97 with a standard deviation of 0.31. However, the mean differences between the NMIMT reference mill and the NASA mill varied from 0.52 to 2.29. As a result of this test NASA considered the KSC field mills to be correctly calibrated in an absolute sense but the validity of individual site measurements suspect due to either maintenance problems or exposure variations.

## 1980

Prior to STS 1 NASA and ESMC begin sharing field mill signals but there are two parallel real-time processing systems, the NASA RDS 500 system in the LCC and the ESMC Cyber 74 at CCAFS. The Cyber 74 was now running a version of the NOAA contour plotting software which was displayed on a Tektronix terminal in the CCFF. The CCFF also has two eight pen strip chart recorders and could plot in real-time data from up to 16 mills selected using front panel switches on the console corresponding to the PCM channel IDs.

Agreement was reached between NASA and ESMC that all real-time field mill processing would move to the ESMC system as soon as it could be verified and proved reliable. Much effort was spent in comparing the outputs of the two different mill types and the two different display systems.

Also in 1980 Phil Krider recommended NASA and USAF implement a real-time delta E plotting capability to locate the charge centers and work begins to code this into the ESMC software. By this time, Krider had also recommended that the protection diodes on the input of the field mill electronics be removed because they introduced an irritating non-linearity in the field mill output that had to be corrected with software. By this time, Krider had also recommended that the protection diodes on the input of the field mill electronics be removed because they introduced an irritating non-linearity in the field mill output that had to be corrected with software.

## 1981

Operation and maintenance responsibility for the 25 NASA field mills was transferred to ESMC who in turn tasked Pan Am World Services to perform this function under the Range contract. Efforts continue to improve/debug software, refine the calibration of the instruments, and develop procedures to prevent bad instruments from contaminating the real-time display products.

A new site identification scheme was adopted for both NASA and ESMC field mills (34 sites in total)

## 1984

ESMC adds another 8 pen strip chart recorder to the CCFF, can now display 24 of 30 field mills

## 1985

The eight AFETR field mills were deactivated and replaced by NASA Phase I Field Mills at all CCAFS sites to form a network of 30 mills.

The original NASA instrument design was modified by replacing the multiple circuit cards in the second box with a single output card in the motor enclosure. Now analog data from each mill (including those on CCAFS) to the LCC where they were digitized and PCM coded in Room 2P19. In Room 2P20 they PMC data could be recorded on analog tape and sent to the NASA RDS 800 computer for processing and display. PCM data for all mills was also transmitted to the Central Computer Complex at CCAFS where the data was converted by the ARMS buffer into Cyber format for processing and display.

The instrument modifications were made by the ESMC (Pan Am) and documented in Range Engineering Drawings 403147 through 403153 but there was no record of a formal test since the LPLWS was not covered by ESMCR 80-4 (the Range configuration management policy at the time).

NASA (I think Phil Krider) observes high amplitude, high frequency noise on the field mill output during rain. Subsequent investigation finds the low pass filter in the original NASA design was not retained in the Range modification performed the prior year when the network was consolidated. All field mills were then modified to add a low pass filter in 1986 and 1987. These modifications are documented in ESMC engineering drawings 404232 through 404237.

When I joined Pan Am, I was told to go to the junk yard where the electronics enclosures from the original NASA design were dumped, remove the low pass filters, and then solder them into each mill per the engineering drawings. I remember doing these modifications during the summer of 1986.

In December 1986 I leave Pan Am but before I leave I give them a report on the calibration interval of the LPLWS instruments. During the summer when I was assigned to the LPLWS cleaning crew I was astounded to find many instruments seriously out of calibration after only a week of exposure (after prior cleaning, which I performed). I started to record the before and after cleaning calibration data and plotted the change in calibration as a function of time for all of the instruments over a six month period. The data confirmed the Phase I field mills must be cleaned and checked weekly to maintain 10% accuracy, particularly those within a few kilometers of the coast. I also recommend all instruments be cleaned and checked on F-1 day and the before and after calibration data be recorded.

1987
On March 26 an Atlas Centaur is launched into precipitating, disturbed, debris clouds with high electric fields at ground measured by LPLWS and initiates a lightning flash which leads to the vehicle being destroyed.

1988
Electric Field Mill Training Handbook written by Launa Maier of NASA and Capt Tom Strange of Det 11 2 WS is approved by Bob Sieck and John Madura and issued (January 1988). Still a great document!

LPLWS Requirements (NASA Document 79K32611) released by the joint NASA/ESMC LPLWS Task Team. The Phase I requirements were:

Add new field mill sites to provide coverage of all areas of interest

Conduct survey to verify location of all field mills

Replace the aging Tektronix monitor and strip chart recorders in the CCFF

Improve exposure of all field mills by clearing sites per the standard included in the requirement

Conduct an Operational Acceptance Test of the LPLWS to be overseen by a Detachment 11, 2nd Weather Squadron (Det 11 2 WS) observer

Establish configuration management and control of LPLWS. The configuration of the LPLWS was frozen for the OAT in December 1987 although three sites were not certified

Establish appropriate configuration identification for all aspects of the LPLWS

Begin a LPLWS replacement project (Phase II) where NASA would design and provide improved field mill sensors and ESMC would provide improved processing and display capability

I recall the goal was to certify LPLWS Phase I as operational prior to the return to flight of Space Shuttle in October 1988. I think this was accomplished but I can't find the official documentation or date, this was a NASA/Det 11 2 WS led effort.

## 1990

Concerned that the NASA/ESMC Task Team assumed the LPLWS calibration was accurate and development of the new instruments appeared to be dragging out, Launa and I flush mounted an LPLWS field mill next to the operational mill at Site 29 in September. We ensured the flush mounted mill was absolutely calibrated using two different plate spacing with a really large plate and collected simultaneous data over a wide range of fields. We found the operational LPLWS output was well correlated with the flush mount instrument but the operational instrument output (electric field) was systematically 40% too high. Test was repeated at another site (Site 11) and with other instruments and confirmed to be consistent. Error appears to be due to a change in the calibration plate spacing sometime between 1976 and 1990. The plate spacing in use in 1990 was 9.4 cm and probably should have been 10.0 cm. We reported the results to Det 11 2 WS and KSC personnel but there was no support to change the calibration procedure of the old instruments since the system had just been officially accepted and the expectation was the new Phase II instruments would be fielded shortly and these would be correctly calibrated by MSFC upon delivery.

## 1992

On 22 January the launch of STS 42 was delayed for one hour by high fair weather electric fields exceeding the +1000 V/m launch commit criteria. After the launch we reminded everyone the LPLWS calibration was 40% too high and the true fields probably did not exceed the launch commit criteria. Lots of meetings and finger pointing, I remember being scolded by Det 11 2 WS for not doing a better job explaining to them the 1990 findings about the calibration error and Jack Ernst for not following "proper procedures" for reporting the results.

Immediately following the STS 42 launch ESMC and KSC agreed to allow CSR to change the calibration procedure for the LPLWS based on the 1990 flush mount calibration data but Jan Zysko was still not convinced we knew what we were doing. Jan brought in Charlie Moore and Steve Hunyady from New Mexico Tech to conduct an independent verification of the LPLWS calibration using their reference mills. Their report to NASA on 30 May 1992 confirms our recalibration was indeed accurate and this calibration was retained until the Phase I instruments were decommissioned.

I consider it a major accomplishment that NASA and the USAF didn't need to bring Charlie Moore back in 2008 to perform yet another independent test of the field mill calibration, something that had previously been required every 16 years. I note with some sadness his passing last week; he was a person to be admired for many reasons.

## 1997

December – Operational acceptance of LPLWS Phase II and decommission of the Phase I instruments and processing systems.

## 2010

NASA/KSC and the 45th Space Wing are discussing ways of funding upgrades or replacements for the Eastern Range CGLSS, LDAR and LPLWS.

# Appendix VIII Weather Instrumentation at KSC/CCAFS

As noted in Section 1.2, the Eastern Range benefits from an extensive suite of weather instrumentation including surface electric field mills, weather radar systems, surface meteorological towers, rain gauges, weather balloon systems, Doppler wind profiling radars and more. Each of these systems will be discussed separately here.

*A8.1 Systems for Atmospheric Electricity: Electric Field Mills (LPLWS), CGLSS, LDAR and NLDN*
The Eastern Range operates three major lightning-related instrumentation systems: electric field mills (called the Launch Pad Lightning Warning System or LPLWS), the Cloud to Ground Lightning Surveillance System (CGLSS) and the Lightning Detection and Ranging (LDAR) system. CGLSS and LDAR are sometimes referred to in combination by the Range as the 4-Dimensional Lightning Surveillance System (4DLSS). In addition, the Range receives data from the commercial National Lightning Detection Network (NLDN). Each of these is described below.

*A8.1.1 Electric Field Mills (LPLWS)*
The LPLWS is basically a large-area network of electrostatic field sensors (field mills) that measures the vertical component of the electrostatic field, E, produced by cloud charges aloft, even if those clouds are not producing lightning. The sensors are termed 'mills' because they employ a rotating and grounded metallic shutter to alternately cover and uncover a set of insulated stators that respond to E. The amplitude and polarity of the current flowing to and from the stators is proportional to the amplitude and polarity of the local electric field. When an electrified cloud forms overhead or moves into the region from somewhere else, the E field will increase and often change polarity. In most cases, E will be large close to the cloud charges and small farther away, so 2-dimensional maps of E will show approximately where the cloud charges are located.

If the cloud electric field is measured in conjunction with a weather radar that senses precipitation, the onset of the electric field can be compared with the onset and type of precipitation, the rate of echo growth, and the time-evolution of the cells, thereby improving both the nowcasting (i.e. detection) and forecasting of high fields aloft. When lightning discharges occur, maps and analyses of the lightning-caused changes in E or $\Delta E$ can be used to determine an approximate location (in 3-dimensions) of the change in the cloud charge (Maier and Krider, 1986; Koshak and Krider, 1989; Krider, 1989).

The unique features of the LPLWS are its large area (approximately 20 x 30 square kilometers), the number of field mills (25 to 33), and the fact that each sensor is mounted (and calibrated) in the same way on uniform sites that are cleared of vegetation. A network like this minimizes the frequency of false alarms, such as might be caused by a single sensor having an incorrect reading, and also any 'failures-to-warn' that might occur if the positive and negative cloud charges are not vertically aligned or if the field seen by a single sensor is masked by intervening space charge. Further details on the implementation and calibration of the LPLWS at the KSC-ER are given in Appendix VII by Michael W. Maier.

Among the first scientific studies that were based on the field mill network were analyses of the electrostatic field changes produced by Florida lightning by Jacobson and Krider (1976) and the overall behavior of E under both small and large storms at KSC by Livingston and Krider (1978). These studies showed that the charge centers inside Florida storms are located at altitudes where the air temperatures are below freezing, i.e. where the environmental temperatures are between about -10 °C and -20 °C, and that the time-average values of E are often surprisingly small when averaged over five minute intervals, even under active storms. The finding that the lightning charges are located at subfreezing temperatures lent support to the idea that a non-inductive ice-ice collision process is the dominant microphysical mechanism in cloud electrification (Saunders, 1988; 2008), and the LLCC still rely on this assumption.

**Figure A8.1.1-1  The KSC/CCAFS Launch Pad Lightning Warning System (LPLWS)**

Note:  Each surface field mill is depicted as a green square. Figure courtesy of William Roeder, 45WS.

*A8.1.2 Cloud To Ground Lightning Surveillance System (CGLSS)*
KSC has made major contributions to the development of two complementary systems for detecting and locating lightning, the Cloud-to-Ground Lightning Surveillance System (or CGLSS) and the Lightning Detection and Ranging (or LDAR) system. The CGLSS utilizes a network of gated, broadband electric and magnetic field sensors (Krider and Noggle, 1975) to detect the waveform signatures that are characteristic of return strokes, the high-current components of CG flashes (Krider et al, 1976; Herrman et al., 1976). When a proper signature is detected (in the time-domain) at two or more known locations (the antenna sites), the coincident times-of-arrival and magnetic directions can be used to compute the points where return strokes strike the ground (Krider et al., 1980; Cummins et al., 1998; 2006).

**Figure A8.1.1-2  An electric field mill**

According to a time-domain antenna theory developed by M. A. Uman and his collaborators shortly after the Apollo 12 incident (see for example Uman et al., 1975), the initial peak of the electromagnetic pulse that is radiated by a return stroke is proportional the peak current in the stroke, multiplied by the speed of the stroke up the leader channel, and divided by the distance to the stroke. [Note: this theory is sometimes called the simple 'Transmission-Line Model' or TLM because it assumes the current pulse propagates up a straight channel, without distortion, and at a constant speed.] Since the CGLSS measures the peak field and can compute the stroke location, and since the stroke velocities are known and roughly constant, the CGLSS can also provide an estimate of the peak current in the stroke and its polarity.

The first CGLSS system was installed at the KSC-ER between 1 June 1979 and 12 July 1979. It was a prototype consisting of three medium-gain magnetic direction-finders (DFs) and was installed as part of the Federal Evaluation of Lightning Tracking Systems (FELTS). This system was subsequently purchased in February 1981 with joint funding provided by NASA and the Air Force. In August 1983, a low-gain system was added to obtain more accurate locations near the KSC-ER launch complexes. By February 1984, the system contained two low-gain direction-finders (DFs) that were located at the Ti-Co Airport (28.5N 80.8W) and on Merritt Island (28.4N 81.3W), and three medium-gain sensors, one co-located with the low-gain sensor on Merritt Island, and the other two were at the Orlando and Melbourne Airports (Harms, et al., 2001). After 1984, the system continued to be developed and evaluated and was eventually accepted into the ER inventory as a fully certified system on 24 July 1989. Between 1989 and 1994, the CGLSS was upgraded to a network of five LLP Model 141 Advanced Lightning Direction Finders (ALDFs), and during 1995-1998, the system was converted to a 6-station, short-baseline network of medium-gain IMPACT (IMproved Accuracy from Combined Technology) sensors (Cummins et al., 1998).

The present CGLSS system has an effective range of about 100 km and covers the KSC-ER launch and operations areas with good accuracy and high detection efficiency (Ward et al, 2008). The CGLSS is operated 24 hours a day, 7 days a week by the Range Technical Services (RTS) Contractor, and the data are sent in real-time to the 45 WS for use in operations.

**Figure A8.1.2-1  4DLSS sensor locations**

Note:   L denotes LDAR II sensors.  C denotes CGLSS sensors. Figure courtesy of William Roeder, 45WS.

*A8.1.3 Lightning Detection and Ranging (LDAR)*
The first LDAR system was developed at KSC by Carl Lennon and associates after a design described by Proctor (1971). It contained seven broadband VHF radio receivers that were deployed at the sites shown in Figure A8.1.3-1and were precisely time-synchronized, initially using microwave communications links and later GPS timing.

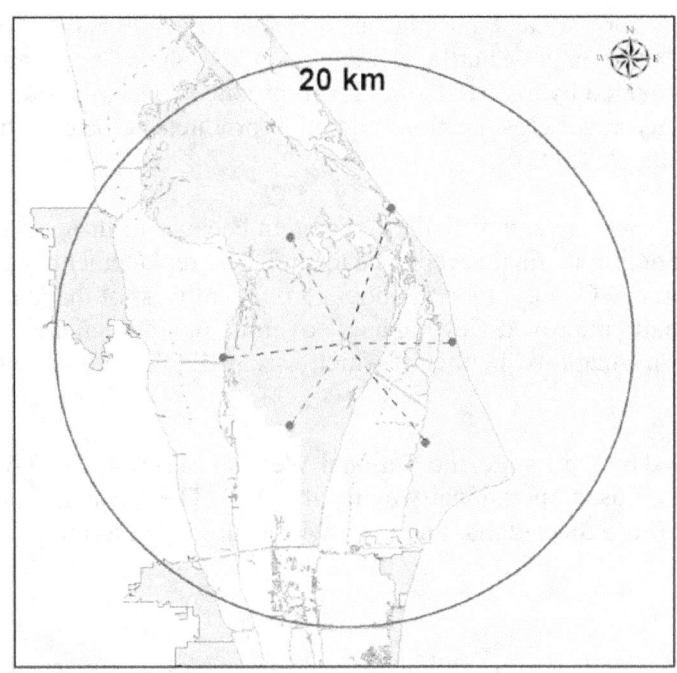

**Figure A8.1.3-1  Locations of the original KSC LDAR sensors**

Note:  Figure courtesy of Jennifer Wilson.

Each site received VHF radiation at 66 MHz, logarithmically amplified the signal, and then transmitted the time and key signal parameters to a central station where the source locations were computed (Lennon and Maier, 1991; Maier et al., 1995). The 3-D locations of the sources of lightning VHF pulses are computed using the differences in the times-of-arrival of the signals detected at the different receiver sites. Since the main sources of VHF radio emissions are the processes associated with air breakdown, the LDAR system detects primarily the in-cloud portions of CG flashes, leader processes, and intracloud discharges. Today each LDAR receiver site operates automatically and is powered by batteries that are recharged by solar panels. The first LDAR system was developed by the KSC Instrumentation and Measurements Branch and was operated and maintained by a NASA contractor. The current system, which is owned and operated by the Eastern Range, is a commercial version of LDAR called LDAR II marketed by Vaisala, Inc. The commercialization was done under a Space Act Agreement between NASA/KSC and a company called Global Atmospherics, Inc. which was subsequently acquired by Vaisala. The 45 WS receives and evaluates the LDAR data 24 hours a day, 7 days a week. For further details on the evolution of the LDAR system and other instrumentation at the KSC-ER, see Boyd et al. (1995), Harms, et al.(1997; 1998; 2001) and Roeder et al.(1999).

*A8.1.4 U.S. National Lightning Detection Network (NLDN)*
Data from the U.S. National Lightning Detection Network (NLDN) (Cummins et al., 1998, 2006; Cummins and Murphy, 2009) have been used to detect and track CG lightning flashes beyond 100 km since the early 1990s. The NLDN sensors are similar to those in the CGLSS except that they have higher gains and larger distances between the sensors. The coincident data from two or more sensors are collected and processed in real-time by a network control center in Tucson, Arizona, and the GPS time, location, and polarity of each lightning stroke, together with an estimate of its peak current, are provided to the KSC-ER in real-time. For further information about the NLDN and its history, see Cummins and Murphy (2009).

*A8.2 Weather Radar*

The operational weather radar for the Eastern Range as of December 2009 is the WSR-74C, a 5 cm conventional weather radar based on pre-Shuttle technology. WSR-74C radar products are generated using proprietary IRIS software provided by SIGMET, Inc. It can produce constant altitude (CAPPI) displays, composite and max reflectivity, radar cross sections and other products at the direction of the user. (Eastern Range Instrumentation Handbook, 2005).

The WSR-74C has become expensive and difficult to maintain because of its age. As this is being written, a modern replacement is undergoing its final acceptance testing. The replacement is a Radtec model TDR-43-250 5 cm radar located in Orange County, Florida about 25 mi southwest of the launch complexes. The new location provides a substantially improved viewing angle over the operational area. The new radar has both Doppler and dual-polarization capability, neither of which was available on the previous Eastern Range weather radars.

In addition to the radar owned by the Range, the National Weather Service NEXRAD 10 cm Doppler radar at Melbourne, Florida (KMLB) is used operationally by the 45th WS. There is a dedicated NEXRAD terminal in Range Weather Operations. It is expected that this radar will be upgraded to include dual-polarization capability in 2011.

*A8.3 Surface Systems*

Surface systems include a variety of sensors and observations, most of which are of the kind that appear in standard weather reports. These include measurements of wind direction and speed, temperature, barometric pressure, relative humidity, visibility and rainfall. These measurements are made by human observers and automated systems as described below.

*A8.3.1 Meteorological Towers*

The current Eastern Range tower network has 46 towers including one 150m and three 62m towers. They cover an area of about 122 km$^2$ for an average station density of one tower every 27 km$^2$. Figure A8.3.1-1 shows the current tower locations on the Eastern Range. The identification numbers are based on location. The first two digits of the 4 digit ID (including a leading zero when necessary) correspond to the distance in nautical miles west of the coastline. The last two digits are the distance (n. mi.) north of Port Canaveral. Towers south of the Port have a leading digit of 9.

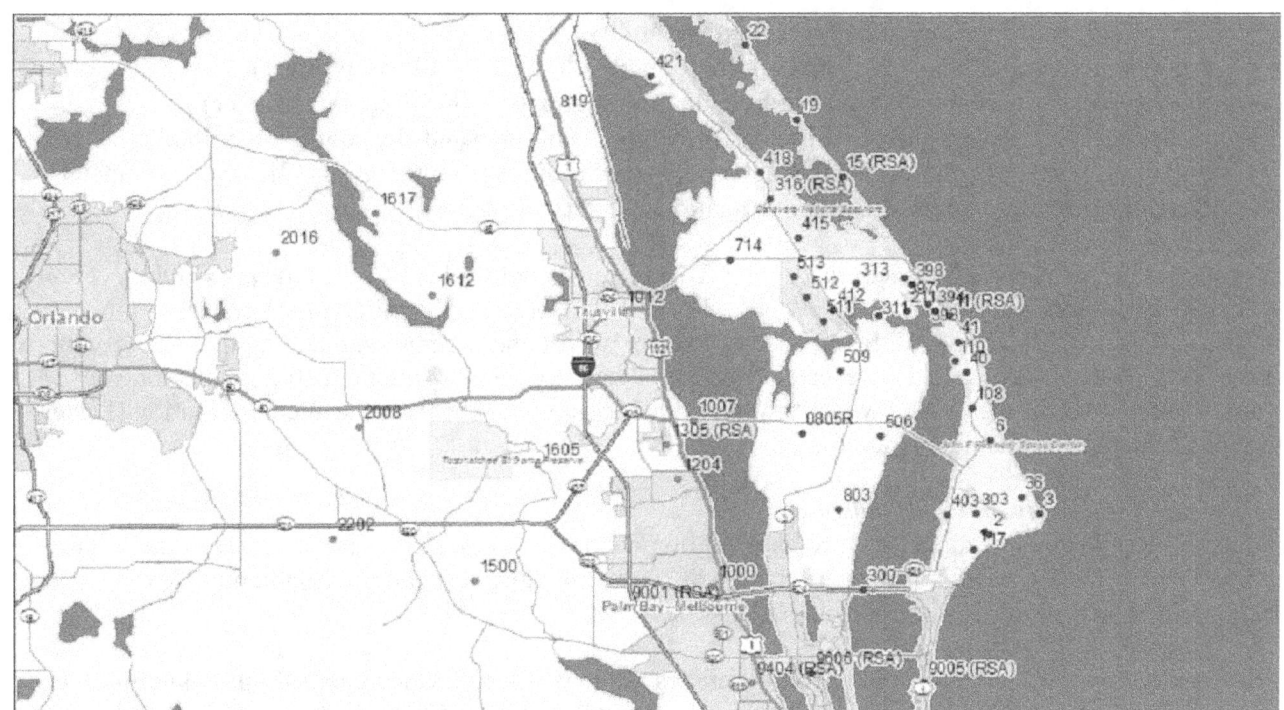

**Figure A8.3.1-1  Location of surface meteorological towers operated by the Eastern Range**

Note:  Towers 511, 512 and 513 in the grey polygon (center, right side of the figure) are located respectively at the south, center and north of the Shuttle Landing Facility. In this figure, several towers are associated with launch complexes (LC) of the same number as follows: 17 (Delta II, SE shore of the Cape), 36 (Atlas, just north of the tip of the Cape), 40 and 41 (formerly Titan, now Atlas V and Falcon, north of #36) . LC 37 (Delta IV) is not shown but is between LC 36 and LC 40. Tower 398 on this map is located near LC39, the Shuttle launch complex.

Wind speed and direction are measured with propeller vane anemometers with an accuracy of about 0.3 m/s and 3 °C respectively. Air temperature is measured with platinum RTD sensors and RH with capacitance-type humidity sensors. Towers designated "launch critical" or "safety critical" are mechanically aspirated. The remaining towers are naturally aspirated. Temperature accuracy is better than 0.5 °C RMS and RH accuracy is better than 3% RMS (Eastern Range Instrumentation Handbook, 2005). Measurements are collected at one second intervals and reported at five minute intervals. For each five minute interval the average wind speed and (vector average) wind direction as well as the one second peak wind speed and direction of that peak are reported. The temperature and RH values are 5 minute averages.

*A8.3.2. Manual Observations*
Surface weather observations at the Eastern Range are made by personnel of the Range Technical Services Contractor for use by the 45th Weather Squadron, Range customers and the Air Force Weather Agency located at Offutt AFB, Nebraska. The information in this subsection is taken from the Eastern Range Instrumentation Handbook, July 2005 revision.

The official observing site, designated station KTTS, is located at KSC in room 4024 of the Air Traffic Control Tower at the Shuttle Landing Facility. Observers submit routine aviation weather reports (METAR) as well as non-routine aviation reports (SPEC/LOCAL) on a round-the-clock basis with additional special observations for Shuttle launch and landing operations. The routine observations include the following:
- Wind speed and direction (reported by three 10m towers along the SLF)

- Maximum wind speed and wind direction variability
- Ambient air temperature and dewpoint (reported by the center SLF wind tower)
- Station pressure (as altimeter setting from an on-site ML-658/GM digital barometer)
- Observed weather (e.g., thunder, precipitation type, visibility) and sky conditions (cloud type, height and coverage) (based on personal observations and readings from three CT-12K laser ceilometers)
- Remarks providing any additional information for clarity or safety.

The observations are reported in accordance with Air Force Manual 15-111, *Surface Weather Observations* and supplemented by the *Federal Meteorological Handbook* (FMH-1). Visibility, cloud cover, weather events, and obstructions to visibility are all based on the observer's skill and judgment at making these observations, and are not provided by sensors. However, for visibility, they use known landmarks (e.g., launch towers, Vehicle Assembly Building [VAB], water towers) with known distances from the Weather Station (Eastern Range Instrumentation Handbook, 2006).

*A8.3.3 Ceiling, Visibility and Soil Moisture*
The Eastern Range deploys three CT-12K Laser Ceilometers. One is located near the center of the Shuttle Landing Facility (SLF), another is located in the KSC industrial area near the Environmental Health Center, and the third is located north of the SLF at Field Mill site 2.

NASA deploys five soil moisture and visibility sensors to the west of KSC to support Shuttle landings. The concern is that post-dawn landings could be subject to a "sunrise surprise" in which undetected ground fog in the St. Johns river basin in Orange County could advect over the SLF reducing visibility at landing below acceptable levels. If the de-orbit decision is made before dawn, there is no ground or satellite-based remote sensing technique that can reliably detect ground fog, so in-situ sensors must be deployed.

*A8.3.4 Rainfall*
At each of the 31 field mill sites described in section 1.2.6 above, a tipping bucket rain gauge is installed. These gauges were funded by the Tropical Rainfall Measuring Mission (TRMM) to provide ground truth for the TRMM satellite launched 27 November 1997. The data are available on the KSC weather data archive website <http://trmm.ksc.nasa.gov/>. Tipping bucket counts, each indicating 0.01 inches (.254 mm) of precipitation, are recorded once per second but only hourly totals are archived.

*A8.3.5 Offshore Data Buoys*
NASA funds the deployment and operation of two buoys by the National Data Buoy Center. One of these is located 20 nautical miles east of Cape Canaveral and the other is located 120 nautical miles east of Ponce De Leon Inlet in Florida. These buoys provide the only available measurements of off-shore conditions affecting the safety of operations by the Shuttle booster recovery ships, and the only upstream surface data available to 45WS forecasters during easterly flow conditions. The buoys report hourly observations via satellite of the following:
- Wind speed and direction
- Barometric pressure
- Air and water temperatures
- Rainfall
- Wave height and period

(Source: NASAfacts, Space Shuttle Weather Launch Commit Criteria and KSC End of Mission Weather Landing Criteria, FS-2006-06-020-KSC, Rev. Nov. 2006)

## A8.4 Upper Air Systems

Upper air measurements have been made using weather balloons for more than a century. Within the last several decades, wind profiling radars have come into common operational use. The Eastern Range and KSC benefit from both kinds of upper air sensors.

### A8.4.1 Balloons

The Eastern and Western Ranges use a weather balloon system called the Automated Meteorological Profiling System (AMPS). AMPS balloons come in two types: low resolution (lo-res) and high resolution (hi-res). Wind measurements from balloons assume the balloon follows the wind and derives the horizontal wind vector from the time derivative of the balloon's horizontal position. Thermodynamic measurements are made *in-situ* using sensors carried by the balloon.

The lo-res AMPS uses a standard latex weather balloon with a GPS-tracked expendable sonde or "flight element" that contains sensors for temperature (thermistor) and relative humidity (carbon film hygristor). Altitude is directly measured from the GPS data and pressure is calculated from the hydrostatic equation using the altitude and thermodynamic data for the trajectory of the sonde as it rises. The balloon rise rate is about 1000 ft/min (5.5 m/s) and the maximum altitude usually exceeds 100 kft (33 km). The wind measurement accuracy is about 1 m/s RMS with a vertical resolution of about 350m. The air temperature and RH RMS accuracies are respectively 0.2 °C and 5%. The barometric pressure accuracy is about 0.5 hPa (mb).

Data are downlinked from the sonde on one of 16 channels near a frequency of 403 MHz GPS and meteorological data are transmitted separately. At the ground, the received signals are processed in an AMPS signal processing system and the results are provided to an Upper Air Systems Controller who can perform manual quality control before the data are released to MIDDS (see description in section A8.5.1 below) for dissemination to the users.

The hi-res AMPS uses a special balloon which is identical to radar tracked "Jimsphere" balloons except that they are clear because they do not have the aluminized coating used for radar tacking. The balloon is 2m in diameter with surface roughness elements that control boundary layer separation to reduce self-induced balloon motions. It rises at about 1000 ft/minute to a maximum altitude of between 55 and 60 kft (17 – 18 km). The flight element for these sondes is smaller and lighter because it does not carry thermodynamic sensors. The wind measurement accuracy and vertical resolution of the AMPS hi-res balloons is about the same as that of Jimspheres (a radar-tracked predecessor to the hi-res AMPS), less than 1 m/s RMS and 150m respectively (Eastern Range Handbook, rev July 2005; Wilfong et al., 1997). Data transmission and processing uses the same system as the AMPS lo-res packages.

### A8.4.2 Doppler Radar Wind Profilers

The Eastern Range and Western Ranges operate networks of 915 MHz boundary layer profilers and the Kennedy Space Center operates a 50 MHz tropospheric wind profiler. Their locations are shown in Figure A8.4.2-1. These systems have much in common, and these common elements will be described first. The system-specifics for each type follows. Data from all of these profilers is fed into MIDDS (see section A8.5.1 below) for dissemination to the users.

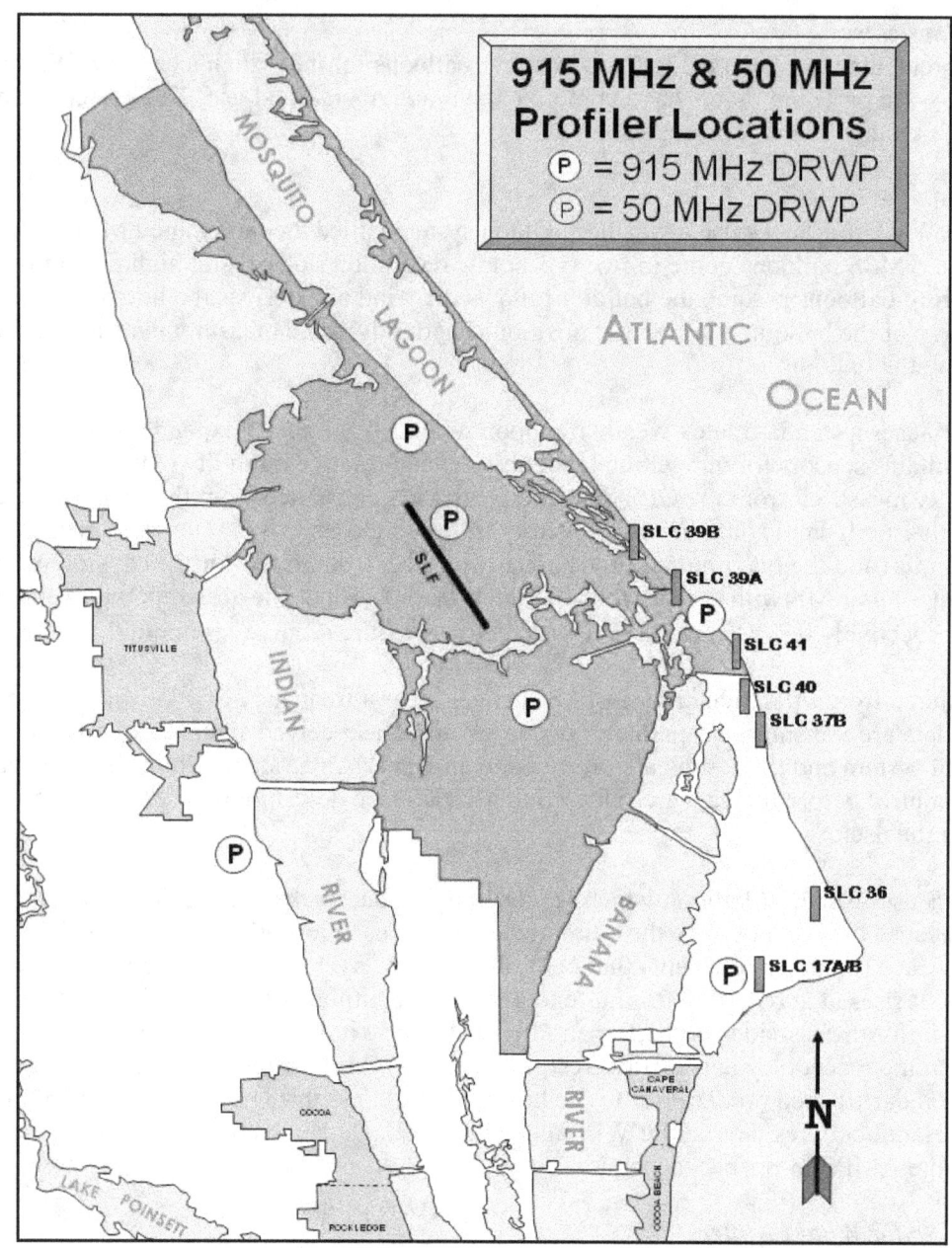

**Figure A8.4.2-1  Locations of the Eastern Range and KSC wind profilers**

Note:  The Shuttle Landing Facility (SLF) and various space launch complexes (SLC) are also shown. The KSC 50 MHz profiler is the (blue) one immediately adjacent to the north end of the SLF. The remaining five profilers (yellow) constitute the ER 915 MHZ network. (Figure courtesy of William Roeder, 45WS).

The Doppler radar wind profilers (DRWP) transmit radio pulses sequentially along three beam directions. One beam, called the vertical beam, is directed straight upward. The remaining two beams, called oblique beams, are directed 15 degrees off of the vertical at 90 degrees with respect to each other in the horizontal. The pulses are backscattered from fluctuations in the radio index of refraction in the atmosphere and received by the radar as a function of time after transmission. From the round trip time between transmission and reception, the range to the scattering volume can be readily calculated. The signals are integrated over range bins (or gates) on the order of 300 ns duration or about 100m length in the atmosphere. The scattered signal is Doppler shifted by an amount proportional to the component of the wind velocity parallel to the beam direction. Both the in-phase and quadrature components of the reflected signal are measured from which the beam-aligned wind component can be calculated. The full wind vector is then calculated from the measured components along the three non-co-planer radar beam directions. The scattered signal comes primarily from atmospheric fluctuations with scales near half of the wavelength of the transmitted radar signal.

The KSC tropospheric profiler is designed to support launch operations and must provide reliable measurements to a height of 18 km. In order to reach this altitude dependably it must operate at a wavelength long enough for there to be significant atmospheric turbulence at that scale in the upper troposphere and have a very large power-aperture product. The KSC profiler operates at 49.75 MHz (6 meter wavelength) with a peak power of 250 kW and a physical aperture of 15,000 square meters. The range gates are 150m providing continuous coverage from 2 to 18 km altitude above ground level. The RMS accuracy is about the same as the best wind-finding balloons (Jimspheres), 1 m/s, and its effective vertical resolution is Nyquist limited at 300m (Schumann et al., 1999; Merceret, 1999).

Normally the KSC profiler operates unattended 24/7/365. For day of launch applications, additional manual quality control is provided by the Eastern Range Technical Services Contractor. This QC process involves examining the Doppler spectra in each beam and, if necessary, adjusting certain variable parameters of the Median Filter First Guess signal processing algorithm (Schumann *et al.*, 1999).

The Ranges profiler networks are used primarily to support Range Safety requirements to identify potential toxic and blast hazards in the event of a processing or launch accident. Wind data are needed at multiple locations in the boundary layer, but not to altitudes in the upper troposphere. Boundary layer turbulence contains significant energy at smaller scales which permits operating the radars at smaller wavelengths. The ER and WR profilers operate at 915 MHz (33 cm wavelength). They are standard commercial instruments (Vaisala model LAP-3000). Range gates of 100m are used to cover the altitude range from about 100m to 5 km. Wind measurement accuracy is thought to be on the order of 1 m/s RMS (Eastern Range Handbook, Rev January 2006; Lambert et al., 2003).

The boundary layer profilers have an additional capability called Radio-Acoustic Sounding System (RASS). The RASS transmits acoustic pulses upward and tracks them with the radar. Since the speed of sound depends only on the virtual temperature, the upward velocity of the sound pulse can be converted to a vertical profile of vertical temperature. In the author's [FJM] experience, this capability has proved neither reliable nor useful.

*A8.4.3 Weather Reconnaissance Aircraft*
Weather reconnaissance aircraft (Wx Recon) are deployed for Space Shuttle launches and landings and for launches of expendable launch vehicles to assist the weather support team in evaluating operational weather constraints. These aircraft can more precisely determine the exact location, height and thickness of clouds than can radar, and can also determine whether the clouds are optically transparent, and the presence and type of precipitation. All of these cloud properties are important for determining whether launch commit criteria or landing flight rules are violated.

For the unmanned vehicles, the Eastern Range currently relies on aircraft operated by commercial contractors. Only pilot or trained observer reports are provided. There is no on-board instrumentation beyond the aircraft navigation system (Eastern Range Instrumentation Handbook, 2004). The aircraft is typically a business jet in the LearJet 35 class with the capability to remain on station for up to 5 hours operating out of small local commercial airports (*ibid*). Pilots and observers receive annual training on the lightning LCC from the 45th Weather Squadron.

Wx Recon support to the Space Shuttle Program is provided by a NASA T-38 jet trainer and the Shuttle Training Aircraft, both flown by astronauts to observe actual conditions for evaluation by the Shuttle Weather

Officer at the launch site and the Spaceflight Meteorology Group (SMG) at Johnson Space Center. The Shuttle Weather Officer (not to be confused with the Shuttle *Launch* Weather Officer provided by the 45WS) is a weather support astronaut located in the firing room at KSC with the Mission Management Team. (Source: NASA Facts "Space Shuttle Weather Launch Commit Criteria and KSC End of Mission Weather landing Criteria"). Like the contract aircraft used for the unmanned operations, the Shuttle Wx Recon aircraft do not carry meteorological instrumentation.

## A8.5 Additional Observation and Forecast Assets

### A8.5.1 Data Integration and Display Systems
The Eastern Range collects and displays much of its meteorological data on the Meteorological Interactive Data Display System (MIDDS). Additional MIDDS users include the Applied Meteorology Unit (AMU, see section 5.2.3) co-located with Range Weather Operations, the National Weather Service Office at Melbourne, the Spaceflight Meteorology Group (SMG) at Johnson Space Center in Houston, the Natural Environments Branch at Marshall Spaceflight Center in Huntsville, and the KSC weather data archive <http://trmm.ksc.nasa.gov/> which obtains much of its data from the Range through exchange of MIDDS files.

Local data acquired by MIDDS includes the surface tower data and upper air data described in previous sections as well as some local radar imagery. Other data includes satellite images and output from national models. MIDDS provides the ability to overlay images and to generate products automatically through a scripting language called McBasi. The 45th Weather Squadron uses a variety of custom scripts, many developed by the AMU.

### A8.5.2 Forecast Tools and Indices
At the Eastern Range the information provided by instrumentation and models is augmented by a variety of software tools. Many of these tools were developed by the Applied Meteorology Unit (AMU) co-located with the 45thWS at Range Weather Operations. The AMU is a joint NASA/Air Force/National Weather Service undertaking to develop, evaluate and transition weather technology to operations in support of America's space program. More information about the AMU may be found in Sections 1.3 and 5.2.3 above and at <http://science.ksc.nasa.gov/amu/>.

The following list of operational tools is neither detailed nor complete. The tools used are rapidly changing through updates and new developments. This list is designed to show the general types of tools in routine use either for daily support to ground operations or for specialized support to launches or landings. Where applicable reports are publically available, they are cited.

1. Severe weather tools
Several severe weather tools are available on the MIDDS system (see section A8.5.1 above) for the forecasting of severe weather in general (Bauman et al, 2005), microbursts (Wheeler, 1996), and fair weather waterspouts.

2. Peak wind tools
Separate tools are available to assist in the forecast of peak winds by daily forecasters (Barrett and Short, 2008) and Launch Weather Officers (Lambert, 2002).

3. Lightning forecast tools
A probability of lightning tool (Lambert, 2007; Lambert and Wheeler, 2005) assists in the issuance of daily lightning forecasts

4. Daily forecast tools
Tools are available for forecasting routine daily weather phenomena such as low-level convergence (Bauman, 2006) and fog (Wheeler, 1994)

5. Tools relating to launch constraints
Tools targeted at specialized day of launch requirements or constraints include an anvil forecast tool for assessing the likelihood of violating the anvil rules of the LLCC (Lambert, 2000), a Shuttle Low Temperature Launch Constraint recovery tool, and a Shuttle launch imaging tool that assesses the likelihood that the required three independent views of the flight path will be obscured by cloud cover (Short and Lane, 2005).

6. Tools relating to the effective application of specific systems
Specialized scan strategies are used with the WSR-74C radar system (see ITEM II above) to maximize the ability of the Launch weather team to observe parts of the atmosphere critical to spaceflight operations (Short, 2008; Taylor, 1994). A unique Median Filter First-Guess signal processing package is used on the KSC 50 MHz Doppler radar wind profiler (see Section A8.4.2. above). A set of specialized menus simplifies the use of the MIDDS system.

*A8.6 Numerical Weather Prediction (computer models)*
Although a locally run numerical weather prediction model is not used operationally by the 45th Weather Squadron, Eastern Range Safety operates an Eastern Range Dispersion Assessment System (ERDAS) which contains a version of the Regional Atmospheric Modeling System (RAMS) to generate wind fields for predictive dispersion modeling. This is used to assess toxic and blast hazards. Some details of the model and its use as well as an evaluation of its performance may be found in Case (2001). On the other hand, the National Weather Service Office at nearby Melbourne, Florida uses high resolution numerical weather prediction extensively. The material in this subsection was provided by David Sharp, the Science and Operations Officer at the NWS Office in Melbourne. His assistance is deeply appreciated. The National Weather Service Forecast Office in Melbourne (MLB), FL, has been involved in both individual and collaborative numerical weather prediction endeavors since the late 1990s (Case, 1999). Specific emphasis has been placed upon the assimilation of unique observational data sets (Lazarus et al., 2006; Zavodsky et al., 2004;) and the generation of both high-resolution automated analyses (Blottman et al., 2001 ) and model forecasts (*e.g.*, Watson et al., 2007) over the Florida Peninsula and adjacent coastal waters. The commitment has been to improve warnings and forecasts for East Central Florida for short-term (00-24 hour) operations (Sharp et al., 2002).

Guided by mutual operational priorities, several initiatives have resulted in community-wide benefits. The most notable initiative involves the provision of real-time 3-dimensional diagnostics delivered to forecasters every 15 minutes at 4 km resolution. Rapid-refresh analyses of countless environmental parameters are made available through an optimized Advanced Regional Prediction System (ARPS) Data Assimilation System (Xue et al., 2003). This diagnostic system (Case et al., 2002) assimilates observational data from surface observing systems and area mesonets, from wind profilers and tower networks at the Kennedy Space Center, from environmental satellites and weather radar, and more, to foster a detailed monitoring of the evolving local atmosphere. It has been successfully used to track the advance of freezing (0 °C) and hard freezing temperatures (-2 °C) for area growers, to monitor the evolution of buoyancy and shear in both pre-storm and near-storm environments associated with severe and tornadic convection, to identify favorable areas of low-level wind shear for take-off/landing aviation traffic, to assess the length and movement of wind shift lines and sea breezes for incident commanders, to detail areas of long duration relative humidity below 35 percent for (wild) fire fighters, etc. Output is available on forecaster AWIPS/D2D (Advanced Weather Interactive Processing System/Display 2-Dimensions) workstations for visualization and further use within gridded short-term forecasts as persistence grids and forecast validation grids. Output is also made available on MLB web pages <http://www.srh.noaa.gov/mlb/ARPS.html> for partners to use. These partners include State and County Emergency Management and the United States Air Force's 45th Weather Squadron.

# References

ABFM Analysis Group, 1991: Airborne Field Mill project operational analysis final report for the summer 1990 deployment, June, 1991, 137 pp.

ABFM Analysis Group, 1992a: Airborne Field Mill project operational analysis final report for the winter 1991 deployment, Feb. 25, 1992, 133 pp.

ABFM Analysis Group, 1992b: Airborne Field Mill project operational analysis final report for the summer 1991 deployment, March 11, 1992, 111 pp.

ABFM Analysis Group, 1992c: Airborne Field Mill project operational analysis final report for the winter 1992 deployment, Oct. 8, 1992, 122 pp.

Anderson, R.V., and J.C. Bailey, 1987: Vector electric fields measured in a lightning environment, NRL Memo. Rept. 5899, 101pp., Naval Research Laboratory, Washington, DC, April 7, 1987.

Andrus, P.G. & Walkup, L.E., 1969: Electrostatic Hazards During Launch Vehicle Flight Operations, NASA Contract Number NASw-1146, Battelle Memorial Institute, Columbus Laboratories, Columbus OH 43201 12 Feb. 1969, p.3.

Arabian, D.D., 1976: Summary of lightning activities by NASA for the Apollo-Soyuz test project, Supplement No. 1 to Apollo-Soyuz Mission Evaluation Report, NASA –TM-74755, NASA Lyndon B. Johnson Space Center, Houston, Texas, July 1976, 38 pp.

Bailey, J.C. and R.V. Anderson, 1987: Experimental calibration of a vector electric field meter measurement system on an aircraft. Memorandum Report 5900, Naval Research Laboratory, Washington D.C.

Bailey, J., J.C. Willett, E.P. Krider, and C. Leteinturier, 1988: Sub-microsecond structure of the radiation fields from multiple events in lightning flashes, 8th International Conference on Atmospheric Electricity, Uppsala, Sweden, June 13-16, 1988.

Barnes, A.A. Jr., and J.I. Metcalf, 1988: "Summary of the triggered lighting workshop held at Cape Canaveral AFS, Florida, 17-19 February, 1988," AFGL Tech, Memo. No. 151, 29 April 1988.

Barret, L., 1986: Campagne Foudre RTLP 86, Note Technique STT/ASP 86-12/LB-mA, Laboratoire D'Applications Speciales de la Physique, 85 X., 38041 Grenoble Cedex, France, December, 1986.

Barret, L., 1986: Campagne Foudre RTLP 86 (in French), Note Technique STT/ASP 86-12/LB-mA, Laboratoire D'Applications Speciales de la Physique, 85 X., 38041 Grenoble Cedex, France, 1 December, 1986.

Barret, L., A. Eybert-Berard, J.P. Berlandis, and G. Terrier, 1990: Campagne Foudre RTLP 1989 (in French), Note Technique STT/LMSP 90-57/LB/AEB/JPB/GT-mtC, Centre D'Etudes Nucleaires de Grenoble, 85 X., 38041 Grenoble Cedex, France, June, 1990.

Barrett, J. and D.A. Short, 2008: Peak Wind Tool for General Forecasting, Ensco, Inc., NASA Contractor Report 2008-214743, 59 pp.

Bateman, M. G., M. F. Stewart, R. J. Blakeslee, S. J. Podgorny, H. J. Christian, D. M. Mach, J. C. Bailey, and D. Daskar, 2007: A low-noise, microprocessor-controlled, internally digitizing rotating-vane electric field mill for airborne platforms, *J. Atm.& Ocean. Tech.* **24**, 1245–1255. DOI: 10.1175/JTECH2039.1

Bauman, W.D., 2006: Forecasting Low-Level Convergence Bands Under Southeast Flow, Ensco, Inc., NASA Contractor Report 2006-214210, 40 pp.

Bauman, W.H., W.P. Roeder, R.A. Lafosse, D.W. Sharp and F.J. Merceret, 2004: The Applied Meteorology Unit - Operational Contributions to Spaceport Canaveral, 11th Conference on Aviation, Range and Aerospace Meteorology, American Meteorological Society, Hyannis, MA, 4-8 October 2004.

Bauman, W.D., M.M. Wheeler and D.A. Short, 2005: Severe Weather Decision Aid, Ensco, Inc., NASA Contractor Report 2005-212563, 41 pp.

Bazelyan, E. M., and Yu P. Raizer. 2000: *Lightning Physics and Lightning Protection*. Bristol: Institute of Physics Publishing, 325 pp.

Bazelyan, E.M, N.L. Aleksandrov, Yu. P. Raizer, and A.M. Konchakov, 2007: The effect of air density on atmospheric electric fields required for lightning initiation from a long airborne object, *Atmospheric Research,* *86*, 126–138, doi:10.1016/j.atmosres.2007.04.001

Beard, K. V., 1987: Cloud and Precipitation Physics Research 1983-1986, *Rev. Geophys.*, 25(3), 357–370.

Bellue, D.G., B.F. Boyd, W.W. Vaughan, T. Garner, J.W. Weems, J.T. Madura, and H.C. Herring, 2006: Shuttle Weather Support – An Overview. AIAA Aerospace Sciences, Reno, 9 - 12 Jan 2006, Paper AIAA-2006-0684, 34pp.

Benson, C.D. and W.B. Faherty, 1978: *Moonport: A History of Apollo Launch Facilities and Operations*, NASA SP4204, 636 pp., Available from the Government Printing Office.

Blottman, P.F., S.M. Spratt, D.W. Sharp, A.J. Cristaldi, J.L. Case, and J. Manobianco, 2001: An Operational Local Data Integration System (LDIS) at NWS Melbourne, Preprints, 18th Conference on Weather Analysis and Forecasting and the 14th Conference on Numerical Weather Prediction, Amer. Meteor. Soc., Ft. Lauderdale, FL, J135-138.

Bondiou, A., P. Laroche, and I. Gallimaberti, 1994: Modeling of the positive leader development in the case of long air gap discharge and triggered lightning, 22nd International Conference on Lightning Protection, Budapest, Hungary, September 19-24, 1994.

Boulay, J.L., J.P. Moreau, A. Asselineau, and P.L. Rustan, 1988: Analysis of recent in-flight lightning measurements on different aircraft, presented at the Aerospace and Ground Conference on Lightning and Static Electricity, Oklahoma City, April 19-22, 1988.

Boyd, B., W. Roeder, J. Lorens, D. Hazen, and J. Weems, 1995: Weather Support to Pre-launch Operations at the Eastern Range and Kennedy Space Center, Preprints Sixth Conference on Aviation Weather Systems, Dallas, TX, Amer. Meteor. Soc., pp 135-140.

Boyd, B.F., J. W. Weems, W. P. Roeder, C. S. Pinder, and T. M. McNamara, 2003: Use of Weather Radar to Support America's Space Program – Past, Present and Future, Preprints, The 31st International Conference on Radar Meteorology, 6 – 12 Aug 03, Seattle, WA, Paper 11B.7, 815-818, 4pp.

Breed, D.W., J.E. Dye, J.J. Jones and G.M. Barnes, 1992: Aircraft observations of electrification in CaPE, *Proc. 9th Internat. Conf. Atmosph. Electr.,* St. Petersburg Russia, June 15-19, 1992, Vol. III, 706-709.

Bringi, V.N., K. R. Knupp, A. Detwiler, L. Liu, I. J. Caylor, and R. A. Black, 1997: Evolution of a Florida thunderstorm during the Convection and Precipitation Electrification Experiment: The case of 9 August 1991. *Mon. Wea. Rev.,* **125,** 2131–2160.

Brook, M., G. Armstrong, R.P.H. Winder, B. Vonnegut, and C.B. Moore, 1961: Artificial initiation of lightning discharges, *J. Geophys. Res.,* **66,** 3967-3969.

Brook, M., C. R. Holmes, and C. B. Moore, 1970: 'Lightning and Rockets; Some Implications of the Apollo 12 Lightning Event,' *Naval Research Reviews,* **23,** No. 4, pp.1-17.

Case, J., 1999: Simulation of a real-time Local Data Integration System over east-central Florida. NASA Contractor Report CR-1999-208558, Kennedy Space Center, FL, 46 pp.

Case, J., 2001: Final report on the Evaluation of the Regional Atmospheric Modeling System in the Eastern Range Dispersion Assessment System, Ensco, Inc., NASA Contractor Report CR-2001-210259, 130pp.

Case, J.L., J. Manobianco, T.D. Oram, T. Garner, P.F. Blottman, and S.M. Spratt, 2002: Local Data Integration over East-Central Florida Using the ARPS Data Analysis System, *Wea. Forecasting,* **17,** 3-26.

Chauzy, S., J.-C. Médale, S. Prieur, and S. Soula, 1991: Multilevel measurement of the electric field underneath a thundercloud 1. A new system and the associated data processing, J. Geophys. Res., 96, 22,319-22,326.

Christian, H.J., K. Crouch, B. Fisher, V. Mazur, R.A. Perala and L. Ruhnke, 1987: The Atlas-Centaur 67 incident. Report of the Atlas/Centaur-67/FLTSATCOM F-6 Investigation Board, 10 pp., available from the Public Affairs Office, NASA, Kennedy Space Center, FL 32899.

Christian, H. J., V. Mazur, B. D. Fisher, L. H. Ruhnke, K. Crouch, and R. P. Perala, 1989: The Atlas/Centaur Lightning Strike Incident, *J. Geophys. Res.,* **94 (D11),** 13,169-13,177.

Clark, J.F., 1957: Airborne measurement of atmospheric potential gradient, *J. Geophys. Res.,* **62,** 617-628.

Clifford, D. W. and H. J. Kasemir, 1982: Triggered Lightning, *IEEE Trans. EMC, 24* (2), 112-122.

Cobb, W.E., 1975: Electric Fields in Florida Cumulus, paper M33, American Geophysical Union Fall Meeting, San Francisco, CA, 8-12 December, 1975.

Columbia Accident Investigation Board, 2003: Columbia Accident Investigation Board Report, Volume 1, August 2003, 248 pp. Available from the Government Printing Office, Washington, D.C.

Committee on Science, Space and Technology, 1987: The Atlas/Centaur 67 Launch Mishap, Hearing Before the Subcommittee on Space Science and Applications of the Committee on Science, Space and Technology, House of representatives, 100th Congress, 1st Session, 4 August 1987, 108 pp. Available from the Government Printing Office.

Cotton, W. R., 1979: Cloud Physics: A Review for 1975-1978 IUGG Quadrennial Report, *Rev. Geophys. Space Phys.*, **17 (7)**, 1840-1851.

Cummins, K.L., M.J. Murphy, E.A. Bardo, W.L. Hiscox, R.B. Pyle, and A.E. Pifer, 1998: A combined TOA/MDF technology upgrade of the U.S. National Lightning Detection Network, *J. Geophys. Res.*, **98**, 9035-9044.

Cummins, K. L., J. A. Cramer, C. Biagi, E. P. Krider, J. Jerauld, M. A. Uman, and V. A. Rakov, 2006: "The U.S. National Lightning Detection Network: Post-Upgrade Status," 86th Annual AMS Meeting, Atlanta, GA, 29 January-2 February 2006.

Cummins , K.L. and M.J. Murphy, 2009: An Overview of Lightning Locating Systems: History, Techniques, and Data Uses, With an In-Depth Look at the U.S. NLDN, *IEEE Trans. on EMC, Vol. 31* (3), pp. 499-518.

Depasse, P. , 1994: Statistics on artificially triggered lightning, *J. Geophys. Res.*, **99**, 18,515-18,522.

Dolezalek, H. and R. Reiter, Editors. 1977: *Electrical Processes in Atmospheres*, Proceedings of the Fifth International Conference on Atmospheric Electricity held at Garmisch-Partenkirchen (Germany), 2-7 September 1974, Dietrich Steinkopff Verlag, Darmstadt, 865 pp.

Durrett,W.R. , 1976: "Lightning—Apollo to Shuttle", Proceedings 13th Space Congress, Cocoa Beach, FL 7-9 April 1976, p 4-27.

Dye, J. E., C. A. Knight, V. Toutenhoofd and T. W. Cannon, 1974: The mechanism of precipitation formation in NE Colorado cumulus. Part III: Coordinated microphysical and radar observations and summary. *J. Atmos. Sci.*, **31**, 2152-2159.

Dye, J.E., W. P. Winn, J. J. Jones, and D. W. Breed, 1989: The electrification of New Mexico thunderstorms. 1. Relationship between precipitation development and the onset of electrification. *J. Geophys. Res.*, **94,** 8643–8656.

Dye, J.E., D.W. Breed, G.M. Barnes, J.J. Jones, R.C. Solomon, 1992: The co-evolution of precipitation and electric fields in Florida cumuli during CaPE, *Proc. 9th Internat. Conf. Atmosph. Electr.,* St. Petersburg Russia, June 15-19, 1992, Vol. I, 179-184.

Dye, J.E., S. Lewis, M.G. Bateman, D.M. Mach, F.J. Merceret, J.G. Ward and C.A. Grainger, 2003: Final Report on the Airborne Field Mill Project (ABFM) 2000-2001 Field Campaign, NASA Technical Memo 2004-211534, Kennedy Space Center, FL, 126 pp.

Dye, J.E., M.G. Bateman, H.J. Christian, E. Defer, C.A. Grainger, W.D. Hall, E.P. Krider, S.A. Lewis, D.M. Mach, F.J. Merceret, J.C. Willett, P.T. Willis, 2006: Electric Fields, Cloud Microphysics, and Reflectivity in Anvils of Florida Thunderstorms, *J. Geophys. Res.*, **112**, D11215, doi: 10.1029/2006JD007550.

Dye, J.E. and J.C. Willlett, 2006: The enhancement of reflectivity and electric field in long-lived Florida anvils, *Mon. Weath. Rev.*, **135**, 3362-3380.

Dye, J.E. M.G. Bateman, D.M. Mach, C.A. Grainger, H.J. Christian, H.C. Koons, E.P. Krider, F.J. Merceret, J.C. Willett, 2006: The scientific basis for a radar-based lightning launch commit criterion for anvil clouds, *Amer. Meteor. Soc. Conf. on Aviat., Range and Aerosp. Meteorol.,* Feb. 2006, Atlanta Georgia, paper 8.4.

Dye, J.E., E. P. Krider, F. J. Merceret, J.C. Willett, M.G. Bateman, D.M. Mach, R.M. Waltersheid, T.P. O'Brien, H.J. Christian, 2008: Analysis of proposed 2007-2008 revisions to the lightning launch commit criteria for United States space launches, *13th Conf. Aviat., Range and Aerosp. Meteorol., Amer. Meteorol. Soc.* Jan. 2008, New Orleans, LA, paper 8.2.

*Eastern Range Instrumentation Handbook*, various years, Prepared by Computer Sciences Raytheon for 45RMS/RMSS, Air Force Space Command, Patrick AFB, Florida, updated as required.

Eybert-Berard, A., L. Barret, and J.P. Berlandis, 1986: Campagne Foudre aux Etats-Unis Kennedy Space Center (Florida) Programme RTLP 85 (in French), Note Technique STT/ASP 86-01, Centre D'Etudes Nucleaires de Grenoble, 85 X., 38041 Grenoble Cedex, France, April, 1986.

Eybert-Berard, A., L. Barret, and J.P. Berlandis, 1988 : Campagne d'Experimentations Foudre RTLP 87 (NASA-Kennedy Space Center, Floride USA) (in French), Note Technique STT/LASP 88-21/AEB/LB/JPB-pD, Centre D'Etudes Nucleaires de Grenoble, 85 X., 38041 Grenoble Cedex, France, July, 1988.

Eybert-Berard, A., L. Barret, J.P. Berlandis, and G. Terrier, 1989: Caracterizations des Decharges Electromagnetiques Provoquees Campagne RTLP 1988 (in French), Note Technique STT/LASP 89-21/AEB/LB/JPB/GT-pD, Centre D'Etudes Nucleaires de Grenoble, 85 X., 38041 Grenoble Cedex, France, June, 1989.

Fisher, R.J., and G. Schnetzer, 1991: Damage to metallic samples produced by measured lightning currents, presented at the 1991 International Aerospace and Ground Conference on Lightning and Static Electricity, Cocoa Beach, FL, April 16-19, 1991.

Fitzgerald, D.R., 1965: Measurement techniques in clouds, in *Problems of Atmospheric and Space Electricity*, ed. S.C. Coronti, 199-212.

Fitzgerald, D.R., 1970: Aircraft and rocket triggered natural lightning discharges, 1970 Lightning Static Elec. Conf., AFAL Rep. 1970.

Fitzgerald, D.R., 1977: Electrical structures of large overwater shower clouds, in *Electrical Processes in Atmospheres*, H. Dolezalek and R. Reiter eds. 362-367.

Foote, G.B., Ed., 1991: CaPE Scientific Overview and Operations Plan, National Center for Atmospheric Research, Boulder CO 80307, June 1991, 145 pp.

Gaskell, W, A. J. Illingworth, J. Latham, and C. B. Moore, 1978: Airborne studies of electric fields and the charge and size of precipitation elements in thunderstorms, *Quart. J. R. Met. Soc.,* **104**, 447-460.

Gaskell, W. and A. J. Illingworth, 1980: Charge transfer accompanying individual collisions between ice particles and its role in thunderstorm electrification, *Quart. J. R. Met. Soc.,* **106**, 841-854.

Giori, K.L. and R. Harris-Hobbs, 1991: Electrification aspects of the Launch Commit Criteria: Case studies, Task B Final Report SRI Project 1449, April 1991, available from SRI International, 333 Ravenswood Ave. Menlo Park CA 94025-3493.

Glickman, T.S., Ed., 2000: *Glossary of Meteorology*, 2d Ed., American Meteorological Society, Boston, MA, 850 pp.

Golde, R. H., Editor, 1977: *Lightning, Vol. 1 Physics, Vol 2 Lightning Protection*, Academic Press, London.

Grimwood, James M., Barton C. Hacker and Peter J. Vorzimmer, 1969: *Project Gemini Technology and Operations; A Chronology,* NASA SP-4002 (Washington, 1969), pp. 154 and 209.

Hallett, J., 1983: Progress in Cloud Physics 1979-1982, *Rev. Geophys. Space Phys., 21* (5), 965-984.

Hallett, J., R.I. Sax, D. Lamb and A.S. Ramachandra, 1978: Aircraft measurements of ice in Florida cumuli, *Quart. J. Royal Meteorol. Soc., 104,* 631-651.

Harms, D. E., B. F. Boyd, M. S. Gremillion, M. E. Fitzpatrick, and T. D. Hollis, 2001: Weather Support To Space Launch: "A Quarter-Century Look At Weather Instrumentation Improvements," 11th Symposium on Meteorological Observations and Instrumentation, 14 – 19 Jan 2001, Albuquerque, NM, Amer. Meteor. Soc., pp. 259-264.

Harms, D. E., B. F. Boyd, R. M. Lucci, M. S. Hinson, and M. W. Maier, 1997: "Systems Used to Evaluate the Natural and Triggered Lightning Threat to the Eastern Range and Kennedy Space Center, Preprints 28th Conference on Radar Meteorology, Austin, TX, Amer. Meteor. Soc., pp. 240-241.

Harms, D. E., B. F. Boyd, R. M. Lucci, and M. W. Maier, 1998: "Weather Systems Modernization To Support the Space Launch Program at the Eastern Range and Kennedy Space Center," Preprint, 10th Symposium On Meteorological Observations and Instrumentation, Phoenix, AZ, 11-16 January, Amer. Meteor. Soc., pp 317-322.

Harris-Hobbs, R., K. Giori, M. Bellmore, and A. Lunsford, 1994:Application of Airborne Field Mill Data for Use in Launch Support, *J. Atmos. Ocean Tech.,* **11,** 738–750.

Heritage, H., Ed., 1988: "Launch Vehicle, Lightning/Atmospheric Electrical Constraints, Post-Atlas/Centaur 67 Incident"; Program Group, Aerospace Report #TOR-0088 (3441-45)-2, The Aerospace Corporation, El Segundo CA 90245; 31 August 1988.

Herrman, B.D., M.A. Uman, R.D. Brantley, and E.P. Krider, 1976: Test of the principle of operation of a wideband magnetic direction finder for lightning return strokes, *J. Appl. Meteorol.,* **15 (4),** 402–405.

Hiscox, W.L., E.P. Krider, A.E. Pifer, and M.A. Uman, 1984: A systematic method for identifying and correcting "site errors" in a network of magnetic direction finders, in Proceedings of the 1984 International Aerospace and Ground Conference on Lightning and Static Electricity, pp. 7-1 - 7-15, Orlando, FL, June 26-28, 1984.

Holitza, F.J. and H.W. Kasemir, 1974: Accelerated decay of thunderstorm electric fields by chaff seedings, *J. Geophys. Res., 79,* 425-429.

Hubert, P., and B. Hubert, 1986: Some results of observations on triggered lightning at Kennedy Space Center in July, August 1995, Centre d'Etudes Nucleaires de Saclay Technical Report DPHG/SAP/86-19R/PH/np, Saclay, France, pp., 17 January 1986.

Idone, V.P., The luminous development of Florida triggered lightning, 1992: *Res. Lett. Atmos. Electr.,* **12,** 23-28.

Idone, V.P., and R.E. Orville, 1988: Channel tortuosity variations in Florida triggered lightning, *Geophys. Res. Lett.,* **15,** 645-648.

Idone, V.P., A.B. Saljoughy, R.W. Henderson, P.K. Moore, and R.B. Pyle, 1993: A reexamination of the peak current calibration of the national lightning detection network, *J. Geophys. Res.*, **98**, 18,323-18,332.

Illingworth, A. J. and P. R. Krehbiel, 1981: Thunderstorm Electricity, *Phys. Technol.*, **12**, 122-139.

Illingworth, A., 1985: Charge Separation in Thunderstorms: Small Scale Processes, *J. Geophys. Res.*, **90** (D4), 6026-6032.

Imyanitov, I. M., Ye. V. Chubarina, and Ya. M. Shvarts, 1971: NASA Tech Translation TT F-728: Electricity of Clouds, Translation of "Elektrichestvo oblakov," Hydrometeorological Press, Leningrad, 1971, 122 pg.

Jacobson, E.A., and E.P. Krider, 1976: Electrostatic field changes produced by Florida lightning, *J. Atmos. Sci.*, **33 (1)**, 103–117.

Jameson, A.R., M. J. Murphy, and E. P. Krider, 1996: Multiple parameter radar observations of isolated Florida thunderstorms during the onset of electrification. *J. Appl. Meteor.*, **35,** 343–354.

Johnson, D.L., 1976: Atmospheric Environment for ASTP (SA-210) Launch, NASA technical Memorandum X-64990, 33pp.

Jones, J.J., W.P. Winn, S.J. Hunyady, C.B. Moore, J.W. Bullock and P. Fleischhaker, 1990: Aircraft measurements of electrified clouds at Kennedy Space Center, Final Report: Part I 1988 Flights (April 27, 1990).

---------Final Report: Part II Case Study Nov. 4, 1988 (April 27, 1990).

---------Final Report: Part III, 1989 Flights (Aug. 21, 1990).

Jones, J.J., W.P. Winn and F. Han, 1993: Electric Field measurements with an airplane: problems caused by emitted charge, *J. Geophys. Res.*, **98**, 5235-5244.

Kanter, L.E., 1975: The last of the Apollos, *NOAA Magazine, 5 No. 4*, October 1975, 12-17.

Kapryan J.W., 1978: "STS Lightning Launch Commit Criteria" KSC/VO Memo to JSC/Program Operations Office (Dec 1978).

Kasemir, H.W., 1964: The cylindrical field mill, *Tech. Rep.*ECOM-2526, U.S. Army Electronics Command, Fort Monmouth, New Jersey.

Kasemir, H.W., 1972: The cylindrical field mill, *Meteorol. Rundsch., 25,* 33-38.

Kasemir, H.W., F.J. Holitza, W.E. Cobb, and W.D. Rust, 1976: Lightning suppression by chaff seeding at the base of thunderstorms, *J. Geophys. Res., 81*, 1956-1970.

Koons, H.C., and R. Walterscheid, 1996: Lightning launch commit criteria, Aerospace Report No. TR-95(5566)-1, 1 February 1996.

Koshak, W.J., and E.P. Krider, 1989: Analysis of lightning field changes during active Florida thunderstorms, *J. Geophys. Res.*, **94 (D1)**, 1165–1186.

Kositsky, J. K.L. Giori, R.A. Maffione, D.H. Cronin, J.E. Nanevicz and R. Harris-Hobbs, 1991a: Airborne Field Mill (ABFM) system calibration report. SRI project 1449, Task A, Final Report, January 1991, available from SRI International, 333 Ravenswood Ave. Menlo Park CA 94025-3493.

Kositsky, J. and J.E. Nanevicz, 1991: Scale-model charge transfer technique for measuring enhancement factors. *Int. Conf on Lightning and Static Electricity,* Cocoa Beach, FL.

Kositsky, J. and K.L. Giori, 1991: Airborne Field Mill (ABFM) special studies. SRI project 1449, Task C, Final Report, Available from SRI International, 333 Ravenswood Ave. Menlo Park CA 94025-3493.

Krider, E.P., R.C. Noggle, M.A. Uman, and R. E. Orville, 1974: Lightning and the Apollo17/Saturn V exhaust plume, *J. Spacecraft and Rockets,* **11 (2)**, 72–75.

Krider, E.P., and R.C. Noggle, 1975: Broadband antenna systems for lightning magnetic fields, *J. Appl. Meteorol.,* **14 (2)**, 252–256.

Krider, E.P., R.C. Noggle, and M. A. Uman,1976: A gated, wideband magnetic direction finder for lightning return strokes, *J. Appl. Meteorol.,* **15 (3)**, 301–306.

Krider, E.P., R.C. Noggle, A.E. Pifer, and D.L. Vance, 1980: Lightning direction-finding systems for forest fire detection, *Bull. Amer. Meteorol. Soc.,* **61 (9)**, 980–986.

Krider, E. P. and R. G. Roble, Editors, 1986: NAS Study in Geophysics, *The Earth's Electrical Environment,* National Academy Press, Washington, D.C., 263 pp. <http://www.nap.edu/catalog.php?record_id=898>

Krider, E.P., 1988: Spatial distributions of lightning strikes to ground during small thunderstorms in Florida, in Proceedings of the 1988 International Aerospace and Ground Conference on Lightning and Static Electricity, pp. 318-323, U.S. Dep. of Commer., Oklahoma City, Okla., April 19-22, 1988.

Krider, E. P., 1989: Electric field changes and cloud electrical structure, *J. Geophys. Res.,* **94 (D11)**, 13,145–13,149.

Krider, E.P., C. Leteinturier, and J.C. Willett, 1992: Sub-microsecond field variations in natural lightning processes, *Res. Lett. Atmos. Electr.,* **12**, 3-9, 1992.

Krider, E.P., H.C. Koons and R. Walterscheid, W.D. Rust, and J.C. Willett, 1999: Natural and triggered lightning launch commit criteria (LCC), Aerospace Report No. TR-99(1413)-1, 15 January 1999.

Krider, E.P., J.C. Willett, G.S. Peng, F.S. Simmons, G.W. Law, and R.W. Seibold, 2006: Triggered lightning risk assessment for reusable launch vehicles at the Southwest Regional and Oklahoma Spaceports, Aerospace Report No. ATR-2006(5195)-1, 30 January 2006.

Krider, E.P. H.J. Christian, J.E. Dye, H.C. Koons, J. Madura, F. Merceret, W.D. Rust, R.L. Walterscheid, and J.C. Willett, 2006: Natural and triggered Lightning Launch Commit Criteria, *Conf. on Aviat., Range and Aerosp. Meteorol.,* Feb. 2006, Atlanta Georgia, paper 8.3.

Lalande, P., and A. Bondiou-Clergerie, 1997: Collection and analysis of available in-flight measurement of lightning strikes to aircraft, Report AI-95-SC.204-RE/210-D2.1, ONERA (France) Transport Research and Technological Development Program DG VII, 1997.

Lalande, P., A. Bondiou-Clergerie, and P. Laroche, 1999: Studying aircraft lightning strokes, Aerospace Engineering (publisher: SAE Aerospace), 39–42.

Lambert, W.C., 2000: Improved Anvil Forecasting: Phase I Final Report, Ensco, Inc., NASA Contractor Report 2000-208573, 23 pp.

Lambert, W.C., 2002: Statistical Short-Range Guidance for Peak Wind Speed Forecasts on Kennedy Space Center/Cape Canaveral Air Force Station: Phase I Results, Ensco, Inc., NASA Contractor Report 2002-211180, 33 pp.

Lambert, W.C., 2007: Objective Lightning Probability Forecasting for Kennedy Space Center and Cape Canaveral Air Force Station, Phase II, Ensco, Inc., NASA Contractor Report 2007-214732, 57 pp.; Lambert, W.C. and M.M. Wheeler, 2005: Objective Lightning Probability Forecasting for Kennedy Space Center and Cape Canaveral Air Force Station, Ensco, Inc., NASA Contractor Report 2005-212564, 46 pp.

Lambert, W.C., F.J. Merceret, G.E. Taylor and Jennifer G. Ward, 2003: Performance of Five 915-MHz Wind Profilers and an Associated Automated Quality Control Algorithm in an Operational Environment, *J. Atm. & Ocean. Tech.*, **20(11)**, 1488-1495.

Laroche, P., 1986: Airborne measurements of electrical atmospheric field produced by convective clouds. *Rev. Phys. Appl.*, **21**, 809-815.

Laroche, P., A. Eybert-Berard, and L. Barret, 1985: Triggered lightning flash characterization, 10th International Aerospace and Ground Conference on Lightning and Static Electricity, Paris, France, June 10-13, 1985.

Laroche, P., A. Delannoy, and H. Le Court de Béru, 1989: Electrostatic field conditions on an aircraft stricken by lightning. Paper presented at the Internat. Conf. on Lightning and Static Electricity, September 26–28, 1989, University of Bath, UK.

Laroche, P., A. Eybert-Berard, L. Barret, and J.P. Berlandis, 1988: Observations of preliminary discharges initiating flashes triggered by the rocket and wire technique, 8th International Conference on Atmospheric Electricity, Uppsala, Sweden, June 13-16, 1988.

Laroche, P., A. Bondiou, A. Eybert-Berard, L. Barret, J.P. Berlandis, G. Terrier, and W. Jafferis, 1989: Lightning flashes triggered in altitude by the rocket and wire technique, 1989 International Conference on Lightning and Static Electricity, Bath, UK, September 26-28, 1989.

Laroche, P., V. Idone, A. Eybert-Berard, and L. Barret, 1991: Observations of bi-directional leader development in triggered lightning flash, 1991 International Aerospace and Ground Conference on Lightning and Static Electricity, Cocoa Beach, FL, April 16-19, 1991.

Latham, J., 1981: The Electrification of Thunderstorms, *Quart. J. R. Met. Soc.*, **107** (452), 277-298.

Latham, J., and J. Dye, 1989: Calculations on the Electrical Development of a Small Thunderstorm, *J. Geophys. Res.*, **94(D11)**, 13141-13144.

Lazarus, S.M., C.G. Calvert, M. Splitt, P. Santos, D. Sharp, P. Blottman, and S.M. Spratt, 2006: Multi-Platform Real-Time Sea Surface Temperature Analysis for the Initialization of Short-Term

Operational Forecasts. Preprints, 10th Symposium on Integrated Observing and Assimilation Systems for the Atmosphere, Oceans, and Land Surface (IOAS-AOLS), Amer. Meteor. Soc., Atlanta, GA, CD-ROM 6.12.

Lennon, C. and L. Maier, 1991: Lightning Mapping System, 1991 International Aerospace and Ground Conference on Lightning and Static Electricity, NASA Conference Pub 3106, Vol II, pp 89-1 to 89-10.

Les Renardières Group, 1977: Positive discharges in long air gaps at Les Renardières: 1975 results and conclusions, *Electra*, **53**, 33-153.

Les Renardières Group, 1981: Negative discharges in long air gaps at Les Renardières: 1978 results, *Electra*, **74**, 70-216.

Leteinturier, C., C. Weidman, and J. Hamelin, 1990: Current and electric field derivatives in triggered lightning return strokes, *J. Geophys. Res.*, **95**, 811-828.

Leteinturier, C., J.H. Hamelin, and A. Eybert-Berard, 1991: Sub-microsecond characteristics of lightning return-stroke currents, *IEEE Trans. EMC*, **33**, 351-357.

Le Vine, D.M., 1980: Sources of the strongest RF radiation from lighting, *J. Geophys. Res.*, **85**, 4091-4095.

Le Vine, D.M., J.C. Willett, and J.C. Bailey, 1989: Comparison of fast electric field changes from subsequent return strokes of natural and triggered lightning, *J. Geophys. Res.*, **94**, 13,259-13,265.

Lhermitte, R., Advancements in Remote Sensing of the Atmosphere, 1979: *Rev. Geophys. Space Phys.*, **17** (7), 1833-1840.

Lhermitte, R. and P. R. Krehbiel, 1979: Doppler Radar and Radio Observations of Thunderstorms, *IEEE Trans. Geosci. Electronics*, **17 (4)**, 162-171.

Lhermitte, R. and E. Williams, 1983: Cloud Electrification, *Rev. Geophys. Space Phys.*, **21(5)**, 984-992.

Livingston, J.M., and E.P. Krider, 1978: Electric fields produced by Florida thunderstorms, *J. Geophys. Res.*, **83 (C1)**, 385–401.

MacGorman, Donald R., and W. David Rust, 1998: *The Electrical Nature of Storms*. Oxford Univ. Press, ISBN 0-19-507337-1

Mach, D.M., 2009: Reanalysis of the ABFM-I Dataset Final Report, Marshall Space Flight Center Final Report for Interagency Agreement No. DTRT57-08-X-70053 between Volpe Center, the FAA Office of the Associate, June 30, 2009.

Mach, D.M., J.C. Bailey and H.J. Christian, 1992: Electrification of stratiform clouds near Cape Canaveral Florida USA, *Proc. 9th Internat. Conf. Atmosph. Electr.*, St. Petersburg Russia, June 15-19, 1992, Vol. I, 175.

Mach, D. M. and W. J. Koshak, , 2007: General matrix inversion technique for the calibration of electric field sensor arrays on aircraft platforms, *J. Atmos. Ocean. Tech.*, **24**, pp. 1576-1587, DOI: 10.1175/JTECH2080.1.

Maier, L.M., and E.P. Krider, 1986: The charges that are deposited by cloud-to-ground lightning in Florida, *J. Geophys. Res.*, **91 (D12)**, 13,275–13,289.

Maier, M.W., 1991: Preliminary evaluation of National Lightning Detection Network performance at Cape Canaveral during August 1990, CSR-322-0007, Instrumentation Systems Evaluation Test Report (CRDL 137A2), 24 pp., October 1, 1991.

Maier, M.W., and W. Jafferis, 1985: Locating rocket triggered lightning using the LLP lightning locating system at the NASA Kennedy Space Center, in *Proceedings of the 10th International Aerospace and Ground Conference on Lightning and Static Electricity*, pp. 337-345, Paris, France, June 10-13, 1985.

Maier, L., C. Lennon, and T. Britt, 1995: "Lightning Detection and Ranging (LDAR) System Performance Analysis," *Preprints, Sixth Conference on Aviation Weather Systems*, Dallas, TX, 15-20 January, Amer. Meteor. Soc., pp. 305-309.

Marshall, T., W. Rust, M. Stolzenburg, W. Roeder, and P. Krehbiel, 1999: A study of enhanced fair-weather electric fields occurring soon after sunrise, *J. Geophys. Res.*, **104(D20)**, 24455-24469.

Martner, B.E., B.W. Bartram, J.S. Gibson, W.C. Carroll, R.F. Ranking and S.Y. Matrosov, 2002: An overview of NOAA/ETL's vertical-profiling cloud radar and radiometer package, *Preprints, 6th Symp. On Integrated Observing Systems,* Orlando, FL, Amer. Meteor. Soc., pp 41-43.

Mason, B.J. and C.B. Moore, 1976: Theories of Thunderstorm Electrification, *Quart. J. R. Met.* Soc., **102**, 219-240.

Mazur, V., 1989: A physical model of lightning initiation on aircraft in thunderstorms, *J. Geophys. Res.*, **94**, 3326-3340.

Melander, B.G., R.T. Hasbrouck, and J. Johnson, 1988: Lightning test results for LLNL's lightning invulnerable device system using both simulated and triggered lightning, presented at the Aerospace and Ground Conference on Lightning and Static Electricity, Oklahoma City, April 19-22, 1988.

Merceret, Francis J., 1999: The Vertical Resolution of the Kennedy Space Center 50 MHz Wind Profiler, *J. Atm. & Ocean. Tech.,* **16(9)**, 1273-1278.

Merceret, F.J., M. McAleenan, T.M. McNamara, J.W. Weems and W.P. Roeder, 2006: *Implementing the VAHIRR Launch Commit Criterion using existing radar products*, Paper 8.7, Twelfth AMS Conference on Aviation and Range Meteorology, Atlanta, GA, 29 January - 2 February 2006.

Merceret, F.J., D.A. Short and J.G. Ward, 2006: Radar Evaluation of Optical Cloud Constraints to Space Launch Operations, *J. Spacecraft and Rockets*, **43 (1)**, 243-251.

Merceret, F.J., J.G. Ward, D.M. Mach, M.G. Bateman and J.E. Dye, 2008: On the Magnitude of Electric Fields Near Thunderstorm Associated Clouds, *J. Appl. Meteor. & Climatol.*, **47**, 240-248.

Mo, Q., A.E. Ebneter, P. Fleischhacker, and W.P. Winn. 1998. Electric field measurements with an airplane: A solution to problems caused by emitted charge. *J. Geophys. Res.* 103, 17,163–17,173.

Winn, W. P., Aircraft measurement of electric field: Self-calibration, 1993: *J. Geophys. Res.,* **98**, 6351-6365.

Nagler, K.M., 1966: Weather Support Problems in the Gemini and Apollo Programs, AMS/AIAA Paper No. 66-334, AMS/AIAA Conference on Aerospace Meteorology, Los Angeles, CA, March 28-31, 1966, 11 pp.

Nanevicz , J.E., 1973: Advanced Materials and Concepts for the Development of antistatic Coatings for Aircraft Transparencies, SRI project 2393, Final Report, Contract, F33615-72-C-2113, Stanford Research Institute, Menlo Park, CA.

NASA, 1970: Analysis of Apollo12 Lightning Incidents, MSC-01540, Feb 1970.

NASA, 1991: Mishap Investigation Team Report, "Mighty Mouse Mishap, 2.75 Inch FFAR Inadvertent Launch on August 26, 1991," KSC Mishap No. 91-0245, October 10, 1991.

National Research Council, 1988: *Meteorological Support for Space Operations, Review and Recommendations.* Panel on Meteorological Support for Space Operations. National Research Council. National Academy Press. Washington DC 1988.

Newman, M.M., 1958: Lightning discharge channel characteristics and related atmospherics, in *Recent Advances in Atmospheric Electricity*, L.G. Smith, ed., Pergamon Press, New York, pp. 475-484.

Newman, M.M., J.R. Stahmann, J.D. Robb, E.A. Lewis, S.G. Martin, and S.V. Zinn, 1967: Triggered lightning strikes at very close range, *J. Geophys. Res.*, **72**, 4761-4764.

O'Brien, T.P., and R. Walterscheid, 2007: Supplemental statistical analysis of ABFM-II data for lightning launch commit criteria, Aerospace Report No. TOR-2007(1494)-6, 15 June 2007.

Orville, R.E., Meteorological Applications of Lightning Data, 1987: *Rev. Geophys. Space Phys.*, **25 (3)**, 411-414.

Orville, R.E., 1991: Calibration of a magnetic direction finding network using measured triggered lightning return stroke peak currents, *J. Geophys. Res.*, **96**, 17-135-17,142.

Orville, R.E., 2008: Development of the national lightening detection network, *Bull. Am. Met. Soc.*, pp. 178-190, February, 2008.

Parker, L.W. and H.J. Kasemir, 1982: Airborne Warning Systems for Natural and Aircraft-Initiated Lightning, *IEEE Trans. EMC*, **24 (2)**, 137-159.

Petterson, B.J., and W.R. Wood, 1968: Measurements of lightning strokes to aircraft, Sandia Laboratories Report SC-M- 67-549, Albuquerque, NM (also Report DS-68-1 of the Department of Transportation, Federal Aviation Administration, Washington, DC, 10590).

Pierce, E.T., 1971: Triggered lightning and some unsuspected lightning hazards, presented at the 138th Annual Meeting of the AAAS, Philadelphia, 27 December, 1971.

Pierce, E.T., 1976: The Thunderstorm Research International Program (TRIP) – 1976, *Bull. Am. Meteor. Soc.*, **57 (10)**, 1214-1216.

Pifer, A.E., and E.P. Krider, 1972: The optical temperature of the Apollo 15 exhaust plume, *J. Spacecraft and Rockets*, **9 (11)**, 847–848.

Pigini, A., G. Rizzi, R. Brambilla, and E. Garbagnati, 1979: Switching impulse strength of very large air caps, presented at the Third International Symposium on High Voltage Engineering, Milan, August 28-31, 1979.

Pitts, F.L., R.A. Perala, T.H. Rudolph, and L.D. Lee, 1987: New results for quantification of lightning/aircraft electrodynamics, *Electromagnetics, 7*, 451–485.

Pitts, F.L., B.D. Fisher, V. Mazur, and R.A. Perala, 1988: Aircraft jolts from lightning bolts, *IEEE Spectrum,* **25**, 34–38.

Plumer, J.A., 1992: Aircraft lightning protection design and certification standards, *Res. Lett. Atmos. Electr.,* **12**, 83-96.

Proctor, D.E., 1971: A hyperbolic system for obtaining VHF radio pictures of lightning. *J. Geophys. Res.* **76**: 1478-1489.

Rakov, V.A., and M.A. Uman, 2003: *Lightning Physics and Effects*, Cambridge University Press, 687pp.

Ramachandran, R., A. Detwiler, J. Helsdon Jr., P. L. Smith, and V.N. Bringi, 1996: Precipitation development and electrification in Florida thunderstorm cells during CaPE. *J. Geophys. Res.,* **101**(D1), 1599–1619.

Richmond, R.D., 1984: Rocket triggered lightning: a comparison with natural flashes, International Aerospace and Ground Conference on Lightning and Static Electricity, Orlando, FL, June 26-28, 1984.

Roeder, W.P., J.E. Sardonia, S.C. Jacobs, M.S. Hinson, D.E. Harms, and J.T. Madura, 1999: Avoiding Triggered Lightning Threat To Space Launch From The Eastern Range/Kennedy Space Center, 8th Conference on Aviation, Range, And Aerospace Meteorology, 10-15 January, Dallas, TX, Amer. Meteor. Soc., pp 120-124.

Rogers, W.P., Chairman, 1986: Report of the Presidential Commission on the Space Shuttle Challenger Accident, accessible at http://history.nasa.gov/rogersrep/.

Rubinstein, M., M.A. Uman, E.M. Thomson, and P.J. Medelius, 1991: Characterization of vertical electric fields and associated voltages induced on a overhead power line from close artificially initiated lightning, presented at the 1991 International Aerospace and Ground Conference on Lightning and Static Electricity, Cocoa Beach, FL, April 16-19, 1991.

Rubinstein, M., M.A. Uman, E.M. Thomson, P.J. Medelius, and F. Rachidi, 1992: Measurements and characterization of ground level vertical electric fields 500 m and 30 m from triggered-lightning, in *Proceedings of the 9th International Conference on Atmospheric Electricity*, St. Petersburg, Russia, pp. 278-278, 1992.

Rubinstein, M., M.A. Uman, P.J. Medelius, and E.M. Thomson, 1994: Measurements of the voltage induced on a overhead power line 20 m from triggered lightning, *IEEE Trans. Electromagn. Compat.*, **36**, 134-140, 1994.

Rubinstein, M., F. Rachidi, M.A. Uman, R. Thottappillil, R. Rakov, and C.A. Nucci, 1995: Characterization of vertical electric fields 500 m and 30 m from triggered lightning, *J. Geophys. Res.*, **100**, 8863-8872, 1995.

Rust, W.D. and P.R. Krehbiel, 1977: Microwave Radiometric Detection of Corona From Chaff Within Thunderstorms, *J. Geophys. Res.,* **82 (27)**, 3945-3950.

Rustan, P.L., 1986: The lightning threat to aerospace vehicles, *AIAA J. Aircr.*, **23**, 62–67.

Saunders, C. P. R., 1988: Thunderstorm electrification, *Weather*, **43**, 318–324.

Saunders, C. P. R., 2008: Charge separation mechanisms in clouds, *Sp. Sci. Rev.*, **137**, 335-353.

Schumann, R.S., G.E. Taylor, F.J. Merceret and T.L. Wilfong, 1999: Performance Characteristics of the Kennedy Space Center 50 MHz Doppler Radar Wind Profiler Using the Median Filter/First Guess Data Reduction Algorithm, *J. Atm. & Ocean. Tech.*, **16(5)**, 532-549.

Sharp, D.W., S.M. Spratt, P.F. Blottman, and J.L. Case, 2002: Using High-Resolution Diagnostics to Facilitate the Short-Term Threat Assessment of Tornadoes during Tropical Storm Gabrielle. *Preprints, 21st Conference on Severe Local Storms*, Amer. Meteor. Soc., San Antonio, TX, 623-626.

Short, D.A., 2008: Radar Scan Strategies for the Patrick AFB WSR-74C Replacement, Ensco, Inc., NASA Contractor Report 2008-214745, 28 pp.

Short, D.A. and R. Lane, 2005: Effect of Clouds on Shuttle Imaging, Ensco, Inc., NASA Contractor Report 2005-211538, 27 pp.

Soula, S. and S. Chauzy, 1985: Multilevel measurement of the electric field underneath a thundercloud 2. Dynamical evolution of a ground space charge layer, *J. Geophys. Res.*, **96**, 22,327-22,366, 1991.

St. Privat D'Allier Group, 1985: Artificially triggered lightning in France. Applications: possibilities, limitations, presented at 6th Symposium on Electromagnetic Compatibility, Zurich, 5-7 March, 1985.

Standler, R.B., and W.P. Winn, 1979: Effects of coronae on electric fields beneath thunderstorms, *Q. J. Roy. Met. Soc.*, **105**, 285-302.

Taillet, J., 1974: Static Electricity Phenomena Involved in the Europa II F-11 Launch, *Journal of the British Interplanetary Society*, **27**, pp,185-191.

Taylor, G.T., 1994: Report on the Comparison of the Scan Strategies Employed by the Patrick Air Force Base WSR-74C/McGill Radar and the NWS Melbourne WSR-88D Radar, Ensco, Inc., NASA Contractor Report 196291, 32 pp.

Theon, J.S., 1986: Report of the Space Shuttle Weather Advisory Panel to the NASA Associate Administrator for Space Flight, October 1986.

Theon, J.S., 1988: The Space Shuttle Weather Advisory Panel, Paper 88-0489, 26th AIAA Aerospace Sciences Meeting, Reno, NV, 11-14 January 1988.

Thottappillil, R., and M. Uman, 1993: Comparison of Lightning Return-Stroke Models, *J. Geophys. Res.*, **98(D12)**, 22903-22914.

Uman, M.A., D.K. McLain, and E.P. Krider, 1975: The electromagnetic radiation from a finite antenna, *Am. J. Physics*, **43 (1)**, 33–38, 1975.

Uman, M.A., Lightning, 1983: *Rev. Geophys. Space Phys.*, **21 (5)**, 992-997.

Uman, M.A., 1987: *The Lightning Discharge*, Academic Press, Orlando, 377 pp.

Uman, M.A., G.A. Dawson, and W.A. Hoppel, 1975: Progress in atmospheric electricity, *Rev. Geophys. Space Phys.*, **13(3)**, 760-765.

Uman, M.A., W.H. Beasley, J.A. Tiller, Y.T. Lin, E.P. Krider, C.D. Weidman, P.R. Krehbiel, M. Brook, A.A. Few, J.L. Bohannon, C.L. Lennon, H.A. Poehler, W. Jefferies, J.R. Gulick, and J.R. Nicholson, 1978: An unusual lightning flash observed during TRIP-76, *Science, 201*, 9-16.

Uman, M.A. and E.P. Krider, 1982: A Review of Natural Lightning: Experimental Data and Modeling, *IEEE Trans. EMC,* **24 (2)**, 79-112.

Uman, M.A. and E.P. Krider, 1989: Natural and Artificially Initiated Lightning, *Science,* **246 (4929)**, 457-464.

Uman, M.A., and V.A. Rakov, 2003: The interaction of lightning with airborne vehicles, *Prog. Aerosp. Sci.,* **39**, 61-81.

Volland, H., Editor,1982: *Handbook of Atmospherics, Vols. 1 and 2,* CRC Press, Boca Raton, FL.

Walterscheid, R.L., L.J. Gelinas, G.W. Law, G.S. Peng, R.W. Seibold, F.S. Simmons, and P.F. Zittel, 2010: Triggered Lightning Risk Assessment for Reusable Launch Vehicles at Four Regional Spaceports, Aerospace Report No. ATR-2010(5387)-1, in press.

Ward, J.G., K.L. Cummins, and E.P Krider, 2008: Comparison of the KSC-ER Cloud–to–Ground Lightning Surveillance System (CGLSS) and the U.S. National Lightning Detection Network™ (NLDN), 20th International Lightning Detection Conference, Tucson, AZ, 21 April – 23 April 2008.

Watson, L.R., B. Hoeth, P.F. Blottman, 2007: Weather Research and Forecasting Model Sensitivity Comparisons for Warm Season Convective Initiation. *Preprints, 22nd Conference on Weather Analysis and Forecasting & 18th Conference on Numerical Weather Prediction,* Amer. Meteor. Soc., Park City, UT, J11A1.

Wheeler, M.M., 1994: Final Report on the Analysis of Rapidly Developing Fog at the Kennedy Space Center, Ensco, Inc., NASA Contractor Report 195888, 56 pp.

Wheeler, M.M., 1996: Verification and Implementation of Microburst Day Potential Index (MDPI) and Wind INDEX (WINDEX) Forecasting Tools at Cape Canaveral Air Station, Ensco, Inc., NASA Contractor Report 201354, 27 pp.

Willett, J.C., 1992: Rocket-triggered-lightning experiments in Florida, *Res. Lett. Atmos. Electr.,* **12**, 37-45.

Willett, J.C., V.P. Idone, R.E. Orville, C. Leteinturier, A. Eybert-Berard, L. Barret, and E.P. Krider, 1988: An experimental test of the "transmission - line model" of electromagnetic radiation from triggered lightning return strokes, *J. Geophys. Res.,* **93**, 3867-3878.

Willett, J.C., J.C. Bailey, V.P. Idone, A. Eybert-Berard, and L. Barret, 1989a: Sub-microsecond intercomparison of radiation fields and currents in triggered lightning return strokes based on the transmission-line model, *J. Geophys. Res.,* **94**, 13,275-13,286.

Willett, J.C., J.C. Bailey, and E.P. Krider, 1989b: A class of unusual lightning electric field waveforms with very strong high-frequency radiation, *J. Geophys. Res.,* **94**, 16,255-16,267.

Willett, J.C., J.C. Bailey, C. Leteinturier, and E.P. Krider, 1990: Lightning electromagnetic radiation field spectra in the interval from 0.2 to 20 MHz, *J. Geophys. Res.,* **95**, 20,367-20,387.

Willett, J.C., D.A. Davis, and P. Laroche, 1999: An experimental study of positive leaders initiating rocket-triggered lightning, *Atmospheric Research*, **51**, 189-219.

Willett, J.C. and J.E. Dye, 2003: A Simple Model to Estimate Electrical Decay Times in Anvils, *Proceedings, Intern. Conf. on Atmos. Elec.*, Versailles, France, June 2003, 267-217.

Willett, J.C., D.M. Le Vine, and V.P. Idone , 2008: Lightning return stroke current waveforms aloft from measured field change, current, and channel geometry, *J. Geophys. Res.*, **113**, D07305, doi:10.1029/2006JD008116.

Wilfong, T.L., S.A. Smith and C.L. Crosiar, 1997: Characteristics of High-Resolution Wind Profiles Derived from Radar-Tracked Jimspheres and the Rose Processing Program, *J. Atmos. Oceanic Technol.*, **14**, 318 – 325.

Wilford, J.N., 1987: NASA Readies New Assault On Still-Mysterious Lightning, New Youk Times, 14 July 1987 accessed at <http://www.nytimes.com/1987/07/14/science/nasa-readies-new-assault-on-still-mysterious-lightning.html?scp=1&sq=&st=nyt>.

Williams, E. R., Large-Scale Charge Separation in Thunderclouds, 1985: *J. Geophys. Res.*, **90 (D4)**, 6013–6025.

Williams, E. R., 1988: The Electrification of Thunderstorms, *Scientific American*, October, pp. 88-99.

Williams, E. R., 1989: The Tripole Structure of Thunderstorms, *J Geophys. Res.*, **94 (D ll)**, 13,151-13,167.

Willis, P.T., J. Hallett, R.A. Black, and W.Hendricks, 1994: An aircraft study of rapid precipitation development and electrification in a growing convective cloud. *Atmos. Res.*, **33**, 1–24.

Winn, W.P., Aircraft measurement of electric field: Self-calibration, 1993: *J. Geophys. Res.*, **98**, 6351-6365.

Xue, M., D.H. Wang, J.D. Gao, K. Brewster, and K.K. Droegemeier, 2003: The Advanced Regional Prediction System (ARPS), storm-scale numerical weather prediction and data assimilation. *Meteor. Atmos. Physics*, **82**, 139-170.

Zavodsky, B.T., S.M. Lazarus, P.F. Blottman, and D.W. Sharp, 2004: Assimilation of MODIS Temperature and Water Vapor Profiles into a Mesoscale Analysis System. *Preprints, 20th Conference on Weather Analysis and Forecasting and the 16th Conference on Numerical Weather Prediction*, Amer. Meteor. Soc., CD-ROM, 2.5.

www.ingramcontent.com/pod-product-compliance
Lightning Source LLC
Chambersburg PA
CBHW081257170526
45165CB00011B/3322